# Exploring the of Landscape

What have cultural anthropologists, historical geographers, landscape ecologists and environmental artists got in common? Along with eight other disciplines, from domains as diverse as planning and design, the arts and humanities as well as the social and natural sciences, they are all fields of importance to the theory and practice of landscape architecture.

In the context of the EU-funded LE:NOTRE Project, carried out under the auspices of ECLAS, the European Council of Landscape Architecture Schools, international experts from a wide range of related fields were asked to reflect, each from their own perspective, on the interface between their discipline and landscape architecture. The resulting insights presented in this book represent an important contribution to the development of the discipline of landscape architecture, as well as suggesting new ways in which future collaboration can help to create a greater interdisciplinary richness at a time when the awareness of the importance of the landscape is growing across a wide range of subject areas.

*Exploring the Boundaries of Landscape Architecture* is the first systematic attempt to explore the territory at the boundaries of landscape architecture. It addresses academics, professionals and students not just from landscape architecture but also from its neighbouring disciplines, all of whom will benefit from a better understanding of their areas of shared interest and the chance to develop a common language through which to converse.

**Simon Bell** OPENspace Research Centre, Edinburgh College of Art, Edinburgh, Scotland, and Estonian University of Life Sciences, Tartu, Estonia

**Ingrid Sarlöv Herlin** Swedish Agricultural University, Alnarp, Sweden

**Richard Stiles** Vienna University of Technology, Vienna, Austria

# Exploring the Boundaries of Landscape Architecture

Edited by

**Simon Bell**
OPENspace Research Centre, Edinburgh College of Art, Edinburgh, Scotland, and Estonian University of Life Sciences, Tartu, Estonia

**Ingrid Sarlöv Herlin**
Swedish Agricultural University, Alnarp, Sweden

**Richard Stiles**
Vienna University of Technology, Vienna, Austria

LONDON AND NEW YORK

First published 2012 by Routledge
2 Park Square, Milton Park, Abingdon, Oxon, OX14 4RN

Simultaneously published in the USA and Canada
by Routledge
711 Third Avenue, New York, NY 10017

*Routledge is an imprint of the Taylor & Francis Group, an informa business*

© 2012 selection and editorial material, Simon Bell, Ingrid Sarlöv Herlin and Richard Stiles; individual chapters, the contributors

The right of the editors to be identified as the author of the editorial material, and of the authors for their individual chapters, has been asserted in accordance with sections 77 and 78 of the Copyright, Designs and Patents Act 1988.

All rights reserved. No part of this book may be reprinted or reproduced or utilised in any form or by any electronic, mechanical, or other means, now known or hereafter invented, including photocopying and recording, or in any information storage or retrieval system, without permission in writing from the publishers.

*Trademark notice*: Product or corporate names may be trademarks or registered trademarks, and are used only for identification and explanation without intent to infringe.

This project has been funded with support from the European Commission. This publication reflects the views only of the authors, and the Commission cannot be held responsible for any use which may be made of the information contained therein.

*British Library Cataloguing in Publication Data*
A catalogue record for this book is available from the British Library

*Library of Congress Cataloging in Publication Data*
    Exploring the boundaries of landscape architecture / [edited by]
    Simon Bell, Ingrid Sarlöv Herlin, Richard Stiles.
      p. cm.
    Includes bibliographical references and index.
    1. Landscape architecture. I. Bell, Simon, 1957 May 24- II. Herlin,
    Ingrid Sarlöv. III. Stiles, Richard.
    SB472.E88 2011
    712'.6—dc22                            2011005570

ISBN13: 978-0-415-67984-8 (hbk)
ISBN13: 978-0-415-67985-5 (pbk)
ISBN13: 978-0-203-80514-5 (ebk)

Typeset in Univers
by Keystroke, Station Road, Codsall, Wolverhampton

Printed and bound in Great Britain by
TJ International Ltd, Padstow, Cornwall

# Contents

|  |  |
|---|---|
| List of figures and tables | vii |
| Foreword<br>Diedrich Bruns, President of ECLAS | xi |
| Introduction: landscape architecture in a changing world<br>Simon Bell, Ingrid Sarlöv Herlin and Richard Stiles | 1 |

**Part 1 Architecture theory, dendrology, sociology and landscape archeology** — 13

**Chapter 1** Theoretical landscapes: On the interface between architectural theory and landscape architecture — 15
Kari Jormakka, Vienna University of Technology, Austria

**Chapter 2** Trees: the living structure of the landscape: Dendrology, arboriculture and landscape architecture — 41
Gabor Schmidt, Corvinus University, Budapest, Hungary

**Chapter 3** Space, place and perception: The sociology of landscape — 60
Detlev Ipsen, University of Kassel, Germany

**Chapter 4** A prospect of time: Interactions between landscape architecture and archaeology — 83
Graham Fairclough, English Heritage, UK

**Part 2 Art, landscape ecology, historical geography and forestry** — 115

**Chapter 5** Space, place and the gaze: Landscape architecture and contemporary visual art — 117
Knut Åsdam, Norwegian University of Life Sciences, Ås, Norway

**Chapter 6** A shared perspective? On the relationship between landscape ecology and landscape architecture — 131
Bob Bunce, Alterra, Wageningen, the Netherlands

**Chapter 7** The past speaks to the present: Historical geography and landscape architecture — 150
Klaus-Dieter Kleefeld and Winfried Schenk, Rheinische Friedrich Wilhelm University, Bonn, Germany

Contents

| | | |
|---|---|---|
| **Chapter 8** | Trees shaping landscapes: Links between forestry and landscape architecture<br>*João Bento and Domingos Lopes, University of Alto Douro y Tras os Montes, Vila Real, Portugal* | 173 |
| **Part 3** | **Economics, cultural anthropology, regional planning and cultural geography** | 195 |
| **Chapter 9** | . . . and how much for the view? Economics and landscape architecture<br>*Colin Price, Bangor University, UK* | 197 |
| **Chapter 10** | Space, place, site and locality: The study of landscape in cultural anthropology<br>*Robert Rotenberg, De Paul University, Chicago* | 233 |
| **Chapter 11** | Greening planning: Regional planning and landscape architecture<br>*Marco Venturi, University of Venezia, Italy* | 259 |
| **Chapter 12** | The place of landscape: Conversing with cultural geography<br>*Stephen Daniels, University of Nottingham, UK* | 276 |
| **Part 4** | **Conclusions** | 297 |
| **Chapter 13** | Crossing the boundaries?<br>*Maggie Roe, University of Newcastle upon Tyne, UK* | 299 |
| | Index | 316 |

# List of figures and tables

**Figures**

| | | |
|---|---|---|
| **1.1** | A garden maze. | 15 |
| **1.2** | Willow cathedral. | 20 |
| **1.3** | The suit of a gardener. | 23 |
| **1.4** | The optics of the landscape. | 26 |
| **1.5** | The formation of the continents. | 28 |
| **1.6** | Diverse decorations. | 33 |
| **2.1** | Examples of the variety of form and character as shown in an arboretum. | 46 |
| **2.2** | Examples of trees along a street showing the architectural properties of trees in these locations where growth conditions may be difficult. | 48 |
| **2.3** | A variety of trees adds to the visual diversity of urban areas. Here in spring the blossom is particularly beautiful. | 51 |
| **2.4** | Winter-hardiness zones in Europe (after Krüssmann, 1989 and Schreiber 2007). | 53 |
| **2.5** | Pruning and other tree care activities is specialist work needing special equipment. | 55 |
| **3.1** | The concept of landscape. | 62 |
| **3.2a and b** | Landscape as a frame. | 63 |
| **3.3** | The panoramic view. | 66 |
| **3.4** | Types of landscape transition. | 68 |
| **3.5** | Landscape in private and public communication. | 78 |
| **3.6** | The relationship of rural and urban will be the main field in the sociology of landscape. | 79 |
| **4.1** | '. . . to study the whole not just some of the parts . . .': A 'traditional' – 'cultural' – landscape at the settlement village of Blindcrake, at the edge of the Lakes National Park: small scale strip fields reflecting mediaeval land divisions in a distinctive settlement region of NW England (UK). | 86 |
| **4.2** | '. . . the conscious shaping of his external environment . . .': ornamental landscaping at Downton, Herefordshire (UK), an 18th century industrialist's 'castle' and improved 'nature'. | 91 |
| **4.3** | '. . . the past being in the present . . .': the military airbase at Scampton, Lincolnshire (UK), used since 1916, its runway extended for Cold War nuclear bombers and the Roman way | |

## List of figures and tables

|     | | |
| --- | --- | --- |
| | diverting from its 2000-year-old path; historic landscapes are not only ancient. | 95 |
| 4.4 | '. . . products of perception . . .': the Scillies (UK) – St Martins Island and Tresco beyond, looking into the west; a flooded land, today's islands are the tips of hills that were farmed and settled until late prehistory; lost field walls emerge at low tide; an historic seascape as well as an historic landscape. | 99 |
| 4.5 | '. . . transitions and change . . . continuity and survival . . .': in the middle of England, through 18th century fields, mediaeval ridge-and-furrow and buried out of sight but not out of perception prehistoric fields, a communications corridor for thousands of years, Roman road, modern roads, railway, canal all pass south to north within a kilometre or two of 'DIRFT' (Daventry International Rail-Freight Terminal), a landscape marking 21st-century lifestyle of mass consumerism. | 107 |
| 5.1 | Knut Åsdam: The Care of the Self Finally Edit, 1999–2007 – DETAIL Architectural installation with film: Finally (2006). Dimensions variable. 'Night time' park consisting of trees, bushes, grass and undergrowth with paths leading to different areas. The film Finally (2006) projected within one of the 'squares' of the park. | 118 |
| 5.2 | Hamish Fulton: Drepung Kora, 2007. Iris print, 45 × 30 cm | 120 |
| 5.3 | Navy Target 103A, Imperial County, California. | 121 |
| 5.4 | CLUI Field Session at the Desert Research Station. | 126 |
| 5.5 | View of the CLUI Programme: A Tour of the Monuments of the Great American Void. | 128 |
| 6.1 | Landscape structure: An example of a fragmented landscape in Languedoc, in southern France. Small, separate patches of tall scrub mixed with semi-natural grassland, deciduous forest (poplar plantation and oak), pine plantation and pure grassland, with a road running across the centre of the picture. Grass strips connect the patches. | 135 |
| 6.2 | Landscape function: a) Scythe-cut grassland near Poprad, in Slovakia. Traditional management of grassland maintains biodiversity, whereas intensive methods – using herbicides and pesticides – have major adverse impacts. b) Abandonment is now taking place in European hills and mountains, on shallow soils and steep slopes. In the picture, taken in the hills near Castelnaudry in southern France, old machinery has been left on a former meadow, with tall grasses and invasive shrubs taking over. | 138 |
| 6.3 | Landscape Description: A classic, mountain landscape in the High Tatras in Slovakia. The lake, waterfalls, scree and cliffs could be separated or treated as a single landscape. | 141 |
| 6.4 | Biodiversity: A landscape near Clun in Shropshire, central | |

| | | |
|---|---|---|
| | England, with many elements; especially linear features, such as hedgerows; but also grassland, crops and small, deciduous woods. All the different patches need to be sampled in order to assess the biodiversity. | 142 |
| 6.5 | Monitoring: Pre-desert in Almeria, in southern Spain. The procedure described by Bunce *et al.* (2008) is based on life forms such as low evergreen shrubs and palms. | 143 |
| 7.1 | A diagram of historical development based on the concept of the constraints of the solar energy system and the possibilities released by the use of fossil energy sources. | 154 |
| 7.2 | A scheme showing the complex etymology of the word "Landschaft". | 156 |
| 7.3 | This diagram structures the different levels of significance and value of historical elements in the landscape. | 161 |
| 7.4 | A model of the process of understanding, valuing and caring for cultural landscapes. | 162 |
| 7.5 | A model showing different approaches in historical geographical research. | 165 |
| 8.1 | A recreational area inside a forest, one of the main areas where landscape architecture intersects with forestry. | 178 |
| 8.2 | A forest road also acting as a fire break, part of the infrastructure which can have a significant landscape impact. | 182 |
| 8.3 | Pasturage activities in Mediterranean ecosystems – traditional management practices which create important cultural landscapes. | 185 |
| 8.4 | A diagram showing the different aspects which make up the forestry educational curriculum. | 186 |
| 8.5 | Trees defining paths in the landscape, experiences and play of light and shade. | 190 |
| 8.6 | Trees shaping landscapes or trees presented as singular elements? | 193 |
| 9.1 | The Ring of Brodgar, Orkney – a Neolithic ritual landscape wrought with much labour. | 201 |
| 9.2 | Versailles, product of the formal taste in landscape that was superseded by the naturalistic preferences of the Romantic Movement. | 205 |
| 9.3 | Mounds and tree planting designed by Sylvia Crowe partly screens and partly softens, but cannot turn Wylfa Nuclear Power Station on Anglesey in Wales into an aesthetic asset. | 215 |
| 9.4 | Fürst Pückler's Pharaonic aspirations, and their Romantic parkland setting at Branitz, nearly ruined him. | 217 |
| 9.5 | An urban view that, according to an estate agent in 1990, put £30,000 on the price of a house commanding it. But how much for Durham Cathedral, how much for the trees, how much for the topography . . .? | 225 |

**List of figures and tables**

| | | |
|---|---|---|
| **11.1 a and b** | Koper (Capodistria), Slovenia: Proposals for the expansion of the port with environmental and energy arrangements. The port is vast area, dwarfing the old Venetian town. Planning is a vital task. | 263 |
| **11.2 a and b** | Stanezice, Slovenia: A new city centre for 3000 dwellings to be planned within the region and to take advantage of new communication infrastructure. | 266 |
| **11.3 a and b** | Rovigo, Italy: Castle, civic park and system of squares, showing planning and design in the context of an old urban fabric. | 273 |
| **12.1** | *Fu Kei No Zuzogaku* (2001) Japanese edition of *The Iconography of Landscape* (CUP 1988) Tokyo, Chijin Shobo. | 278 |
| **12.2** | Humphry Repton's Trade Card. | 280 |
| **12.3** | Housesteads Roman fort on Hadrian's Wall, Northumbria, UK, now owned and managed by the National Trust. | 283 |
| **12.4** | Eskdale in the Lake District National Park, Cumbria, UK, owned and managed by the National Trust. | 284 |
| **12.5** | The author and Denis Cosgrove leading a group of students in Vicenza in the Veneto, Italy, on one of the field tours described in the text. | 292 |
| **13.1** | Interdisciplinary ways of working. | 300 |
| **13.2** | The Dimensions of Landscape. | 302 |
| **13.3** | Overlapping fields of value and sources of theory in landscape architecture. | 305 |
| **13.4** | Knowledge building in Landscape Architecture and Cultural Geography. | 307 |
| **13.5** | A summer moment, St Abbs Head, Scotland. | 313 |

**Tables**

| | | |
|---|---|---|
| **3.1** | Levels of influence in a regional limited landscape | 64 |
| **3.2** | The empirical field of landscape consciousness | 78 |
| **8.1** | Global distribution of the main types of forest | 174 |

# Foreword

The European Council of Landscape Architecture Schools (ECLAS) exists to foster and develop scholarship in landscape architecture throughout Europe, mainly by strengthening contacts and enriching the dialogue between members of Europe's landscape academic community. For the first time in October 2002 ECLAS obtained European Union funding for a Thematic Network Project[1] in landscape architecture. This was the beginning of the LE:NOTRE Project, 'Landscape Education: New Opportunities for Teaching and Research in Europe'. (The name of the project also pays homage to the famous landscape architect André Le Nôtre.) The first LN Project was followed by others and this book is one of the results of the LE:NOTRE TWO Project.

Since the 2002 beginnings the LN Projects have served as a platform for increased co-operation and intensified exchange between members of the landscape architecture community in Europe and worldwide. Based on surveys of landscape architecture education and research, elements and structures of theory, methodology, taught courses and formal research have been established. Collaboration between universities and research facilities has been developed in teaching, research and community service. A glossary and key word list of landscape architecture terminology have been prepared and made available on the project website (http://www.le-notre.org). This website also includes numerous other resources, such as listings of landscape architecture education and research institutions in Europe (and beyond), and now serves as an interactive tool acting as a centre for the collection and dissemination of information between project partners.

In 2006, and initially funded through the LN Project, ECLAS was able to establish 'JoLA', the Journal of Landscape Architecture. JoLA is an academic journal stimulating scholarly debate; all contributions are blind peer reviewed. The journal is unique in addressing landscape architecture alone with special sections for research-based articles as well as critical readings of landscape architecture projects and 'visual essays'. Another good approach to identify a field is to describe its 'edges'. Review papers on the interface with landscape architecture were commissioned from academics representing a number of *neighbouring disciplines*. This book includes papers from twelve such disciplines.

The rationale to invite colleagues from outside of the field to help describe landscape architecture has several aspects. Two are explained below; one is historic, the other strategic. When, during the nineteenth and until the beginning of the twentieth century, the planning, design and management of landscapes (including urban landscapes and open space) was increasingly perceived as important, available

## Foreword

professionals had received their training at institutes of higher education in fields such as gardening, horticulture, agriculture and forestry, or, alternatively, they were artists, architects or engineers, who had obtained some knowledge about plants, soil, etc., for example by working with gardeners. Thus, to this day, formal programmes of university education and research facilities are still found in art colleges and technical universities, in agricultural universities and horticultural schools, as well as universities of professional education and general purpose or 'classical' universities. Depending on the type of academic setting the educational staff might include not only landscape architects but also represent architecture, planning and the arts, and the humanities, natural and social sciences.

The strategic reason to ask help from the neighbouring disciplines in describing the interface with and linkages to landscape architecture is their collaboration both in practice and in research. In practice landscape architects usually work in teams with architects, artists, engineers and other professionals, be it in site design, regional scale planning or in landscape management. In research landscape architects have developed their own body of knowledge and, at the same time, relate to methods and techniques that were developed outside of the field. Advancing any field requires thinking beyond disciplinary borders, and this is true in particular for a relatively young field such as landscape architecture. Landscape architects must have a holistic knowledge and understanding of landscape in time and space, and the pressures and driving forces to which landscapes are subjected; they involve specialist knowledge from a wide range of disciplines.

Representing European landscape architecture schools I am indebted to colleagues from several of our neighbouring disciplines who have taken the time and energy to share their insights with us. All of these people have experience in working with landscape architects in academic settings, and many also in practice. We have mutual respect and highly regard each other's special expertise. Most of all we share learning experiences where, together, we have achieved much more than we could have individually. We also share a concern with the shaping of landscapes at various scales to create, enhance, maintain and protect places so as to be functional, aesthetically pleasing, meaningful and sustainable and appropriate to diverse human needs and goals.

This book is, I hope, the first of many to be published with an ECLAS 'branding' and it reflects not only the aims carried in its title but also the increased activity and self-confidence of landscape architecture as a discipline which feels comfortable exploring widely and undertaking self-reflection and criticism. I hope it will be a valuable contribution to the corpus of material of use by academics and professionals alike in their striving to improve the landscape for everyone's benefit.

Dr Diedrich Bruns, Landscape Architect.
ECLAS President
January 2011

## Notes

1. Thematic Network Projects are part of the European Union's Socrates-Erasmus Programme (The Lifelong Learning Programme, LLP, is the successor to Socrates).

# Introduction: landscape architecture in a changing world

*Simon Bell, Ingrid Sarlöv Herlin and Richard Stiles*

Landscape is 'an area, as perceived by people' according to the definition in the European Landscape Convention.[1] But landscape is more than that. As well as being both physical space and mental perception, landscape is increasingly also an arena within which a wide range of different, and often contrasting, academic disciplines from both sides of C.P. Snow's 'Two Cultures' divide encounter each other (Snow, 1959/1993). One of these many disciplines, perhaps one of the few for which the landscape lies at the very centre of its concerns, is landscape architecture.

The task of landscape architecture is the conservation and development of natural and cultural landscape resources, together with their associated meanings and values, for the benefit of current and future generations. It operates by means of planning, design and management.

The main aims of the European Landscape Convention (ELC) are defined as being 'to promote landscape protection, management and planning', and at first sight the way in which all of these activities are defined as 'action' in the ELC has suggested to some people that the ELC was really all about landscape architecture. But to believe this is to miss the point entirely, and not only because the ELC was conceived and drafted without any known contributions from landscape architecture; but because it even emphasises the need for multidisciplinary approaches.

Any semblance of support for such a landscape architecture-centric view of the ELC is further undermined by the fact the word 'landscape' is to be found in the names of a growing range of disciplines, including landscape ecology, landscape archaeology and landscape urbanism. What is more, in a number of countries the use of the term 'architect' is legally protected and cannot currently be used by landscape architects who are forced to work under another professional title, while in yet other countries the profession itself is still not even recognised as such. Furthermore, it is not even a question of 'which profession?' as the ELC also stresses the importance of the role of the wider public alongside that of professionals and specialists in shaping policy on landscape conservation and development.

Seen from this perspective, the ELC can in no way be regarded as a reason for other disciplines to defer to the pre-eminence of landscape architecture as the key to achieving the aims of landscape protection, management and planning or conservation, but rather as a clarion call to landscape architecture to re-evaluate its old assumptions and above all to engage more closely and creatively with the many other disciplines who also have an interest in understanding and shaping our common landscape.

One way to look at this book, therefore, is to see it as an initial response to this call. Thus seen, it can be viewed as a metaphorical first voyage of discovery, venturing forth from landscape architecture's safe shores of familiar past certainties and striking out towards the unknown frontiers of the discipline.

This is not to suggest that landscape architects have not been working in teams with other professionals for a long time, but such collaborations tend to differ from the approach behind this book in two important ways. First, they tend to be with the other environmental professions, above all architects, civil engineers, urban and regional planners, whereas this book aims to cast its net more widely, and at the same time seek to explore the relationships with neighbouring academic disciplines. Second, inter-professional collaboration has always tended to take place on a pragmatic and *ad hoc*, project-by-project basis, whereas here for the first time an attempt has been made to initiate a more or less 'systematic' survey of the 'neighbouring lands' bordering on landscape architecture.

## Planning the voyage: the origins of the book project

When the first ideas for this expedition to the borders of the discipline of landscape architecture were formulated, the world looked rather different from how it does today. The European Landscape Convention had only just been opened for signature, and the idea for a Science Policy Briefing on landscape was not yet even a 'twinkle in the eye' of the European Science Foundation. Nevertheless, the potential benefits of a comprehensive re-examination of the discipline had already been recognised. The question was how to achieve this? The possibility for fundamentally re-evaluating any discipline does not come along very often; it is not sufficient to have the insight that it might be worthwhile, there is also the need for a certain structure and resources for it to be feasible.

In the case of landscape architecture, the opportunity came with a chance for the European Council of Landscape Architecture Schools (ECLAS) to apply for a European Union funded Thematic Network Project.[2] According to the European Union, Thematic Networks 'were created to deal with forward looking, strategic reflection of the scientific, educational and institutional issues in the main areas of higher education'.

'Strategic reflection' was exactly what was called for, and as part of this, during the development of the proposal for what came to be the 'LE:NOTRE Project' in 2001 and 2002, the idea of commissioning review papers from leading members of neighbouring disciplines with the aim of exploring the interface between their discipline and landscape architecture, was defined as one of the goals for the project.

Following the success of the proposal, the project was faced with two questions:

- which disciplines should be considered – what are the 'neighbouring disciplines' of landscape architecture; and
- which individuals from these disciplines should be approached?

The answer to the first was based on the principle that as wide a range of disciplines as possible should be selected, in order to reflect the scope and breadth of landscape architecture and the way in which it crosses many of the recognised domains of knowledge. In fact C.P. Snow's 'two cultures' proved not enough to encompass these and the choice ended up spanning not just the arts and the sciences, but the humanities and social sciences as well.

In response to the second question – which individual academics should be chosen to discuss the disciplines concerned – two important criteria were defined. Because the Thematic Network was a Europe-wide project, one of the key aims was to involve academics from a wide range of different countries in order, as far as possible, to gain a European perspective (in fact in total eleven countries are represented in the authorship of this book, including the editors). However, it was also important that the individuals chosen had an appropriately international overview of their discipline and so were able to write not just from a national viewpoint.

The other key criterion was that each of them should have some degree of familiarity with landscape architecture, such that they were in a position to write about the relationship between the two disciplines from a point of view of understanding and so not have to rely on poorly informed stereotypes. The corollary to this was, of course, that they should also not be so close to landscape architecture that they risked having 'gone native', as it was also thought necessary that they should be able to take an unbiased view of the discipline from the 'outside looking in'. With the approval of the LE:NOTRE TWO Project and its subsequent extension beyond Europe through 'LE:NOTRE Mundus'[3] it became possible to expand the initial choice of disciplines, not just to include six new ones, but also to extend the geographic domains from which the authors came, while the brief remained the same.

The initial brief reflected the intention that the review papers would be treated as internal documents for reference purposes within the context of the Thematic Network Project and that they would be used as raw material to develop the outputs planned for it. Thus they would initially be made available to the members of the network via the project website, but would not be distributed more widely.

The brief for the papers suggested they begin with a broad introductory presentation of the discipline in question and to reflect upon its current and potential future relationship with the discipline of landscape architecture. This was to be accompanied by some thoughts of the author on the interface between the disciplines on which they were based.

Two other specific issues were to be covered, the first being the basic education of landscape architects: 'which are the main subject areas from your discipline of which landscape architecture students should gain a basic appreciation . . . (in) their introductory education?'; while the second dealt with the potential for research: 'what are the main research approaches and methods associated with your discipline with which landscape architecture research students should be familiar in order to be able to engage in collaborative research programmes and projects?'

In order to ensure that the papers covered the subjects in a broadly comparable manner a common structure was proposed, largely to help make the papers easier to work with within the context of the project. Although the possibility of some form of publication was considered, this was initially not seen as feasible given the wider constraints under which the LE:NOTRE Project had to operate.

However, a number of things happened to change this situation. First, it became increasingly clear that the individual quality of most of the papers, as well as their overall range, meant that not to publish them more widely would deprive both the discipline and the wider landscape academic community of a potentially very valuable resource. Second, the way in which the wider context of landscape architecture was continuing to change, partly in response to the success of the European Landscape Convention, could not have been foreseen even at the start of the project only a few years earlier.

## A changing world

That the world around landscape architecture's borders is changing was witnessed, not just by the drawing up of the European Landscape Convention – although the significance of this in raising the overall profile of landscape cannot be over-estimated – but also by the speed with which the Convention has been accepted and ratified by a significant majority of Council of Europe member states, including most European Union countries. The ELC was opened for signature at about the same time that the ECLAS decided to investigate the possibilities of applying for funding for a Thematic Network Project. By the second year of the project and completion of the first round of neighbouring disciplines papers, the ELC had already received the necessary ten signatures and ratifications required for it to come into force. Four years later, by end of the LE:NOTRE TWO Project in September 2009, when the second set of papers was completed, it had come into force in 30 countries, including 22 of the 27 EU member states.

This demonstration of the way in which landscape is becoming a mainstream political concern is made explicit in the preamble to the ELC. Passages such as:

> Noting that the landscape has an important public interest role in the cultural, ecological, environmental and social fields, and constitutes a resource favourable to economic activity and whose protection, management and planning can contribute to job creation

emphasise the wide range of policy areas within which it has a role to play, as well as highlighting the broad scope of academic disciplines with which there could be an interface.

As if to emphasise this fact, following close on the tenth anniversary of the Landscape Convention, a new joint Science Policy Briefing 'Landscape in a Changing World', prepared on behalf of the European Science Foundation and COST,[4] also stressed the extent to which landscape has moved towards the centre of both political attention and academic concern:

> Many of the social, economic and environmental decisions facing Europe and the wider world concern the cultural uses and meanings of land. Their spatial dimensions can be addressed through the idea of landscape, which comes into being whenever land and people come together.

Both these passages make clear the broad range of disciplines with which landscape architecture needs to engage in order to respond adequately to the growing recognition of the significance of landscape as a potentially unifying force in policy and research.

'Landscape in a Changing World' is of key importance for landscape architecture. It was prepared through the first collaboration between the humanities domain committees of COST and the European Science Foundation.[5] Major inputs were provided by historians, geographers and archaeologists, as a result of which: 'The briefing's key proposal is to integrate the human(ities)' perspective fully, and in its own right into landscape research as the starting place for reaching out to all other landscape research areas.' It is clear that the landscape, as an arena for interdisciplinary discourse, is becoming increasingly populated with a broad range of disciplines, and that the potential for meetings of minds across the cultural divide is greater now than it has ever been.

However, just because landscape straddles the traditional divide between the 'natural scientists' and the 'literary intellectuals', about which C.P. Snow expressed his concerns more than a half-century ago, does not mean that such disciplines which meet each other in the 'landscape arena' will be able to communicate meaningfully. Simply by virtue of landscape's location on an academic and intellectual fault line between 'two cultures' need not in itself necessarily result in any integration, even if some meaningful communication does occur. Nevertheless, the transdisciplinary potential of landscape architecture has to be realised if it is to have any impact in the new conditions arising from these changes and, in order to achieve this, considerable theoretical and practical effort is required. While many disciplines should contribute to this process, landscape architecture has perhaps more reasons than many to take the lead, both because landscape is its *raison d'être* and not least because it has the pragmatic need to integrate theoretical considerations in a way which can give rise to new approaches to practical action.

## Landscape architecture and 'the third culture'

So, for landscape architecture, the 'survey of the lands beyond' represented by this book is an important first step, but it is not enough. There is a need to both analyse and synthesise in order to be able to benefit from what are only the preliminary results of this project and so to build a functional bridge between the 'two cultures'.

In the decades since the publication of The Two Cultures, two different responses can be identified, but neither of them really provides a viable model for landscape and landscape architecture. The first, as exemplified by The Third Culture (Brockman, 1995) suggests that there is no longer the need for communication across the old fault line, as a new breed of articulate scientists has arisen, who are able to explain and popularise their work to a wider audience, including decision makers and opinion formers, even if most of these were still educated on the 'other' side of the divide.

The second version of the 'third culture' would appear at first sight to be of more use to landscape architecture. This comes from the suggestion that in addition to the classic two cultures of the sciences and the humanities, there is a third culture which is 'design' itself (Cross, 2007). The problem with this approach is that the argument for recognising design as a third culture does not derive from any suggestion that it is somehow in a position to bridge the divide between the other two but that it stands as an equal alongside both of them. So, this version of the 'third culture' not only fails to provide the bridge between the first two, but instead generates the need for two further bridges!

So where does this leave landscape architecture? Attention was drawn to the existence of a clash of value systems leading to potential fault lines within the discipline by Ian Thompson in his book 'Ecology, Community and Delight' (Thompson, 1999). Here he poses the question: 'Is it [landscape architecture] primarily concerned with making beautiful places, helping people, or saving the planet from ecological catastrophe?' From his analysis it would appear that the fault lines dividing the main domains of knowledge, across which landscape and the discipline of landscape architecture are positioned, continue to run right through the discipline itself. Surely this is reason enough to have a very strong self-interest in forging a common understanding between the cultures, or at least in acting as a facilitator in instigating a new dialogue.

As a planning and design discipline, landscape architecture has perhaps long felt itself closer to the natural sciences than the humanities. Possibly inspired by the work of Ian McHarg (McHarg, 1969/1992), ecology and soil science, geomorphology, climatology and hydrology have tended to be the areas where landscape architects have sought a more in-depth understanding of the 'layers of the landscape' over recent decades. However, while landscape architecture has long had interests in these areas of the natural sciences, the concern has not always been mutual. By contrast, over recent decades there has been a growing interest in landscape from the other side of the 'two cultures' divide, namely on the part of many humanities disciplines, as evidenced, amongst other things, by the ESF's Science Policy Briefing. This interest has been a more mutual one, something which can partly be seen as a product of a certain waning of confidence on the part of landscape architecture in the positivism of the natural sciences in the wake of the post-modern and environmental revolution following the 1970s.

As a result of this, the challenge now faced by landscape architecture in the changing world in which it finds itself is to find how the ideas about and approaches to landscape which all these varied disciplines embody can be integrated in an operational way that can form the basis of a renewed culture of landscape planning, design and management.

## The structure of the book

While seriously addressing this very ambitious challenge is clearly outside the scope of this book, commissioning and collecting together the review papers presented here represents an important first step. But can it go beyond a mere survey and presentation of resources and ideas from these 'new found lands'? Is there a chance to make some first tentative steps in the direction of analysis if not perhaps the opportunity for a synthesis that cuts across the old domain boundaries? These were the very questions which we addressed when considering the structure for the book. What implications did they have for the arrangement of the contributions from such a varied range of disciplines?

Three options appeared possible. First and most obviously would have been to structure the book according to the classic domains of knowledge from which the papers came, and to create three sections, one each for the humanities, the social sciences and the natural sciences. While this would successfully have broken the book down into manageable and broadly homogenous parts, it would also have achieved exactly the opposite effect to our aim of bridging the fault lines. Instead, it would merely have helped to reinforce these age-old divisions.

Option two would have been to reject a structure based on the classic academic domains, and accept that there is no other logical alternative than a simple alphabetical arrangement of the chapters within one continuous section. However this too seemed to risk missing an opportunity to orchestrate, as it were, some greater interaction.

The third option, and the one we selected, was derived by considering the disciplines involved, not just a series of impersonal papers but, instead, in terms of the 'representatives' who had written them. These we envisioned as 'ambassadors', emissaries from the domains of knowledge in question, who were invited to a diplomatic reception. How should this imaginary event be best organised to ensure that it leads to improved diplomatic relations and a better understanding between all concerned? What should the seating plan look like? Who should sit next to whom in order to ensure that the conversation sparkles, or should there instead be a buffet where all can mingle freely?

The first, domain-based, option is one that might have resulted from the event being organised as a buffet. Here individuals with similar backgrounds and interests might naturally have gravitated into three groups based around the conventional domains and spent the rest of the evening exchanging news and 'talking shop'. In the case of the second option, we may imagine a formal meal at one long table, where one's neighbours are the result only of alphabetical chance and, while the result might be a few interesting conversations, it will be a matter of pure luck whether they will sparkle in the desired manner.

Our preferred option is to design the seating plan so as to create provocative juxtapositions of domains and disciplines, of scales of time and space, of levels of theory and practice and even nationalities of authorship, but in groups small enough to be able to engage with each other across round tables. As a strategy this is intended not only to maximise the interest in reading the contributions, but to be as intellectually stimulating and thought provoking as possible, perhaps even prefiguring the first tentative steps in the next part of the process of synthesising these different perspectives. Purposefully bringing together ideas from different fields and engineering the environments in which such potentially meaningful 'chance meetings' can take place is, according to Steven Johnson, an important strategy for generating new and creative ideas (Johnson, 2010).

The first grouping (intimate round table) starts with a look at gardens and plants (architectural theory and dendrology), perhaps a fundamental basis for landscape architecture, and then goes on to consider contemporary society and the ways in which past societies have influenced the landscape (sociology and landscape archaeology). Imagine how each person explains their fields to another and you can see what stimulation could occur! The second group mixes a classic contrast between the arts and the sciences (fine art and landscape ecology) with those involved in both past and contemporary land use (historical geography and forestry). The final group looks at the role of monetary and social values (economics and cultural anthropology) together with functional and symbolic approaches to place (regional planning and cultural geography).

Chapter 1, 'Theoretical landscapes', on the interface between architectural theory and landscape architecture is written by Kari Jormakka, a Finnish architect, historian, critic and pedagogue and currently professor of architectural theory at Vienna University of Technology. Jormakka considers that embedded in design theories, paradigmatic examples, representational methods, work habits, legal and financial structures etc. are ontological assumptions that may limit the range of design options open to landscape architects without ever reaching the level of consciousness. Opening up some of these restricted territories may be the best service theory can offer to landscape architecture.

Chapter 2, 'Trees: The living structure of the landscape' by Gabor Schmidt of the Corvinus University in Budapest, Hungary examines the interface between dendrology and landscape architecture. There are very few landscapes without trees and here at least there is a longer history of engagement between landscape architects and a neighbouring discipline. Some potential areas for collaborative research between dendrology and landscape architecture are the management and evaluation methods for landscape trees and street trees, and green space inventories in the towns and cities – of major importance given the fact that over half the world's population now live in cities and the growing concerns over the urban quality of environment.

Chapter 3, 'Space, place and perception: the sociology of landscape' is written by the late Detlev Ipsen, until his recent death professor of urban and regional sociology at the University of Kassel (Germany). His research focus was on sustainable urban and regional development and the impact of migration and cultural diversification. Ipsen explores sociological aspects of landscape, a field that few have previously dealt with. He suggests that landscape is today becoming a social problem in Europe but can also offer a fruitful approach for solving a series of imminent problems such as how to organise new peri-urban landscapes and megaurban landscapes, which are correlated with the urbanisation of the world. The socio-cultural aspects of landscape can be an important tool to develop spatial identity and personal identification in an increasingly urbanised world.

Chapter 4, 'A prospect of time: interactions between landscape architecture and archaeology' is written by Graham Fairclough, an archaeologist who has worked with English Heritage for many years, mainly in the field of landscape, historic characterisation and heritage management. He has written widely on these topics, is closely associated with the implementation of the ELC, and was recently involved in the preparation of the ESF/COST Science Policy Briefing 'Landscape in a Changing World'. He is also a member of the advisory board of the LE:NOTRE project. He shows how different branches of landscape archaeology recognise the need to look at the whole, to work at 'landscape scale' and are concerned with understanding how people have made our landscape over time, and for understanding our own present-day landscape. Both landscape architecture and landscape archaeology benefit from understanding the origins and history of landscape and the complex of historic and social processes that have created it.

Chapter 5, 'Space, place and the gaze' is on fine art and is written by Knut Åsdam, a Norwegian artist who has contributed to the international art scene with exhibitions, publications and broadcasts for over fifteen years. Identity, space and place are recurring concerns in Åsdam's work. His investigation of the use and perception of public urban spaces takes the diverse form of audio, film, video, photography and architecture. He has been working with landscape architects in various installations, photography and other wide-ranging works. Åsdam thinks that that there are many similarities and important crossovers between visual art and landscape architecture. One particular discussion in the fine art context is the analysis of space and place, installation art, film/video art and audio art. Land art, which emphasises process and impermanence, is an obvious development that is relevant to landscape architecture. There are several areas for students of landscape architecture to engage in the discussion of contemporary visual arts.

Chapter 6, 'A shared perspective? On the relationship between landscape ecology and landscape architecture' looks at landscape ecology and is written by Bob Bunce, a senior internationally recognised landscape ecologist with over thirty years of experience of land use research, currently at Alterra Research Institute in the Netherlands, formerly at the Institute of Terrestrial Ecology in the UK. Bunce was also for many years the president of the International Association for Landscape Ecology. Landscape ecology is a relatively young discipline which concerns the holistic understanding of the interactions between ecological components of landscapes. There is a high degree of overlap between landscape ecology and landscape architecture. The participation of landscape architects in many landscape ecology meetings shows recognition of shared objectives, and there is a strong base for overlapping in teaching.

Chapter 7, 'Past meets present' examines historical geography and is written by Klaus-Dieter Kleefeld and Winfried Schenk, of the Rheinische Friedrich Wilhelm University, Bonn, Germany. Historical geography studies human activities and their resulting spatial structures in a historical perspective in order to deduce laws of temporal spatial differentiation. This means describing, differentiating and explaining the scale and quality of economic, social, political, demographic and natural processes. The method of cultural landscape analyses involves research into the cultural landscapes development, and the formulation of guiding principles and development purposes for future planning and management of the cultural landscape for inhabitants and politicians.

Chapter 8, 'Links between forestry and landscape architecture: Trees shaping landscape' is a study from a forestry perspective written by João Bento and Domingos Lopes of the University of Alto Douro y Tras os Montes, Vila Real, Portugal. They demonstrate that forests are major elements of land cover and the basic climax natural land cover of most of Europe, although cleared from many areas. The strongest connections between forestry and landscape architecture are issues such as recreation, nature tourism, forest landscape planning and design, management of animal resources and nature conservation. Important challenges for joint research in the future are, for example, the maintenance of an increasing forest cover in Europe as a result of afforestation, fragmentation and decline of natural habitats. The authors predict an increased need for urban or peri-urban forest in order to fulfil multiple environmental, social and economic functions.

Chapter 9, '. . . and how much for the view? Economics and landscape architecture' is written by Professor Colin Price of Bangor University, Wales. He is a well-known expert on forest, landscape and environmental economics, and has often written about landscape from an economist's perspective, including in his pioneering book *Landscape Economics* (Macmillan, 1978). In his chapter he remarks on his involvement in economics which arose via forestry, an interest which in turn was sparked by concern over the aesthetic impacts of afforestation projects on the British rural landscape. He also remarks on his verbal sparring with the well-known landscape architect and advisor to the British Forestry Commission, Dame Sylvia Crowe, who associated economists with philistines. This has clearly affected his views of landscape architects and his perception of how they view economists – part of the *raison d'être* of this volume. The chapter takes its starting point in economics with emphasis on aspects of the subject that have particular importance for landscape architects.

Chapter 10, 'Space, place, site and locality: the study of landscape in cultural anthropology' is written by the only contributor from outside Europe: Robert Rotenberg who has taught at De Paul University Chicago, USA since 1979. His research interests focus on peoples' lives in big cities, and on space and place. In the chapter, Professor Rotenberg summarises fifty years of research on a complex facet of the human experience: the meaning people derive from their experience with specific spaces and places. He argues that the landscape designer decides how much of the design responds to the issues and concerns in the professional community or the community of users, and how much derives from creativity. He sees clear possibilities for cooperation between the two fields. If landscape architects could take the time to engage in ethnographic research themselves, their designs become more deeply rooted to the locality.

Chapter 11, 'Greening Planning: Regional planning and landscape architecture' explores regional planning and is written by Marco Venturi, of the Istituto Universitario di Architettura di Venezia (IUAV), Italy. Professor Venturi has a long experience of teaching and research at a number of university departments in Italy, Germany and Algeria, as well as in public service and professional practice with planning projects in a range of countries. In this chapter, Venturi starts by describing the challenges to planning since a majority of the world's population now lives in urban areas, and with new forms of urbanisation, including infrastructure, covering a large part of Europe. This inversion of the relationship between open and built-up space is creating a series of dichotomies that are central to the future of physical planning at different scales, according to Venturi. He describes two great tendencies that seem to generate the biggest transformations: homogenisation, with diffusion of infrastructure everywhere, and the concentration of environmental impact in some limited areas.

Finally Chapter 12, 'The place of landscape: conversing with cultural geography' is written by Stephen Daniels, professor of cultural geography at the University of Nottingham, UK. His research interests include landscape in eighteenth–nineteenth century Britain; geography, art and literature; and history of geographical education. Daniels considers that the conversation between cultural geography and landscape architecture is potentially a highly creative one but so far limited to research in terms of cultural geography's concerns with the theory and history of landscape, and more recently urban space and mobility. What would enhance the exchange between cultural geography and landscape architecture is more joint participation collaborative projects, in which polite conversation could be sharpened by practical experience of the limits as well as opportunities for interdisciplinary work.

The final part is the concluding chapter, 'Crossing the boundaries?' written by Maggie Roe, a landscape architect based in the School of Architecture, Planning and Landscape at the University of Newcastle upon Tyne, UK. Maggie is also editor of the journal *Landscape Research* which itself is multi-disciplinary. Her aim is to reflect on the positions presented by the authors of the foregoing chapters and to identify some of the key messages of value relating to the reappraisal of the way landscape architects can work with other discipline areas. Her conclusions synthesise these messages into some tentative conclusions for future working for the profession.

We hope you, the reader, will find the contents stimulating and for those of you who are teachers we believe that there is much to put to use in courses on landscape architecture theory. If you are a student we very much hope you will be challenged and inspired in your struggle to find your own path in this complex yet rewarding field.

## Notes

1 The full text of the European Landscape Convention can be found here: http://www.coe.int/t/dg4/cultureheritage/heritage/landscape/default_en.asp

2   The LE:NOTRE Project began formally in October 2002 with funding approved for one year, and since then has been the subject of seven further successful funding applications. LE:NOTRE is an acronym standing for: Landscape Education: New Opportunities for Teaching and Research in Europe.
3   This was funded under the European Union's ERASMUS Mundus programme and made it possible to extend the Thematic Network beyond the boundaries of Europe between 2007 and 2009.
4   COST is a European organisation supporting 'Cooperation in Science and Technology' see: http://www.cost.esf.org/
5   It is perhaps also interesting to note that the preparation of this Science Policy Briefing was the first time that the European Science Foundation had acknowledged the existence of the discipline of landscape architecture, which otherwise receives no mention in its scheme of academic domains and the structure of committees which are responsible for them.

## References

Brockman, J. (1995) *The Third Culture*, Simon & Schuster, New York.
Cross, N. (2007) *Designerly Ways of Knowing*, Birkhäuser Verlag, Basel.
European Science Foundation (2010) *Landscape in a Changing World*, ESF, Strasbourg.
Johnson, S. (2010) *Where Good Ideas Come From*, Riverhead Books (Penguin USA), New York.
McHarg, I. (1969/1992) *Design with Nature*, John Wiley and Sons, New York.
Snow, C. P. (1959/1993) *The Two Cultures*, Cambridge University Press, Cambridge.

# Part 1

# Architecture theory, dendrology, sociology and landscape archeology

Chapter 1

# Theoretical landscapes

## On the interface between architectural theory and landscape architecture

*Kari Jormakka*

### Introduction

Summarizing the remote pages of Pierre Boitard's *Manuel de l'architecte des jardins*, Gustave Flaubert divides gardens into:

(a) *the melancholy and romantic*, distinguished by immortelles, ruins, tombs, and 'a votive offering to the Virgin, indicating the place where a lord has fallen under the blade of an assassin'

**Figure 1.1**
A garden maze. From Batty Langley, Practical Geometry, applied to the Arts of Building, Surveying, Gardening, and Mensuration. London: W. & J. Innys, J. Osborn and T. Longman, B. Lintot *et al.*, 1726.

(b) *the terrible*, featuring overhanging rocks, shattered trees and burning huts
(c) *the exotic*, with Peruvian torch-thistles that 'arouse memories in a colonist or a traveler'
(d) *the grave*, offering a temple to philosophy
(e) *the majestic*, populated by obelisks and triumphal arches
(f) *the mysterious*, composed of moss and grottoes
(g) *the dreamy*, centered on a lake
(h) *the fantastic*, where the visitor, after meeting a wild boar, a hermit and several sepulchres, will be taken in an empty barque into a boudoir to be laved by waterspouts. (Flaubert 1971: 59–60).[1]

In mapping a site streaked with aporias and ruptures, Flaubert's categories capture the heterotopic logic that characterizes gardens and, by extension, landscape architecture in general. This lack of hierarchical order may be the reason why many writers, such as John Dixon Hunt, have felt that landscapes have not been as fully theorized as architecture (Hunt 1992). What I attempt to suggest below is that heterotopic thinking, far from being a handicap, is in fact one of the major resources of landscape architecture and something that also enriches architectural theory.

In the *Order of Things*, Michel Foucault appropriates the medical concept of 'heterotopia' to describe contradictory conceptual schemes which make it impossible to name this or that thing because they tangle common names and destroy syntax in advance, 'and not only the syntax with which we construct sentences but also that less apparent syntax which causes words and things to hold together'. In their madness, however, heterotopias also have the more positive function of exposing the equal relativity and arbitrariness of every alternative classification (Foucault 1970). In the essay 'Of other spaces', Foucault approaches the term 'heterotopia' from another angle. Instead of conceptual schemes, he claims to address the real physical environment. Heterotopias are described as real existing places that are 'formed in the very founding of society', as part of the presuppositions of social life. They are 'counter-sites, a kind of effectively enacted utopia in which the real sites, all the other real sites that can be found within the culture, are simultaneously represented, contested, and inverted'. Heterotopias must therefore be contrasted with the ordinary, dominating, real sites but also distinguished from utopias, places which also represent society in perfected or inverted form but do not actually exist. Furthermore, heterotopias have the curious property of being in relation to all the other sites in such a way as to suspect, neutralize or invert the set of relations that they happen to designate, reflect or mirror (Foucault 1993: 422).

## Heterotopias

Among his countless examples of heterotopias, Foucault includes the oriental garden because it represents another site in miniature. Other reasons could also be named for seeing gardens as heterotopias: they combine elements of nature and architecture only to undermine both. A familiar strategy of garden architects involves subverting the natural characteristics of things by forcing natural materials into shapes and states they would not normally take. In the garden of Villa Garzoni at Gollodi, for example, cypresses 'twisted and stretched, now jesting, now serious', taking the form of a tower, a ship, a pear, or an angel (Ponte 1991: 182).[2] The intersections of Platonic forms with ephemeral nature, animals with plants, eternity with transience, figurality with abstraction allow for no totalizing discourse. Fountains are another case in point: the most primitive of all natural elements, water, is harnessed to deny the most basic fact of earthly existence, gravity. Elaborate hydraulic contrivances at the Villa d'Este, Tivoli, Italy, even transform water into music, the most unnatural of all cultural creations of man. Every garden, no matter whether formal or informal, no matter how conventional or unspectacular, is ultimately an artifact and as man-made nature an implicit criticism of the role of the creative subject itself, man as *natura naturans* collapsed with *natura naturata*.

On another scale, too, the culture of gardens is one of inversions. By letting an Egyptian pyramid in the New Garden at Holy See in Potsdam serve as an ice cellar and housing the kitchen in a Roman temple, Carl von Gontard and Carl Gotthard Langhans were not only being sacrilegious but also revealed the fragility of architectural theory, of assignments of programme, of typological categories and of fictions about functional form. Such questioning results in a heterotopic universe that exists in another place and time than ours. It is not without justification that Yves Bonnefoy describes the Désert de Retz as a place of mysteries, 'at the antipodes of ordinary life' (Hunt 2008: 244).

Similarly, the garden of Stowe, with its thirty-eight monuments ranging from a temple of Bacchus to Gothic churches, inspired one visitor to state that 'the owner and the creator of this superb solitude have even had ruins, temples and ancient buildings built here, and *times as well as places* are brought together in the splendor that is more than human' (Baltrusaitis 1995: 212).

However, horticultural heterotopias are not limited to the assortments of oddities in Romantic gardens. Renaissance gardens occasionally question fundamental temporal divisions. The garden of the Villa Lante at Bagnaia, for example, can be read as a narrative that relates the carefully orchestrated transformation of trees into columns and the materialization of voids as columns. Simultaneously old and new, originary and derivative, the elements in the garden assert and deny their own rhetorical mode. Gardens can also undermine the authority of architecture. In Vaux-le-Vicomte, Le Nôtre achieved the astounding feat of destabilizing the very image of stability, the castle, which apparently changes location as new terraces are gradually revealed to the visitor.

## Contradictions

In such exercises in contradiction, garden designers have often followed the credo of Antoine de Ville who argued in 1666 that the gardener must work like God, 'who has ordered and arranged things quite contrary to their qualities in such proportion that they continue without destroying each other'. This way, gardens also reveal and celebrate the essential nature of the world, composed as it is 'of opposing parts, without which nothing can survive'. According to de Ville, trees and plants are jealous one of the other, as are rational and animal creatures, in the same way as the parts of a machine or of a system of fortification. This vital principle of 'contrariety' or 'discord' must not be suppressed but rather regulated by art (Vérin 1991: 136).

Contrariety may not impossibly be one of the essential characteristics of gardens. At the most basic level, this applies to the relationship between the garden and its surroundings. As a utopian ideal, paradise has always been represented by that which was scarce or absent, the other. In the south, the paradise was a grove or an oasis, while in the north, where forested landscapes existed, the exceptional scene was the absence of trees, as evidenced by the fact that the Anglo-Saxons used the word for meadow to signify paradise. Thus, as J. B. Jackson has pointed out, there are in the west two distinct garden traditions, that of the enclosure and the artificial forest and that of the clearing and architectural forays (Jackson 1980). If the former is reflected in the Edenic myths of Gilgamesh or Adam, the latter is portrayed in the creation myths of the north, such as the Finnish *Kalevala*.

The contrariety of garden architecture leads to and results from a certain multiplicity in the corresponding theory. In the third part of *Aesthetics*, G. W. F. Hegel declared that the art of garden design, like dancing, is an 'incomplete art'. The incompleteness of gardening derives from its effortless combination of various arts and sciences, such as botany, engineering, architecture, zoology, hydraulics and musical theory, resulting in the irregular multiplicity of mazes and bosques, bridges over stagnant water, surprises with Gothic chapels, temples, Chinese houses, hermitages and so on. Hegel compares such hybrid assemblages to hermaphrodites, cross-breeding and amphibians which only manifest the impotence of nature to maintain essential border lines. He insists that while gardening can deliver pleasant, graceful and commendable effects it will always fall short of actual perfection (Hegel 1955: 262).

Furthermore, it is not only that the sources of garden theory are heterogeneous, to say the least, but its objectives are often subversive and sometimes outright negative. In his moral essay dedicated to Lord Burlington, Pope advised English gardeners:

> He gains all points who pleasingly confounds,
> Surprises, varies, and conceals the bounds.
> (Pope 1826: 61)

However, not only was illusion the preferred medium of the horticulturalists but the principle was applied to the art of gardening itself: it was to vanish together with the natural boundaries of the site. Thus, the Duke of Harcourt opened his essay on the informal landscape with the paradoxical epigraph: *ars est celare artem*, 'art lies in the concealment of art' (Teyssot 1991: 363).[3]

Others took this doctrine, despite its classical origins, as involving not so much a paradox as an actual contradiction. Antoine Chrysostome Quatremère de Quincy argued that all arts, including architecture and the art of gardening, are based on Aristotelian imitation that does not mean a simple production of a formal likeness. Rather, 'to imitate . . . is to produce a resemblance of a thing, but in some other thing which becomes the image of it. It is precisely the fictitious and incomplete within each of the arts that constitutes them as arts' (Quatremère 1977: 120, 149). However, he observed that in the informal or English garden, the desired image of nature is simply nature itself – which is contradictory: 'The medium of this art is reality . . . Now, nothing can pretend to be at the same time reality and imitation' (Teyssot 1991: 369).

It seems clear enough that nothing can be at the same time reality and its image – unless it is the origin of the opposition and thus capable of occupying either side of the equation. Indeed, the garden can be seen as the *chora*, the unnameable container existing before categories such as truth/illusion and reality/imitation, or the separatrix between city and nature. Through the separation, it constitutes our physical environment either as 'nature' or as 'culture,' in both cases through opposition. Hence, gardens are not only constructed through the architectural act of building a wall but they themselves enact an analogous separation on a conceptual level.

In this sense, the garden is a fluid signifier without a signified, or in itself an abstract machine that can take on different roles in different contexts, constituting the perpetual other. Its matter (from Latin *mater* or 'mother', hence Mother Nature) is dependent on its *haecceitas*; its function is neither semiotic nor physical, neither expression nor content, but a pure function that informs both the expression-form of the discourse on architecture and the content-form of the city. Horace Walpole said that while Mahomet imagined an Elysium, William Kent created many of them (Baltrusaitis 1995: 207). In Deleuzean terms, every garden is a relocation of the Garden of Eden and a deterritorialization of nature but also its reterritorialization within the regime of signs, the necessary counterpart to architectural signifiers.

## Origins

For most theorists of architecture until recent times, nature was the mother of all architecture from ornament to urban design. Vitruvius derived the Ionic capital from the leaves of a tree and attempted to return the Doric order to the simple botany of Arcadia on the Peloponnesos. Analogously, Goethe, Schlegel, Coleridge and Chateaubriand likened the Gothic cathedral to a petrified forest, a conjecture spectacularly demonstrated to the Royal Society of Edinburgh by Sir James Hall who in 1792 planted sixteen trees in his garden in the form of a Latin cross pavilion. In merely six years, the branches (which had been tied together) had grown to form a characteristic Gothic vaulting. Even larger architectural ensembles have been traced back to a vegetative paradigm. Abbé Laugier not only subscribed to the Vitruvian notion of the primitive hut but also insisted that the city be modeled after a forest (Laugier 1977: 128).[4]

An alternative vision permeates the roughly contemporaneous *Encyclopédie* which defines the city as 'an enclosure surrounded by walls', containing several districts, streets, public squares and other buildings, like an enclosed garden (Diderot and d'Alembert 1751–1772). The idea may ultimately come from the Bible. As Abraham Cowley wrote, 'God the first garden made, and the first city, Cain,' after His example (Rabreau 1991: 305). The first of all gardens, the Biblical Paradise, was a site of perfect order, rich in gold, bdellium and onyx, bursting with fruit trees and

# Kari Jormakka

**Figure 1.2**
Willow cathedral. From Sir James Hall, "Essay on the Origin and Principles of Gothic Architecture." *Transactions of the Royal Society of Edinburgh*, Vol. IV, Part II, ii. Edinburgh: T. Cadell Jun., W. Davies, J. Dickson, E. Balfour, 1798, pp. 3–28.

animals, intersected by four rivers, bound by a square enclosure of jasper, husbanded by an immortal man. It was created by the gods, *elohim,* only to be lost by accident, a regrettable oversight, the unaccountable malice of a snake, the gullibility of a woman. But if we may believe Jean-Jacques Rousseau, Paradise already embodied the Fall. He famously writes:

> The first man who, having enclosed a piece of ground, bethought himself of saying 'This is mine', and found people simple enough to believe him, was the real founder of civil society. From how many crimes, wars, and

murders, from how many horrors and misfortunes might not any one have saved mankind, by pulling up the stakes, or filling up the ditch, and crying to his fellows: 'Beware of listening to this impostor; you are undone if you once forget that the earth belong to us all, and the earth itself to nobody.'

(Rousseau 1986: II, 17)

What Rousseau is implying is that Paradise marked the end of the natural state of humankind and the beginning of the unnatural establishment of private property. Surrounded by a wall, the twelve gates of which were guarded by Cherubim with flaming swords, Paradise was the first gated community separated from the natural world of Eden.

The word 'paradise' comes from the Avestan *pairi-daeza*, which means 'a walled enclosure'; the word 'garden' has a similar origin. It comes from the vulgar Latin phrase *hortus gardinus* or 'enclosed garden' where the adjective *\*gardinus*, 'enclosed', derives from prehistoric German *\*gardon* – whence also comes the English word 'yard'. *Gardon* comes from Indo-European *\*ghorto-* which also produced Latin *cohors*, 'court', and *hortus*, 'garden', and ultimately from *\*ghar* which means 'to seize, to enclose'. This is also the root for several architectural words, such as the Indoeuropean *grhár*, 'house', the Old Slavic *grad*, 'castle, city, garden', the Russian *gorod*, 'town', as in Petrograd, and the Czech *hrad*, 'castle'. A garden, thus understood, is the embodiment of the basic architectural gesture of spatial separation and the root of all later architecture.[5]

## Labyrinths

Illustrating one of his *Advertisements for Architecture* with an image of the garden labyrinth at Château de Villandry, Bernard Tschumi exclaims: 'behind every great city there is a garden'. The claim is not unreasonable: already the ancient Greeks traced the origin of architecture back to Daidalos' labyrinth. Unlike the Cretan prison for the Minotaur, however, a garden labyrinth does not celebrate the sordid origin of architecture as the container and veil of sin. Instead, according to Tschumi, it stands for the sensual experience of architecture (Tschumi 1994: 43–44). He claims that such sensual architectural reality cannot be experienced as an abstract object but rather as an immediate and concrete human activity, or a praxis with all its subjectivity. Unhistorically, perhaps, Tschumi further proposes that gardens are built exclusively for delight, pleasure and eroticism. The pleasure of gardens derives ultimately from their uselessness which suggests to him the general principle that 'the necessity of architecture may well be its non-necessity' (Tschumi 1994: 49–50, 86, 88).

Tschumi's theory owes a lot to Georges Bataille for whom a labyrinth represented excess (Bataille 1970). No prison, a labyrinth was in his view 'composed uniquely of openings, where one never knows whether they open to the inside or the outside, whether they are leaving or entering'. For Bataille, the labyrinth stands for counter-reason: like many sciences, it usually employs geometry – the art which Dante described as 'lily-white, unspotted by error and most certain' (Baxandall 1974: 124) – but it does this not to enlighten but rather to mislead, or to open unfamiliar ways. As a result, 'one can never see it in totality, nor can one express it. One is condemned to it and cannot go outside and see the whole' (Tschumi 1994: 43, 49). Indeed, the most important thread of Ariadne that Bataille discovers in the labyrinth is the rejection of totality.

## Processes

The most extreme formulation of the traditional idea of totality is the concept of an organic unity. Originating in the antiquity, organicism has dominated Western art theory at least from the Renaissance to the modernism of the early twentieth century. Alberti's celebrated definition of beauty as perfect harmony in which every change would be for the worse is merely a paraphrase of the *Poetics*, as is Mies van der Rohe's definition of structure (Alberti 1986: VI.2, I.9, II.3).[6] Despite its popularity, the notion of an organic whole may be contradictory in itself, as Jacques Derrida has attempted to show (Derrida 1987: 54).[7] Ironically enough, the faults of organicism are nowhere as evident as when it is applied to nature.

One of the central issues concerns the individuation of organisms. Instead of conceiving of a plant or an animal as a separate entity as Carl Linné would have done, ecologists usually focus on populations and secondly relate different elements in an ecosystem together. A squirrel could not exist without the plants it eats; these would not grow except for certain minerals, water and air, etc. The ecological point of view entails the concept of an ecological superorganism, a concept proposed by Frederic Clements as early as 1916. It is the idea that different ecosystems are in fact organisms in themselves, with particular emergent properties that their constituent parts, animals and plants, do not have.

The application of similar considerations to architecture and urbanism, as well as landscape design, opens up new avenues of thought. Just as no animal is self-sufficient but rather merely an element interacting with others in a larger ecosystem building is self-sufficient or independent. In cities, buildings tap into the infrastructure of water pipes and sewers, electricity lines and communications, streets etc.; even a hermit's hut mediates between the environment and its resident. In the past two or three decades, this perspective has engendered the hybrid discipline of landscape urbanism.

In the interpretation of Richard Weller, landscape urbanism wants to include within the purview of design all that is out there – buildings, infrastructure, green spaces, unused areas etc. – and conceptualize its object as a chaotic ecology, thereby bridging the divides between landscape design, landscape ecology and landscape planning (Weller 2008). Unlike the pictorial paradigm, the designer sees the environment as a process and works with indeterminate and catalytic strategies, as opposed to formal compositions and master plans.

James Corner explains: 'There is no end, no grand scheme for these agents of change, just a cumulative directionality toward further becoming' (Corner 1997: 81).[8] As Corner admits, such a vision of landscape urbanism is 'paradoxical and complex' (Corner 2003: 58). For architects and architecture theorists, it is a *pharmakon* in the double sense of medicine and poison. On the one hand, it reduces the tendency to totalize and reify the environment; on the other, it tends to naturalize social configurations and legitimize political processes with cloudy references to chaos theory, autopoiesis and the gospel of the virtual.

Still, the processual model harbours great potential in also suggesting a new ethic for designers. Most influential Western moralities – ranging from the Judeo-Christian religious tradition to Kantian deontology – accept the anthropocentric thesis that only human beings can be moral subjects as well as its corollary, the instrumentalization of nature and her subordination to human utility. By contrast, the ecological paradigm (far more pronounced in landscape architecture than in building or urban design) entails that natural organisms have intrinsic, rather than merely instrumental, value. Moreover, it insists upon an unusually long-term perspective which extends the designer's responsibility to future generations (cf. Vesilind 2002). In my view, it is only from such a position that a consistent model of sustainability could be developed. In the end, architecture and landscape are 'not about technology, but about a worldview and ethos in harmony with working principles and laws of earth's biotic mantle, mankind's nurturing natural home' (Callicott 1980).

## Definitions

The rejection of organic wholes and binary oppositions, the emphasis on processual thinking and the hint of non-anthropocentric ethics are some aspects of landscape architecture that continue to inspire architectural theory. At this stage it is fair to ask

**Figure 1.3**
The suit of a gardener. From Nicolas de Larmessin II, Les Costumes grotesques et les métiers. Paris, 1695.

whether there is anything that architectural theory might offer to the discipline and students of landscape architecture. Maybe it is best to begin with a brief analysis of the definition of landscape architecture proffered by the European Council of Landscape Architecture Schools.

The Council declares that 'landscape architecture is the discipline concerned with mankind's conscious shaping of his external environment'. From a theoretical point of view, the definition may not be entirely successful. To begin with, it is not quite clear what it means for a discipline to be 'concerned with' the environment. Can we say that ecology is concerned with the environment, both natural and man-made? Would it also be fair to say that real estate development is concerned with the environment, both natural and man-made? But is real estate a discipline in the sense intended here, or does the word 'discipline' carry the connotations of academia, for example? Is landscape architecture not possible without some connection to a university discipline?

The proposed definition may imply that landscape architecture is not a conventional academic discipline in pursuit of truth or knowledge alone. I take the expression 'conscious shaping' as a suggestion that landscape architecture, similarly to architecture, is not only seeking knowledge about how the environment has been shaped. Instead, landscape architects actively and consciously take part in shaping our environment. But why is it important to emphasize that this creative activity is conscious when the unconscious element is not shunned in many other creative domains?

It may be that the word 'shaping' is meant to refer to all effects of human action, or it could be that it only includes the effects of conscious human action. Here, however, we get into many problems. Does the expression 'conscious shaping' assume that we are conscious of what we are doing right now or that we are conscious of the consequences of our actions as well? Of course, it happens very seldom that a person could be aware of all the consequences of her actions; otherwise we would have been spared many environmental problems. If we insist that landscape architecture is only concerned with 'conscious' shaping, it seems that we are interested in the intentions of landscape architects more than the effects their actions may have on the environment. This reading would place landscape architecture squarely in the vicinity of the arts, understood in particular ways that put the author's concept before the audience's experience or the materiality of the object. Another curiosity is the word 'shaping'. Instead of conceptualizing the environment as a process or a system of functions, for example, the forms seem to be privileged, much like in the tradition of the visual arts.

Finally, I am not sure that this definition manages to separate the territories of landscape architecture and architecture. It may be that the formulation 'external environment' is meant to refer to that part of our surroundings that is not made of buildings or other properly architectural objects, whatever those might be. However, few architects would consent to limiting their activities to the design of buildings alone and would demand the right to work with the spaces between buildings as part of their architectural vision. In other words, we cannot define 'landscape' in 'landscape architecture' as that which is external to or outside of buildings. Instead, we might have to define architecture as that which is not landscape. Intuitively, the concept of landscape seems to contain some idea of naturalness; thus, architecture should imply artifactuality. To gain a better understanding, however, we should again consult the Council.

## Oppositions

ECLAS continues to characterize the discipline of landscape architecture by claiming that

> it involves planning, design and management of the landscape to create, maintain, protect and enhance places so as to be both functional, beautiful and sustainable (in every sense of the word), and appropriate to diverse human and ecological needs. . . The exceptionally wide-ranging nature of the landscape means that the subject area is one of unusual breadth, drawing on and integrating not just material from the two sides of the traditional divide between the creative arts and the natural sciences, but incorporating many aspects of the humanities and technology as well.
>
> (http://www.eclas.org)

As has been recognized for a long time, the discipline of landscape architecture is very complex indeed. It taps into knowledge from a wealth of scientific disciplines, including ecology, biology, botany, geology and so on, but it also has an artistic dimension to it. Here, one can sense a certain tension: as a normative activity, design is a question of right or wrong and not a question of true or false, as the sciences usually claim.

As combinations of art and science, architecture and landscape architecture are actually quite similar. When we try to understand the difference between architecture and landscape architecture, we soon arrive at another traditional opposition: the one between nature and culture. In some ways, it seems reasonable to say that architects articulate interior and exterior spaces by means of particular kinds of artifacts, namely buildings, and that landscape architects are designing those parts of the environment that are not enclosed within buildings and that are somehow not artifactual to the same degree. But it does not take long to realize how mistaken it would be to divide the concept of landscape architecture in two and oppose the 'cultural' activity of architecture (understood as the conscious shaping of the environment – or 'form') to the 'natural' element of landscape (understood as the raw material, or '*materia*' as in *mater natura*, for a cultural transformation). Already the etymology of the word reveals that landscape is not a natural given.

First recorded in 1598, the English word 'landscape' was borrowed from Dutch painters to suggest 'a picture depicting scenery on land'. Before the Renaissance painters established the new genre, the Dutch word *landschap* had meant simply 'region, tract of land'; Old English had the cognate *landscipe*. The Dutch suffix *–schap* or *–scap* corresponds to the English suffix *–ship*, denoting condition, character, office, skill, etc. as in 'township', for example. This suffix, in turn, comes from Old English –*scipe*, which is akin to 'shape'; in Dutch *schip*.[9] If we still want to follow the etymological trail a few steps further, we will see that the verb and the noun 'shape' go back to the Proto-Indoeuropean base *\*(s)kep-* 'to cut, to scrape, to hack' (whence also the word 'shave'). From this root we get in German the verb '*schaffen*' and in Swedish '*skapa*', both meaning 'to create'.[10] From this point of view, then, landscape is no less a cultural creation than a township. And one should add that 'land' did not originally denote a natural condition either. The modern word 'land' comes from Old English *land, lond,* signifying 'a definite portion of the ground, the home region of a person or a people, territory marked by political boundaries', from Protogermanic *\*landom* and ultimately from the Indoeuropean root *\*lendh-* 'land, heath'.

If we pull these etymological roots together, we begin to sense that the word 'landscape' carries not the connotations of nature but rather those of ownership, society and identity. It is a cultural construction with social and political connotations, and a construction that is also contingent on a particular historical technology, the art of painting.

Garden design in particular has followed pictorial models. Although Laugier urges us to 'look at the town as a forest', what he means by a 'forest' rather resembles a Baroque park. He namely continues to say:

> it needs a Le Nôtre to design the plan for it, someone who applies taste and intelligence so that there is at one and the same time order and fantasy, symmetry and variety, with roads here in the pattern of a star, there in that of a *patte d'oie,* with a featherlike arrangement in one place, fanlike in another, with parallel roads further away, and everywhere *carrefours* of different design and shape.
>
> (Laugier 1977: 128)

Soon after Laugier wrote his essay, the poet Thomas Gray wrote on his visit to Derwentwater in the English Lake District in 1769: 'I . . . saw in my glass a picture, that, if I could transmit to you, and fix it in all the softness of its living colours, would fairly sell for a thousand pounds' (Thacker 1983: 142). Like most other travelers of the time, Gray went to see the countryside carrying a Claude glass – a round or oval convex mirror with a dark backing. It would reflect the scenery

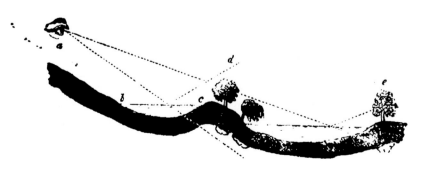

**Figure 1.4**: The optics of the landscape. From Loudon, J. C., *The Landscape Gardening and Landscape Architecture of the Late Humphry Repton, Esq.* London: Longman & Co. and A. C. Black, 1840.

with reduced brightness in a sepia tone, approximating a painting by Claude Lorrain. The design of gardens was even more directly indebted to artistic paradigms. Laugier remarks: 'The picturesque can be found in the embroidering of a parterre as much as in the composition of a painting' (Laugier 1977: 128). Whether French or English, Persian or picturesque, gardens used to be designed in a pictorial fashion as autonomous objects for visual apprehension, or as organic unities in the Aristotelian sense.[11]

No less than a building or a city, a garden is a technological artifact, 'a deviation from nature,' in the words of Sir Joshua Reynolds.[12] In Persian paradises or hunting gardens, the wall protected the flora and the fauna from the erosive effects of farming as well as the merciless hooves of ungulate herds. Without technology, the unnaturally heightened garden landscape would disappear; with the wall, nature survived, if only in effigy or simulacrum. Later manifestations of paradise throughout the Mediterranean were even more graphically artifactual, as the climate necessitated elaborate technologies of irrigation and cultivation. In fact, the four rivers of Eden may simply represent the idea of artificial irrigation.[13]

## Theses

These observations suggest that landscape architecture is not very different from architecture either as an intellectual discipline or as a practical activity. Both architecture and landscape architecture manipulate culturally constructed materials and follow paradigms that in some cases have been appropriated from other fields, in particular from the visual arts. Moreover, both combine research and design, or reflection and creation, with the goal of generating value. Intrinsic values depend on the physical properties of the created objects, such as buildings or landscapes; extrinsic values relate to the objects' 'perceived affordances,' i.e. perceptions of what the object is able to do, as well as relevant metaphors, archetypes, myths, images and many other aspects that affect how we experience the objects. Once we understand that the design element in landscape architecture complicates the research agenda, we have to ask how the criteria of value are determined.

Even though landscape architecture often addresses the natural environment, values cannot be derived from nature. While Aristotle believes in the principle of *ars imitatur naturam* to the extent that he calls the essence of an entity its 'nature', one should not equate the natural with the good. During the Enlightenment, Denis Diderot defined the good as that which comes from nature as opposed to anything devised by man who has been perverted by the wretched customs of society. To skeptics, he explained that

> water, air, earth and fire, everything in nature is good. Even the gale, at the end of autumn, which rocks the forests, beats the trees together, and snaps and separates the dead branches; even the tempest, which lashes the waters of the sea and purifies them; even the volcano, casting a flood of blazing lava from its gaping side, and throwing high into the air the cleansing vapour.
>
> (Thacker 1983: 88)

A little later, however, Giacomo Leopardi was to exclaim in despair: 'Now despise yourself, Nature, you brute force that furtively ordains universal doom and the infinite futility of all existence.'[14] What is natural is sometimes valued as good and sometimes rejected as embarrassing or inferior. But the real problem is that 'nature' is a natural given but rather a contingent construction. Let me give an example.

## Models

In 1681, looking out the window of his quiet study to the broad meadows and blooming fields, the open, smiling pastures, the cooling shadow of the woods, the fresh springs and streams and ponds in which one could swim, Thomas Burnet realized the landscape he saw was nothing but the disheartening wreck of Paradise, a 'Picture of a great Ruin, . . . a World lying in its Rubbish' (Burnet 1726: 148).

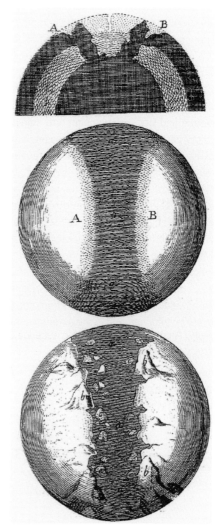

**Figure 1.5**
The formation of the continents. From Thomas Burnet, *The Sacred Theory of The Earth Containing an Account of the Original of the Earth*. London: J. Hooke, 1726.

Comparing the earth to a 'statue, temple, or any work of art', he was disappointed to note that 'there appearing nothing of Order, or any regular Design in its Parts, it seems reasonable to believe that it was not the Work of Nature, according to her first Intention, or according to the first Model that was drawn in Measure and Proportion by the Line and by the Plummet, but a secondary Work, and the best that could be made of broken Materials' (Burnet 1726: 44, 171, 173–174). In his *Sacred Theory of the Earth*, Burnet reasoned that God in His benevolence must have created our world, the 'Darling and Favourite of Heaven,' as a smooth, regular and uniform sphere, endowing the global paradise with the beauty of youth and blooming nature. On the day of the Deluge, the thin crust of the earth must have broken and the waters underground burst free, marring the globe forever with jagged mountains, scarring the perfect surface of the antediluvian orb with deep, uneven blemishes of rivers and seas, and hollowing out 'Holes and Caverns, and strange subterraneous Passages', some 'filled with smoak and fire, some with water, and some with vapours and mouldy Air' (Burnet 1726: 147).

To us, Burnet's claim that the earth does not live up to the standards of nature may sound overzealous but similar conclusions are implicit in many philosophies. The Aristotelian conception of nature, for example, explicitly admits to different degrees of naturalness in nature. Not surprisingly, man is for Aristotle 'the animal of all animals', more natural than other creatures (Aristotle 1912–1931: *Part.An.* 687a22, 686a28, 656a7; *Pol.* 1253a31; *De Inc.* 706a20). This makes man the measure of nature for 'that which is first, most simple and perfect in any gender is the measure of everything that falls in that gender', and man is built to the same specifications as the universe (Aristotle 1912–1931: *Hist. An.* .494a26; *Part.An.* 656a7; *De Inc.* 706b9f). On the other hand, Aristotle also declared that man is the least earthly of all animals, capable of surpassing nature and making it appear as a kind of imperfect artist (Aristotle 1912–1931: *De Gen. An.* 745b17; *Phys.* I99a15; *Pol.* 1337a1; *Protr.*, 9, 44). Indeed, the higher the animal, the more it is characterized by form and less by imperfect, amorphous matter. In architectural theory, the classical notion of man as the origin and measure of all things has been used for centuries to justify all aspects of architectural design governing proportion, shape and organization in plan and section.

Another Aristotelian commonplace states that 'nature does nothing in vain'. Aristotle insists that 'the absence of haphazard and conduciveness of everything to an end are to be found in Nature's works in the highest degree, and the resultant end of her generations and combinations is a form of the beautiful' (Aristotle 1912–1931: *De Caelo* 271a35; *De Part.An.*645a23–26; 639bl9). This notion of nature as a perfectly functioning system formed the basis for the 'Design Argument', used by theological thinkers to prove that the world did not originate by chance but was created by an intelligent being. Also Burnet maintained that every body in the sublunary world is 'perfect and admirable in its kind,' criticizing Epicureans and other believers in aleatorism:

> In the composition of a perfect Animal, there are four several frames or Compages joyn'd together, The Natural, Vital, Animal, and Genital; Let

> them examine any one of these apart, and try if they can find any thing defective or superfluous, or any way inept, for matter or form. Let them view the whole Compages of the Bones, and especially the admirable construction, texture and disposition of the Muscles, which are joyn'd with them for moving the Body, or its parts. Let them take an account of the little Pipes and Conduits for the Juices and the Liquors, or their form and distribution; Or let them take any single Organ to examine, as the Eye, or the Ear, the Hand or the Heart; In each of these they may discover such arguments of Wisdom, and of Art, as will either convince them, or confound them. . .
>
> (Burnet 1726: 73, 410-411)

Contrary to Burnet's thesis, however, a study of any organism will quickly reveal parts for which no adequate explanation can be given. In the human body, for example, the appendix is generally understood as serving no function but it is also quite difficult to determine what the function of the pineal gland is, unless one accepts Descartes' thesis that it is the seat of the soul. Moreover, many forms of cancer, genetically inherited illnesses and defects, and the phenomenon of aging in general suggest that the design of the human body may not be flawless.

The difficulties in providing a complete teleological explanation of any organism (or ecosystem) lend support to the hypothesis that there is no intention behind nature and that there is therefore no reason to expect perfection, closure or finality. Yet, despite the efforts of a century of Darwinism to present the evolution of species as a process involving random mutation of genes and a selective process which only weeds out those mutations that fatally affect populations, organicist writers continue to view evolution as a teleological process of improvement whereby organisms achieve perfect adaptation to their environment and some ecologists to take it as axiomatic that all natural ecosystems are in perfect homeostasis, only upset by the thoughtless interventions of man. The idea of nature as a perfectly functioning, organic and ecological complex can, however, not have been derived from empirical observation but rather from extra-scientific sources. The conception of perfect unity was, indeed, predicated on works of art and books before it was applied to things of nature.

The theological Design Argument rests on the anthropomorphic assumption that the creations of God, including the earth and all living things, were designed in the way human beings produce artifacts. Burnet, for example, made reference to works of art and architecture in specifying what God's original intentions must have been, determining the characteristics of the divine mind after the model of human intelligence and only increasing the values of the parameters. Hence, he insisted that in our world,

> there is more of Art, Counsel and Wisdom shewn, than in all the works of men taken together, or than in all our *Artificial* World. In the construction of the Body of an Animal, there is more of thought and contrivance, more of exquisite invention, and fit disposition of parts, than is in all the

> Temples, Palaces, Ships, Theaters, or any other pieces of Architecture the World ever yet see... Let them take any piece of Humane Art, or any Machine fram'd by the wit of Man, and compare it with the body of an Animal, either for diversity and multiplicity of workmanship, or curiosity in the minute parts, or just connexion and dependance of one thing upon another, or fit subserviency to the ends propos'd, of life, motion, use and ornament to the Creature, and ... in all these respects they find it superior to any work of Humane production...
>
> (Burnet 1726: 410)

In Burnet's conception, then, natural organisms are superior to but not essentially different from human artifacts; nature is seen as the work of a supremely intelligent human being.

Likewise, the earth was judged by Burnet to be merely a ruin because its appearance did not agree with the aesthetics he saw exemplified in the Bible, the other work by 'the Author of Nature' (Burnet 1726: 401). Centuries before nature was described in organicist terms, the Bible was understood as a text where there are no contingencies, where every element is connected to every other element in perfect harmony and where every detail is charged with precious significance.[15] It is not a great conceptual leap to ascribe the same properties to the 'great book which is always open before our eyes (I call it universe)', to use Galileo's words, or to follow Spinoza's dictum *deus sive natura* and turn the author of the book of nature into an ecological superorganism. Ultimately, Biblical exegesis only radicalized the postulates of classical poetics, such as Aristotle's idea that any change in a good work of art would be for the worse, for if Aristotle was correct about the creations of mortal minds, the same must be true to an even higher degree of God's word (Aristotle 1912–1931: *Poet.* 1451a32-35; *N. E.* 1106b10–15).

To sum up, organicist theories read natural organisms in the same way as exegetes read the Bible, postulating universal determination and optimal adaptation as rhetorical axioms of interpretation. This strategy does not leave room for empirical confirmation or refutation. Distinguishing between philosophy and oratory, Burnet explained that 'Orators ... represent Nature with all her graces and ornaments, and if there be anything which is not capable of that, they dissemble it, or pass it over slightly' (Burnet 1726: 147).[16] This is, unfortunately, just what he himself did, along with many architecture theorists from Vitruvius to the present day.

## Constructions

Oscar Wilde explained both 'the secret of Nature's charm, as well as the explanation of Nature's weakness' by reference to aesthetic theories. In the *Decay of Lying*, he suggests that the only effects that she (nature) can show us are effects that we have already seen through poetry, or in paintings.

> Where, if not from the Impressionists, do we get those wonderful brown fogs that come creeping down our streets, blurring the gas lamps and changing the houses into monstrous shadows? ... The extraordinary

> change that has taken place in the climate of London during the last ten years is entirely due to this particular school of art . . . At present, people see fogs, not because there are fogs, but because poets and painters have taught them the mysterious loveliness of such effects.
>
> (Wilde 1923: 47–48, 63)

Once art has established a certain aesthetic effect, parallel phenomena will be foregrounded in the infinite variety of nature. Wilde cautions, however, that such effects do not always have to form a positive value. 'Where she (Nature) used to give us Corots and Daubignys, she gives us now exquisite Monets and entrancing Pisaros (sic)'; however, in imitating art nature can also commit an error and deliver 'on one day a doubtful Cuyp, and on another a more than questionable Rousseau' (Wilde 1923: 48–49). It is fair to conclude that the constitution of nature reflects perceptual habits more closely than aesthetic or other values.

The idea that the world is constructed and not discovered was formulated with more precision by Eddy M. Zemach (1986). To refute the commonplace and Fregean view of language that a name corresponds to an individual or that our vocabulary reflects the individuation of things, Zemach discusses the concept of gelding. Speaking of bovine animals, the English language distinguishes between cows and bulls, oxen and steer: are they different things or the same? Gelding changes the identity of a bull but not that of a cat, for example, for a castrated cat performs the same functions as before but an ox (or a steer) is valuable for other reasons than a bull (an uncastrated bovine) or a cow. Hence, we give them different names and different identities. In a restaurant the functions of transportation and procreation cease and other interests take their place. Consequently, all meat of a *Bos taurus* is called beef, independently of gender or gelding. The concept of 'beef' is relevant for nutritional or culinary interests while 'steer' and 'ox' figure in the general economy of a farm. The identities of things are contingent upon the economical practices of a given linguistic community. Pliny tells us that when Hannibal brought the first elephants to Italy, they were called (Lucanian) 'oxen' whereas in Africa, elephants were called 'bears', a category that also included lions and other dangerous animals (Pliny 1991: VIII.vi.16).

In general, the aspects of nature that do not affect our interests remain in the background and we take notice only of qualities that matter to us. In other words, the individuation of things is value-bound in that objects of thought are constituted in relation to our interests. This means that nature is not a first-level concept capable of grounding a theory of architecture or landscape architecture but merely an after-effect, a secondary concept constituted in a discourse; the relevant questions concerning alternative constitutions of nature have to do with the interests that guide the discourse.

Furthermore, as Donna Haraway has argued, natural sciences do not necessarily get closer to an objective nature to be materially and symbolically appropriated, but are themselves social activities, inextricably bound within the processes that give them birth (Haraway 1989: 12, 257). Now, if she is right and if there can be no pre-discursive encounter with nature, then the natural sciences have no particular privilege over other discourses on nature. We may go as far as to suggest that there may well be no object that would deserve the ambitious name of 'nature', suggesting something that is born, not made.

If the concept of nature is merely a contingent construction then it cannot serve as a foundation: we cannot follow nature because we have to first construct it. In this situation, both architects and landscape architects have to face a few serious questions: How do we decide what to do? How do we know if something is good or bad? What is our expertise?

## Experts

In *The Doctor's Dilemma*, George Bernard Shaw declared that 'all professions are conspiracies against the laity' (Shaw 1927: vi, 32). Because only other colleagues can judge whether a doctor has committed malpractice, the profession is always able to hide its shortcomings. The same may apply to both architecture and landscape architecture.

**Figure 1.6**
Diverse decorations. From Louis-Eustache Audot, *Traité de la composition et de l'ornement des jardins avec cent soixante et une planches.* Paris: Audot, 1839.

In the 1950s, when Jørn Utzon was working in the office of Alvar Aalto, he asked the elder colleague how one knows one is on the right track in designing. Aalto replied: 'you are right when your colleagues, other architects, like your architecture' (Suhonen 1982: 16). The same principle is actually followed in most architectural competitions, where the decisive members of the jury are usually architects. Moreover, the architectural profession often applies complicated language that the uninitiated cannot understand. Of course, brain surgeons use Latin and nuclear physicists speak in Joycean words, such as 'quarks', because everyday language is too imprecise to deal with the issues at hand. The problem is whether or not architecture involves anything like the expertise that nuclear physicists or brain surgeons can demonstrate. Certainly, architects know a lot about materials, structures and building codes, and possess a large library of possible solutions to particular functional programs. However, the debates about architecture seldom concern such issues where the expertise of architects is readily accepted by the public. Beyond such technical, material, structural and organizational issues, there are many questions in architecture that cannot be easily justified by a reference to rationalist criteria. A crucial question to both architects and landscape architects is whether there can be expertise in aesthetic matters, as well.

A classic saying claims that *de gustibus non est disputandum:* there is no disputing about taste. And yet, people clearly do want to debate aesthetic matters even though nobody has ever been able to explicate and fix universally valid aesthetic rules. In response, many architects suppose that architecture is a universal but ineffable quality: it cannot be quantified but it can be objectively discerned by experts and connoisseurs. Moreover, many architects claim to be able to evaluate architecture on the basis of their trained aesthetic sensibility. Even though this particular aesthetic expertise cannot be verbalized or quantified, architects claim it to be real. As evidence they maintain that people always resist great advances in the sciences and the arts first, and only much later praise them.

However, it is very hard to find convincing evidence for the popular claim that the taste of the masses would eventually follow the taste of the avant-garde. To give just one example to the contrary, the harmonic and melodic structure of popular music in the West have remained for a long time similar to those of mid-nineteenth century Romanticism, rather than followed the innovations of impressionistic, dodecaphonic, serial or concrete music by the avant-garde. It is rather the case that contemporary concert music borrows and varies aspects of popular music (e.g. those that do not come from the Western avant-garde tradition but from African or Oriental popular or folk music). In fact, while critics normally condemn banalities oftentimes artists feel stimulated by them.

Are ugliness, kitsch or banality then degenerated forms of beauty, good taste and art or is it rather the other way: is art a perversion of the banal? Ernst Gombrich has demonstrated time and again that art is not born out of inspiration or nature but out of earlier art. Still, the development of art is also controlled by its unspeakable opposites, kitsch and invisible banalities. In architecture, the careers of Peter Behrens, Henry van de Velde and Josef Hoffmann make it clear how much artists fear the banalization of their style. When a style becomes generally accepted

it is also considered vulgar and the avant-garde has to create something radically different not to lose its status. Historically it is often true that kitsch and the anonymous popular culture follow good taste, as personified in the avant-garde – but ontologically the good taste is the parasite: it is a distinction strategy which always has to distance itself from its artless reflection, its banal *Doppelgänger*. In other words, Kitsch is the invisible precondition of art.

## Distinctions

If there is no independent evidence to prove that the taste of experts would be more correct than any other taste, there is no reason to assume that the ineffable quality that experts or members of the avant-garde recognize is objective or universally valid to all people. Rather, it appears that the taste of experts is a means of social distinction by a sub-culture that is systematically opposed to the taste of laymen. Pierre Bourdieu's theory of social fields generalizes this notion and focuses on how cultural activities create values and distinctions on a symbolic level between different groups in society. He argues that aesthetic taste is not universal but it is not arbitrary or irreducibly subjective either. Rather, taste is situational: it is determined by social groupings. To gain cultural capital in the field of architecture, one has to internalize a taste that differs profoundly from the taste of the masses.

Generally, people internalize a taste that improves their standing in what they consider their reference group. Partly because of the educational system (isolation in studio work, oral presentations, intuitive criticism, special jargon), architects tend to have their colleagues as their reference group. Friendships are sought among colleagues; as a result, their taste tends to become rather uniform and specialized. Architects form a small but strongly defined subculture with its own language, taste and values. Some of the values of this subculture need to be different from those of other sections of society. This is more than a problem of language. Ludwig Wittgenstein claims that 'if a lion could speak, we would not understand him', because all meanings are related to the form of life and not to the words themselves or language as an isolated system. This suggests that it would be difficult or impossible for an architect even to understand, let alone design or make value judgements for people from other subcultures.

## Concluding thoughts

Architectural theory is a mixed bag. The term has been used to refer to at least three radically different kinds of writing – and, occasionally, non-verbal projects and buildings. I would characterize these three as design theory, criticism, and the philosophy of architecture. Much of Le Corbusier's literary output can be called design theory: he attempts to formulate new concepts in order to set rules and goals for design. Theory is used as criticism when we attempt to understand what Le Corbusier really has done by comparing his buildings with his writings, or the writings of other architects. Colin Rowe's observations about the resemblance of Corbusian villas to Palladian ones would fall in this category. Finally, architectural theory as the philosophy of architecture investigates the possibility of formulating design theories (the first kind of theory) as well as the relationship between a theory and a building or the intentions of the author and the work (the second type of theory) but there are many other questions as well. It is the third kind of theory that I find to be most useful – but a caveat needs to be added.

In his *Preface to the Works of Shakespeare,* Dr Samuel Johnson cautions his readers not to expect that 'they would in general, by learning criticism, become more useful, happier, or wiser' (Johnson 2008: 453). *Mutatis mutandis* the same holds for architectural theory: it will hardly make a landscape architect happier. Nor will it reveal fundamental truths. As David Lewis has suggested, metaphysical speculation does not so much establish truths as it tries to fix their relative prices (Lewis 1986: 4, 5 et passim).

The definition of truth as *the* goal of philosophical activity was undermined already by Friedrich Nietzsche for whom thinking was a practice of demystification, unmasking and genealogy, ultimately aiming at emancipation. And yet, Nietzsche, according to one of his later admirers, Martin Heidegger, never broke free from what the Greeks called *ousia*, often translated as 'nature' or 'substance' but literally meaning 'the household' or the belongings of a household (Heidegger 2003: 221; McCumber 1999: 21–70). To liberate thought from this residual domesticity may be the real task of theory. Unfortunately, for this demanding purpose, it neither has a method of its own nor a stable object to interrogate. As regards landscape architecture, these lacks may turn out to be a resource, as they indicate fractures in the foundation of the concept and imply a lack of established ousidic structure.

What students of landscape architecture can learn from architectural theory is first to appreciate that everything they are working with is a contingent construction that could be made otherwise if need arises. Second, these constructions are usually neither radically subjective nor random. Instead, they are collective conceptualizations that pertain to particular interests and specific discourses. Third, it is within these discursive contexts with regard to the relevant interests that we can understand the values experts and laymen attribute to the designs. Fourth, different contexts produce different evaluations. While the design of public space is ultimately a matter of managing a conflict of interests, there may be conflicts that cannot be rationally resolved. Fifth, the education of an architect or a landscape architect involves not just the transmission of knowledge and skills but the initiation of the student into a particular value system which – if the Bourdivin view is right – may even be necessarily opposed to that of large sectors of society. Sixth, the disciplinary structures that determine what we understand as a design, what we recognize as a problem to be solved through design, or what we consider to be a good solution, may need to be critically examined. Embedded in design theories, paradigmatic examples, representational methods, work habits, legal and financial structures etc. are ontological assumptions that may limit the range of design options without ever reaching the level of consciousness. Opening up some of these restricted territories may be the best service theory can offer to landscape architecture.

## Notes

1 The inspiration is Pierre Boitard's *Manuel de l'architecte des jardins, ou, L'Art de les composer et de les décorer.* Paris: Roret, 1834.
2 In a poem called *Le Pompe di Collodi* of 1652, Francesco Sbarra describes the Garden of Villa Garzoni at Collodi with the following verses: *In mille guise so contorce e stende / Il bel cipresso hora scherzande, hor grave, / Hor esprime una Torre, hor una Nave, / Hor di Pera, Hor d'angle sembianza prende.* (In a thousand guises the fair cypress twists and stretches, now jesting, now serious. Now it forms a tower, now a ship, now it takes the semblance of a pear, now that of an angel.)
3 The attribution of the phrase *ars est celare artem* to Ovid is not correct. Still, in *Ars amatoria* 2.313, he does say something similar: *'Si latet ars, prodest; adfen deprensa pudorem'* or 'artifice is a fine thing when it's not perceived; once it's discovered, discomfiture follows'. Moreover, in the *Metamorphoses* 10.252, talking about Pygmalion's statue, Ovid comments: *ars adeo latet arte sua*, 'so much does art lie hid by its own artifice'.
4 Laugier writes: 'One must look at the town as a forest. The streets of the one are the roads of the other, and must be cut through in the same way. That which forms the essential beauty of a park is the multiplicity of roads, their width and their alignment.'
5 To be more precise, it should be added that the Biblical account of Paradise allows us to identify both the walled garden and the forest as the origin of architecture, for in the garden of Eden, God planted a number of trees. It is not without interest that the Hebrew word *eden* might be related to the Assyrian *idinu*, 'plateau', or 'field'; to Assyrian-Babylonian *edin*, 'steppe', 'field'; or Sumerian *edinu*, 'desert' or 'steppe'. In a desert landscape, planted trees would stand out as a special condition. The Hebrew word for tree, *ez*, is related to the Sumerian *es* which means temple; in Sumer temples were in fact often called 'trees'. Ishme Dagan, the king of Isin, called the Temple in Lagash the greatest tree in the land of Sumer; King Gudea describes the Temple of Ningirsu in Lagash as a cosmic tree (Hengge 1993: 61–62, 68–69).
6 Mies explains that 'by structure we have a philosophical idea. The structure is the whole, from top to bottom, to the last detail with the same ideas' (Carter 1961: 97).
7 Derrida focuses on the concept of *parergon* in Immanuel Kant's discussion of ornaments in art. *Parergon* is something that exists outside the *ergon*, the work. Kant wrote: 'even what is called ornamentation (*parerga*), i.e. what is only an adjunct, and not an intrinsic constituent in the complete representation of an object, in augmenting the delight of taste does so solely by means of its form. Thus it is with the frames of pictures or the drapery on statues, or the colonnades of palaces.' (Kant 1961: 1, I, 1, §14.) Asking how the frame of a painting actually is related to the painting itself, Derrida points out that if we follow Aristotle in defining a work of art as a whole with a beginning, middle and end, we must then determine how the beginning and the end exist. Their boundaries are obviously marked by something that exists beyond the work. In other words, the unity is defined and made into itself by something it is not, its outside or its other. In this epistemological sense, the *parergon* is the outside that constitutes the inside as the inside. Without the boundary which is the *parergon* we could not recognize the essential (the artwork) as essential. In this sense, the secondary is actually the formal cause for the primary, to use Aristotelian terms. However, the formal cause equals the essence which means that the secondary is essential and therefore primary. Hence, the very concept of an organic unity is self-contradictory.
8 Corner prefaces the quoted statement by claiming that 'The process of which ecology and creativity speak are fundamental to the work of landscape architecture. Whether biological or imaginative, evolutionary or metaphorical, such processes are active, dynamic, and complex, each tending toward the increased differentiation, freedom, and richness of a diversely interacting whole.'
9 The word *scipe*, 'shape, form, mode,' comes from *scepp-an*, 'to shape, make' and is cognate with Icelandic *skapr*, Swedish *skap* or German -schaft, as in Icelandic vin-skapr, Swedisch frändskap or German Freundschaft, i. e. 'friendship'.

10  In more detail, the noun 'shape' comes from Old English *ġesċeap* 'creation, creature, form, pudendum, decree, destiny,' corresponding to Old Saxon *giskapu* 'creatures, decrees,' and Old Norse *skap* 'state, mood condition, pl. fate'. 'Shapely' in the sense of 'well-formed' is recorded from 1382, related to the sense of 'a woman's private parts'. The verb 'shape' is similar, going back to OE *sceppan, scyppan;* corresponding to ON *skepja,* Goth *-skapjan* 'to make'; Protogermanic *\*skapjanan* 'create, ordain', from PIE base *\*(s)kep-* 'to cut, to scrape, to hack'. O.E. *scieppan* survived into M.E. as *shippen,* but *shape* emerged as a regular verb (with pt. *shaped*) in the sixteenth century.

11  Aristotle distinguishes between groups of things which are mere aggregates or heaps and those which are organic wholes. Mere aggregates possess a certain amount of unity because their parts are juxtaposed whereas real wholes have parts which are unified, not only by contiguity but also by form. Moreover, every element is necessary to the whole. As Aristotle explains in the *Poetics,* 'if the presence or the absence of a thing produces no distinguishable difference, that thing is not a part of the whole.' *Poet,* 1451a35. Wholes are the end (*telos* and essence) of their parts; they are both logically and temporally prior to their parts. *Met.* 1016b7–17 1023b26–1024al0. In Aristotelian essentialism, all value comes from the unfolding of the essence. That which does not come from the essence must be against the essence, and must therefore be unacceptable. *De Anima* 407a35–407b1. Hence, organic unity embodies the highest value.

12  Not only the city but also the garden could be defined, in the words of Le Corbusier, as 'the grip of man on nature. It is a human operation directed against nature' (Le Corbusier 1971: I).

13  In the Mesopotamian creation myth 'enuma elish,' the gods create men to water the four regions of the world. In the Palace of Mari, a relief was found showing two goddesses pouring water from vases and creating four rivers.

14  Leopardi, Giacomo, 'A se stesso.' The translation is Jormakka's; the original is as follows: 'Omai disprezza/Te, la natura, il brutto/Poter che, ascoso, a comun danno impera/ E l'infinita vanità del tutto.' Leopardi 1966, 280.

15  However, the genealogy of the notion of the Bible as a perfect unity can be traced from Philo of Alexandria, who maintained that nothing is superfluous in the Scriptures and that every detail is charged with precious significance, to Irenaeus who affirmed that every part of the Scripture is in its perfect place, and harmonizes with the rest; no portion is without significance, nothing is included by accident or without intention. The Calvinists John Henry Heidegger of Zürich, François Turretin of Geneva and Lucas Gernler of Basel took the notion of a perfect text to extremes. In the sixteenth century, some Renaissance humanists had discovered inconsistencies in the *Septuaginta*. To explain these, Johann Reuchlin suggested in 1513 that the Seventy translators had used a Hebrew text without vowels. In the *Arcanum punctationis revelatum,* published anonymously in 1624, and the *Critica sacra* of 1650, Louis Cappel argued that the vowel points did not date from Old Testament times but were a Medieaval invention by Masoretic Jewish grammarians. Jesuits conceded this point and further argued that the Masoretic Jews had added the points out of hatred toward Christians. While the Calvinists John Henry Heidegger of Zürich, François Turretin of Geneva and Lucas Gernler of Basel accepted the Medieaval origin of the punctuation system they still claimed divine authority and inerrancy even for the diacritical points in their notorious *Formula consensus helvetica* of 1675. The reader may find the Calvinist argument absurd but it would be a stranger madness to question the divine authority of the punctuation. Turretin pointed out in 1688 in an argument for the perfectness of the Hebrew Bible and against the Saumur doctrines of Cappel and Moïse Amyraut that if we are allowed to change anything at all in the Biblical text, 'then the determination of the authentic reading will be the work of reason and human judgment, not of the Holy Spirit. Human reason will be enthroned, and, in the Socinian manner, regarded as norm and principle of faith.' (Wood 1967: 25, 28, 27, 29, 30).

16  Describing his own method, Burnet continues: 'But Philosophers view Nature with a more impartial eye, and without favour or prejudice give a just and free account, how they find all the parts of the Universe, some more, some less perfect.'

## Bibliography

Alberti, Leon Battista, *The Ten Books of Architecture*. Tr. Giacomo Leoni, New York: Dover, 1986.
Aristotle, *The Works of Aristotle*. Translated into English under the editorship of D. Ross (and J. A. Smith). Oxford: Clarendon Press, 1912–1931.
Baltrusaitis, Jurgis, 'Jardins et pays d'illusion'. *Aberrations. Les perspectives dépravées*. Paris: Flammarion, 1995.
Bataille, G. 'Le labyrinthe' *Ouvres complètes,* vol. 1, ed. Denis Hollier. Paris: Gallimard, 1970, 433–41.
Baxandall, Michael, *Painting and Experience in Fifteenth Century Italy*. Oxford: Oxford University Press, 1974.
Burnet, Thomas, *The Sacred Theory of The Earth Containing an Account of the Original of the Earth*. London: J. Hooke, 1726.
Callicott, J. Baird, 'The Search for an Environmental Ethic'. In *Matters of Life and Death*. Ed. Tom Regan. New York: MacGraw-Hill Inc., 1980.
Carter, Peter, 'Mies van der Rohe. An appreciation on the occasion this month, of his 75th birthday'. *Architectural Design*. London, 1961, pp. 95–121.
Corner, James, 'Ecology and Landscape as Agents of Creativity'. In *Ecological Design and Planning*, ed. G. Thompson and F. John Steiner. New York: Wiley & Sons, 1997, pp. 80–108.
Corner, James, 'Landscape Urbanism'. In *Landscape Urbanism: A Manual for the Machinic Landscape*. Ed. Mostafavi Mohsen and Ciro Najle. London: AA Publications, 2003, pp. 58–63.
Derrida, Jacques, *The Truth in Painting*. Tr. Geoff Bennington and Ian McLeod., Chicago and London: The University of Chicago Press, 1987.
Diderot, D. and d'Alambert, J. *Encyclopédie ou Dictionnaire raisonné des sciences, des arts et des métiers, par une Société de Gens de lettres*. Paris, 1751–1772.
Flaubert, Gustave, *Bouvard and Pecuchet*, Tr. T.W. Earp and G.W. Stonier. New York: New Directions, 1971.
Foucault, Michel, 'Of Other Spaces: Utopias and Heterotopias'. In *Architecture Culture 1943–1968. A Documentary Anthology*. Ed. Joan Ockman. New York: Rizzoli/Columbia University, 1993, pp. 419–426.
Foucault, Michel, *The Order of Things: An Archaeology of the Human Sciences*. Tr. Alan Sheridan-Smith. New York: Random House, 1970.
Haraway, Donna, *Primate Visions: Gender, Race and Nature in the World of Modern Science*. London: Routledge, 1989.
Hegel, Georg Wilhelm Friedrich, *Ästhetik*. Berlin: Aufbau-Verlag, 1955.
Heidegger, Martin, 'Über Nietzsches Wort: Gott ist tod'. *Holzwege*. Frankfurt: Klostermann, 2003.
Hengge, Paul, *Es steht in der Bibel. Die Bibelkorrektur*. Bielefeld: Verlag Wissenschaft und Politik, 1993.
Hunt, John Dixon, 'Folly in the Garden'. *The Hopkins Review*. 1(2), 2008, pp. 227–250.
Hunt, John Dixon, *Gardens and the Picturesque: Studies in the History of Landscape Architecture*. Cambridge, MA: MIT Press, 1992.
Jackson, J. B., *The Necessity of Ruins*. University of Massachusetts Press, Amherst, 1980.
Johnson, Samuel, 'Preface to the Plays of William Shakespeare' (1765). *The Major Works*. Ed. Donald Greene. Oxford: Oxford University Press, 2008, pp. 419–456.
Kant, Immanuel, *The Critique of Judgement*. Tr. J.C. Meredith, Oxford: Clarendon Press, 1961.
Laugier, Marc-Antoine, *Essay on Architecture*. Tr. Wolfgang and Anni Herrmann. Los Angeles: Hennessey & Ingalls, 1977.
Le Corbusier, *The City of To-Morrow and its Planning*. Tr. Frederick Etchells. London: The Architectural Press, 1971.
Leopardi, Giacomo, *Selected Prose and Poems*. Ed. Iris Origo and John Heath-Stubbs. London: Oxford University Press, 1966.
Lewis, David K., *On the Plurality of Worlds*. Oxford: Blackwell, 1986.
McCumber, John, *Metaphysics and Oppression: Heidegger's Challenge to Western Philosophy*. Bloomington, IN: Indiana University Press, 1999.

Pliny the Elder, *Natural History*. Tr. H. Rackham. Cambridge, MA: Harvard University Press, 1991.

Ponte, Alessandra, 'The Garden of Villa Garzoni at Collodi'. In *The Architecture of Western Gardens. A Design History from the Renaissance to the Present Day*. Ed. Monique Mosser and Georges Teyssot. Cambridge, MA: The MIT Press, 1991, pp. 181–184.

Pope, Alexander, 'Moral Essays. Epistles to Various Persons. Epistle IV to Richard Boyle, Lord of Burlington'. *The Poetical Works of Alexander Pope*. London: Jones & Co. 1826, pp. 60–61.

Quatremère de Quincy, Antoine Chrysostome, 'Type'. *Oppositions*, Spring 1977: 8, pp. 95–II5.

Rabreau, Daniel, 'Urban Walks in France in the Seventeenth and Eighteenth Centuries'. In *The Architecture of Western Gardens. A Design History from the Renaissance to the Present Day*. Ed. Monique Mosser and Georges Teyssot. Cambridge, MA: The MIT Press, 1991, pp. 305–316.

Rousseau, Jean-Jacques, *The First and Second Discourses and Essay on the Origin of Languages*. Ed. Victor Gourevitch, New York: Harper & Row, 1986.

Shaw, George Bernard, *The Doctor's Dilemma, Getting Married, and the Shewing-Up of Blanco Posnet*. New York: Brentano's, 1927.

Suhonen, Pekka, 'Alvar Aalto – mitali Jørn Utzonille'. *Helsingin Sanomat* 7. 8. 1982.

Thacker, Christopher, *The Wildness Pleases. The Origins of Romanticism*. New York: St. Martin's Press, 1983.

Tschumi, Bernard, *Architecture and Disjunction*. Cambridge, MA: MIT, 1994.

Teyssot, Georges, 'The Eclectic Garden and the Imitation of Nature'. In *The Architecture of Western Gardens. A Design History from the Renaissance to the Present Day*. Ed. Monique Mosser and Georges Teyssot. Cambridge, MA: The MIT Press, 1991, pp. 359–372.

Vérin, Hélène, 'Technology in the Park: Engineers and Gardeners in Seventeenth-Century France'. In *The Architecture of Western Gardens. A Design History from the Renaissance to the Present Day*. Ed. Monique Mosser and Georges Teyssot. Cambridge, MA: The MIT Press, 1991, pp. 135–145.

Vesilind, P. Aarne, 'Vestal Virgins and Engineering Ethics'. *Ethics & Environment,* 7 (1), 2002, pp. 92–101.

Weller, Richard, 'Landscape (Sub)Organism in Theory and Practice'. *Landscape Journal,* 27(02), 2008, pp. 247–267.

Wilde, Oscar, 'The Decay of Lying'. *The Complete Works of Oscar Wilde*. Vol V. Intentions. Garden City, NY: Doubleday, Page & Co., 1923.

Wood, A. Skevington, *The Principles of Biblical Interpretation*, Grand Rapids, Michigan: Zondervan, 1967.

Zemach, Eddy M., 'No Identification Without Evaluation'. *British Journal of Aesthetics* 26, 1986, 239–251.

Chapter 2

# Trees: the living structure of the landscape

Dendrology, arboriculture and landscape architecture

*Gabor Schmidt*

## Introduction
This chapter looks at one of the main elements used in landscape architecture – trees and other woody plants. These are used in many ways by landscape architects to give structure to the landscape and usually students have to know a set number of trees and shrubs which they often go on to use in their later work – only knowing a limited number out of many thousands.

## What is dendrology?
Dendrology is the science and study of woody plants (trees, shrubs, lianas and conifers). The name dendrology derives from the ancient Greek δένδρον, *dendron*, 'tree'; and -λογία, – *logia*, science of or study of. In this strict sense dendrology means the science (knowledge) of trees. According to other definitions, dendrology is a specific branch of botany dealing with trees and shrubs.

In practice, however, the term dendrology is used in a much broader sense: It includes not only trees but also all other woody plants including shrubs and the almost herbaceous semi-shrubs (e. g. *Lavandula*, *Santolina*, etc.). Besides their study, practical dendrology also deals with the propagation, breeding, growing, planting and maintenance of these plants (Tóth, 1969).

## Related terms and subjects
### Arboriculture
Arboriculture is the art, science, technology and business of tree care. Arboriculture is practiced by arboriculturalists (arborists in the USA). Arboriculturalists are trained to promote tree health, discern tree problems and take measures to correct them. Certified arboriculturalists are accredited by the International Society of Arboriculture. In most of the landscape schools of the UK or the USA the most essential elements of dendrology (e.g. plant knowledge) are embedded in the subject of arboriculture (Gross, 2002).

### Arboricultura
Arboricultura (as distinguished from the former term arboriculture) is the Latin translation of the word dendrology. In the countries where a Romance language is spoken (Italy, France, Romania, Spain, Portugal, South America, etc.) very often the word 'arboricultura' is used instead of 'dendrology'.

### Urban forestry
Urban forestry is the careful care and management of urban forests, i.e., tree populations in urban settings for the purpose of improving the urban environment. Urban forestry professionals advocate the role of trees as a critical part of the urban green infrastructure. According to another (wider) definition, urban forestry is the management of naturally occurring and planted trees in urban areas (Forrest et al., 1999).

The term urban forestry is used sometimes instead of dendrology or arboriculture by some Scandinavian higher educational establishments, where the teaching of landscape architecture is connected to a forestry department. In these schools, the term is useful although somewhat contradictory to the classical definition of *forest*: a forest (also called a wood, woodland, wild, weald or holt) is generally defined as an area with a high density of trees while the term *forestry* usually refers to the science of planting and caring for forests and the management of growing timber, or in a wider sense, is the art and science of managing forests, tree plantations, and related natural resources.

As seen from the above definitions, dendrology, arboriculture and urban forestry are somewhat overlapping as educational subjects (depending on the form of the particular establishment), but each of them includes elements of the other two: there is no practical dendrology without some knowledge of arboriculture and/or forestry and vice versa.

For the sake of simplicity, and also because of the larger number of agricultural and landscape schools of the world where 'dendrology' is the term most commonly used, this term will be used in the rest of this chapter.

Before proceeding, however, there are three further terms to be clarified: Nursery, Botanical garden, Arboretum.

A *nursery* is a place where plants are propagated and grown. According to the type of plants grown, there are bedding- and balcony-plant nurseries, perennial plant nurseries and woody plant nurseries or tree nurseries. In this chapter, the term nursery always means a nursery for woody plants. According to the size to which the plants are grown we can distinguish between the nurseries which only produce propagation materials and the 'normal' commercial nurseries producing and marketing ready-to-use trees and shrubs. Some nurseries specialize in one type of plant: e.g., conifers, landscape trees, container plants, groundcovers, fruit trees, rock garden plants, etc. (Probocskai, 1969).

A *botanical garden* is a place where plants, especially ferns, conifers and flowering plants, are grown and displayed for the purposes of research and education. This distinguishes them from parks and pleasure gardens where plants, usually with showy flowers or foliage, are grown mainly for amenity. The earliest botanical gardens were founded during the late Renaissance period at the University of Pisa (1543) and the University of Padua (1545) in Italy, for the study and teaching of medical botany (Tar, 2003). Many universities today have botanical gardens for student teaching and academic research, e.g. the Arnold Arboretum of Harvard University, the Bonn University Botanic Garden, Germany, the Cambridge University Botanic Garden, England and the Botanical and Experimental Garden of the Radboud University of Nijmegen in the Netherlands.

An *arboretum* (in Russian-speaking territories: *dendrarium* or *dendropark*) is a specialized kind of botanical garden devoted primarily to trees and to other woody plants, intended at least partly for scientific study. An arboretum specialized in growing conifers is known as a *pinetum*. The term 'arboretum' was first used in English by J. C. Loudon in 1838 in his encyclopaedic book *Arboretum et Fruticetum Britannicum*, but the concept was already long-established by then (Tóth, 1999).

The first arboretum to be designed and planted was the Trsteno Arboretum, near Dubrovnik in what is now Croatia. Its start date is unknown, but it was already in existence by 1492. The prides of the arboretum, two Oriental planes over 500 years old, have survived all disasters unscathed; these ancient trees are both about 45m tall with 5m trunk diameter. (The arboretum is currently owned and managed by the Croatian state).

## The history of dendrology as an independent subject
*Dendrology as section of botany, the first plant collectors*

At the beginning, early in the seventeenth century, dendrology was considered as just one of the numerous fields of botany. Later on, in the age of colonialism of the world a number of devoted botanists (the so-called 'plant hunters') were sent out (and financed) for long expeditions by the great colonizing countries (England, France, Germany, Holland, Portugal, Belgium, Russia) in order to discover, define and possibly to bring home new plants from the new lands. From the sending side, the financing of these expeditions was not simply a gesture of charity but served strict economic purposes: to introduce new trees for timber production (e.g. *Robinia, Eucalyptus, Pseudotsuga*), others for valuable raw materials (e.g. the coffee-tree *Coffea arabica*, the rubber tree, *Howea belmoriana*), or new crops for food production (potatoes, maize), etc.

The first step of introduction was usually the growing of the new plant in a botanical garden. The next, if it seemed worthwhile, was the elaboration of their propagation and growing technology followed by the distribution and mass-production either in the introducing country or on the colonies (Tar, 2003).

Some of the legendary plant hunters of that time include:

- John Tradescant (the younger), 1608–1662, exploring and collecting plants mainly in Virginia, North America. Among his introductions were the Black Locust *Robinia pseudoacacia* and the Tulip Tree *Liriodendron tulipifera*.
- David Douglas, 1799–1831, famous Scottish plant collector in North America, and collector of many trees used in commercial forestry.
- Robert Fortune, 1812–1881, born in Scotland, botanist and gardener, later director of the Chelsea Gardens, London. He undertook several trips to China and was one of the most successful plant collectors; he introduced the tea plant into India.
- Louis Augustin Guillaume Bosc, 1759–1821, French botanist; supervisor at the gardens and nursery of Versailles, and the Paris Botanic Garden; travelled in North America, specialist on *Quercus*.
- Paul Kitaibel, 1757–1811, Hungarian professor of botany in Budapest; explored and collected the vegetation of the Carpathian Region and the West-Balkans.
- Michael Smirnov, 1849–1881, Russian botanist from Odessa; explored the flora of the Caucasus (Krüssmann, 1989).

## Dendrology becoming an independent science and profession

The 'golden age' of plant collections was from the second half of the nineteenth to the first half of the twentieth century. By that time, practically the whole world was 'safe and civilized' (colonized) at least to such an extent that large-scale plant collecting expeditions could be reliably organized. The patrons of these expeditions were, in many cases, wealthy landowners who had large ornamental parks (or nurseries supplying them with plants) and the primary aim of the journeys was to find not so much useful but spectacular and rare new ornamental woody species to make their parks and gardens unique (Lancaster, 1993). The finest of these gardens later became the distribution and breeding centres of many novelties. It was actually at this time that botanists dealing with woody plants started to be called not simply botanists but *dendrologists* and thus dendrology became an individual and well-defined science and profession. The first books (in three volumes) entitled *Dendrology* were written by Karl Heinrich Emil Koch (1809–1871), from Weimar, Germany, professor of botany in Berlin. This represents the first 'official' use of the term in a scientific sense.

According to their background, dendrologists of that (and also of the present) time fall into two main groups: *the botanists* – people of theory, mainly concerned with classification and *the gardeners* – nurserymen, foresters, landscape architects, garden-owners (people of practice and money). There was (and there still is), of course, no strict boundary between the two categories.

Many of the best dendrologists were (and are) not 'full-time' but 'part-time' or amateur botanists. It was they, especially the nurserymen, who mostly did the breeding and commercial distribution of ornamental trees and shrubs for landscape design purposes.

By now, dendrology has became a completely separate profession and (though under various names) an important study subject in the curriculum of agricultural, forestry or landscape architecture educational establishments.

## Basic concepts of dendrology

Like all life sciences, dendrology has two main aspects: a theoretical (in this case botanical) and a practical (cultural, landscape application, etc.) aspect.

From a *botanical aspect*, dendrology is just a specialized field of botany focusing on the classification, and nomenclature of *naturally occurring* woody plants. Besides their geographical area of distribution, botanical dendrology also studies the position and role that woody plants play in natural plant-associations.

From the *practical aspect*, on the other hand, dendrology is not simply a woody plant-orientated field of botany, but also an economically important applied science. It deals with a wide range of *cultivated* woody species and varieties (cultivars) with different emphases depending on their type of utilization, such as:

- horticultural
- silvicultural and
- landscape-orientated dendrology.

*Horticultural dendrology* includes a consideration of all tree and shrub species of commercial horticultural interest and especially their large number of cultivars. The description of plants is usually supplemented with short cultural directions relating to their selection, propagation, growing, marketing, planting and the necessary aftercare (maintenance) (Schmidt and Tóth, 2006).

*Silvicultural dendrology* (forestry dendrology) deals mainly with tree (and some shrub) species, subspecies and cultivars suitable for the purpose of forestry. Silvicultural books contain usually fewer species and cultivars than the horticultural ones but devote much more attention and space to each of them. Besides the usual morphological and botanical description, a large part usually deals with their timber production possibilities, processing characteristics, and short notes regarding their propagation (Gencsi and Vancsura, 1996).

*Landscape-oriented dendrology* is probably the most comprehensive one. It is a synthesis of dendrology from all the botanical, horticultural and silvicultural aspects (see Figure 2.1).

By way of explanation we can look at the differences between botanical and practical dendrology by using the example of the genus *Sorbus*. Botanically, the genus *Sorbus* belongs to the *Rosaceae* family, along with a large number of herbaceous genera *Fragaria* (strawberry), *Filipendula* (Fan-flower), *Alchemilla*, the *Potentillas* (cinquefoils), etc. The *Sorbus* genus is distributed in temperate regions of the northern hemisphere and is represented by over 100 species. Each species has its own natural area of distribution and is subdivided into *naturally occurring* units: subspecies (subsp.) varieties (var.) and/or forms (forma) distinguished from each other by more or less marked morphological and ecological features. *Sorbuses* give rise to a surprisingly large number of natural hybrids which may be stabilized and

**Figure 2.1**
Examples of the variety of form and character as shown in an arboretum. (Photo Gabor Schmidt)

propagate naturally by the way of apomixes (asexual reproduction through unfertilised seeds). These hybrid-originated so-called 'small' *Sorbus* species are a true treasure (and a subject of permanent disputes) for botanists. Most of them, however from the practical point of view, are just interesting rarities.

From the point of silviculture, the *Sorbuses* as trees never form an individual forest but live in a subsidiary position (on the edge of the forest or in lower canopies) of the 'main' forestry trees like oak-woods (*Quercetum*), pine-woods (*Pinetum*) etc. Nevetherless, they are important components of healthy forest ecosystems, and some of them, like the Wild Service Tree (*Sorbus torminalis*) produce a highly valued timber in the wood-industry.

From the horticultural and the landscape design point of view, *Sorbuses* have innumerable forms (cultivars) with high ornamental values which are widely grown in nurseries and used in parks and gardens.

This distinction of the different types of dendrology is, of course, somewhat artificial (and subjective) as there is no clear border between them. Nevertheless, the profile of the educational establishment where a subject under the name dendrology (or arboriculture) is taught and also the textbooks offered for students show a more or less distinguishable profile.

## Education and textbooks in dendrology

In brief, the teaching of dendrology is included into the curriculum of almost all the higher education establishments related to landscape architecture. The subject is usually stronger in university faculties and/or colleges related to plant sciences: botany, agriculture, horticulture, forestry, ecology etc. Landscape architecture students, as noted in the introduction, usually come into direct contact with the subject but not in such a great depth.

In the educational establishments (technical universities and/or colleges) which focus on the built elements of human environment, including landscape architecture, dendrology is also represented in the curriculum. The main difference is only that these (technical or design) schools usually do not have their own departments of dendrology, but take the services from the department of a sister-faculty or employ invited lecturers from other (agricultural) universities or departments.

## *The textbooks*

Owing to the wide interest in the subject by amateur and lay people (compared, say, with sociology or cultural geography) the subject is rich with accessible publications and websites full of illustrations and cultivation instructions. From the point of view of student or practicing landscape architects, textbooks dealing with dendrology fall roughly into four categories:

1   comprehensive textbooks
2   practical textbooks for the everyday work of professionals
3   textbooks for students
4   books for amateur enthusiasts or for very narrowly specialized landscapers.

The reference list at the end of the chapter presents a comprehensive selection of this literature.

Teaching is also very practically orientated, with visits to arboreta or dendroparks, tests on identification, keeping sketchbooks (drawing plants is an excellent way to get to know them) and visits to nurseries. Seeing trees and shrubs during the range of seasons is also important. Practical study of planting techniques, maintenance, pruning, tree health (fungi and disease) and other aspects is also part of the education, more so in some places than others.

In some places students from different study programmes participate in courses together (foresters, horticulturalists, arboriculturalists, landscape architects etc.) which makes for a good start in inter-disciplinarity.

## The intersection between dendrology and landscape architecture

There are few landscapes and almost no gardens without plants of some sort. Woody plants are the most characteristic and the most long-lasting elements of the plant world and play a key role in every type and every scale of landscape. An elementary knowledge of dendrology (as a science of woody plants) is therefore essential for the effective work of landscape architects.

## *The role of woody plants in the landscape*

Woody plants have three main roles in the landscape: *functional, ecological* and *psychological.*

## *The functional role*

The functional role means that woody plants are the 'living construction materials' of landscape design. In parks and gardens they create the green framework of space and provide one means for the natural division or separation of different functional areas (e.g. playgrounds for children, resting places for elderly people, sports fields, picnicking areas, jogging tracks, etc.) (Schmidt, 2003). The use of trees to enclose space, to screen objects from view, to control and create views and to provide contrasts of mass and space, light and shade, height, form and colours is one of the main aspects trees bring to the landscape. The effects differ depending on the configuration – a single tree, a clump, a row, an avenue or a copse, for example.

One of the most characteristic types of functional tree elements are those planted along the edges of roads and streets in cities, towns and villages. Such planting sites are ecologically very difficult for the trees concerned and yet very important for the qualities of the streets at the same time (see Figure 2.2).

When planted at a larger and more extensive scale trees and shrubs can fulfil a range of functions as set out below.

## Shelter belts and wind breaks

Shelter belts are widely used along edges of open spaces and alongside linear features such as highways to provide protection against wind, for trapping particulate pollution and for preventing the dangerous accumulation of the snow in wintertime. In agricultural fields, shelter belts are important for improving the microclimate for crops or livestock by decreasing the speed and reducing the turbulence of the wind, thus, depending on the area and objective, helping to collect rain and snow (more

**Figure 2.2**
Examples of trees along a street showing the architectural properties of trees in these locations where growth conditions may be difficult. (Photo Gabor Schmidt)

humidity in the air and more moisture in the soil), for helping to reduce energy use 50 animals keep warm and therefore maintain weight in winter and, perhaps the most important, for preventing the wind-erosion of soil. For this purpose, usually 12–20 m wide linear plantations of native tree and shrub species are used which provide around 60 per cent canopy density. A foliage which is evergreen (like that of conifers) or is at least retained as dry on the trees through winter (*Carpinus, Quercus*) is normally preferred (Halasz, 2004).

## Greening in industrial areas

In industrial areas, large areas of tree and shrub plantings are valuable to hide visually intrusive objects, and also to reduce adverse effects of dust, air pollutants, etc. In this case, usually hardy and fast-growing native (or naturalised) plants are used in the surroundings (protection and cover), and possibly more decorative species and cultivars in the vicinity of housing areas.

A related application is the *restoration or re-cultivation of disturbed land* resulting from mines, quarries, dumps of pulverised fuel ash from power stations, land fill sites and demolition areas or large-scale construction works. These areas are usually characterized by poor soils or no soils at all and often with toxic residues from former uses. For the protection of such areas, hardiness (tolerance of adverse conditions) and vigour of the plants are the most important aspects. Trees and shrubs of pioneer character like *Salix* spp, *Populus* spp or *Betula* spp in the moist sites and *Pyrus* spp, *Crataegus* spp or *Pinus* spp in dry locations are the most commonly used genera, as well as nitrogen fixing species such as *Robinia* spp or *Alnus* spp (Jambor-Benczur, 2003).

## Landscape design and management under special ecological conditions

The establishment of parks and gardens under unusual ecological conditions such as the flood-plains of lakes and rivers (too much water), or on sandy, rocky or saline soils etc. requires special dendrological knowledge of stress-tolerant plants and their proper application and combination by landscape architects.

## *The ecological role*

Trees and shrubs are perhaps one of the most effective tools for improving the micro- and macroclimate of urban areas. By their respiration they produce oxygen, by evapo-transpiration they increase the humidity and reduce the temperature of the air and by natural shading they reduce excess heat and glare from hard surfaces. These effects are especially important in countries with warmer climates and in the urban environment. The proportion of trees and shrubs is, therefore, relatively high (over 50 per cent) in streets, squares, parks and gardens of southern countries and in the large cities and is relatively low (usually below 50 per cent) in northern cities.

A wide biodiversity of trees and shrubs greatly contribute to *self-sustaining urban ecosystems*: providing natural or nature-like habitats for the other plants, giving food and shelter for small animals (e.g. birds). A park with a high biodiversity is almost always more capable of recovering from stresses than an *alleé* of urban street trees which is practically a monoculture and as such is more vulnerable to new pests and diseases. If the size and function allows, therefore, use as many species as possible in order to keep a particular green space in a state of ecological equilibrium (Buczaki, 1986).

## The psychological role

The psychological roles of woody plants in the landscape seem to be so self-evident that is even difficult to describe. No matter how civilized and urbanized, man is and will always be a part of Nature. Bringing him back into contact with the natural environment (or bringing nature back to artificial human environment) is, subconsciously, always improving the psychological state of people: calming down the nerves, brightening the mind and making the work creative and pleasant (Konijnendijk *et al.*, 2005, Nilsson *et al.*, 2010). There are numerous studies to formulate these effects in figures but the most convincing fact is the instinctive devotion of human beings to trees and shrubs. Although not edible, people spend money to have them in their gardens, lead actions for their protection in parks or streets, and one of the most popular (and relatively cheap) slogans of politicians is that they are 'green in heart' and the promise to give more green spaces for their electors (Radó, 2001).

## The aesthetic role

'*Varietat d'electat*' (variability gives pleasure) is the saying. Woody plants, as key elements in green areas, play a fundamental role through their positive aesthetic value. They give colour, fragrance, diversity of size and shape and they express the seasonal changes of the year (blossom in the spring and summer, spectacular fruits and coloured leaves in the autumn, evergreen foliage and interesting bark in winter, etc.). To have a maximum effect, very deep dendrological knowledge and successful combination of a wide assortment of woody species is required by landscape architects (see Figure 2.3).

## Arboriculture and landscape architecture

Although closely overlapping with dendrology, arboriculture focuses mainly on the planting, care and maintenance of woody plants. In this respect, landscape architects have to keep two things in mind:

1. At the stage of planning, an ecologically appropriate selection of plants (creating almost 'self-sustaining' combinations) results later in cheaper maintenance, as well as the safer and longer life of the whole green element.
2. Later, in the stage of maintaining an already existing park or garden, landscape designers should be familiar with the main principles of tree and shrub care but should never forget that it is an individual and specialized profession needing a high level of knowledge and skills. Not anyone with a chainsaw can or should be allowed to 'prune' or to 'trim' (or more usually to 'butcher') our trees and shrubs. Certified persons have to be employed for this job only: these are the arboriculturists or tree surgeons (Dujesiefkenn 1995; Gross, 2002).

**Figure 2.3**
A variety of trees adds to the visual diversity of urban areas. Here in spring the blossom is particularly beautiful. (Photo Gabor Schmidt)

## Main areas of knowledge of use to landscape architects

Landscape architects have to utilize a very wide range of knowledge, covering almost all fields of life, technical, and socio-ecological sciences. An inevitable question arises: how much and in which areas of these should dendrology cover?

In my opinion, these main areas are as follows.

### *General dendrology*

General dendrology usually begins with the definition(s) and the role of woody plants in landscape architecture. If botany is not taught at the particular school as a specific subject, then the basics of plant-systems and plant morphology should also be included. This should be followed by a *detailed plant-to-plant introduction* of the most important species and cultivars, which forms the most important body of dendrological knowledge needed by landscape architects. This knowledge includes the morphological description, geographic distribution, coenological characteristics and ecological requirements of each tree and shrub, and also how they are used, their application, technological properties related to their availability or special needs for planting, establishment and maintenance. This plant-to-plant introduction in landscape schools is usually done in a plant-systematic order: for example, first the gymnosperms (conifers), and than the angiosperms (broadleaved, woody plants) in accordance with the normal botanical classification (Harlow *et al.*, 1969; Seneta, 1976; Iliev, 2001; etc.).

Gabor Schmidt

## Applied Dendrology

Applied dendrology comes after the basic knowledge of plants. In fact, it is the selection and application of woody plants for different landscape settings and situations. Here the main points of consideration are the *environmental requirements*, the *ornamental value* and the *functional abilities*.

### Environmental requirements

Matching the environmental requirements of trees and shrubs to the conditions of a particular landscape element is essential for their long-term survival and good development. Key aspects are the soil, water, light and temperature (Buczaki, 1986). Therefore information about these aspects for each species is important knowledge. However, because trees are used all over the place and because trees planted in one place may have been grown in a nursery in a different country, problems arise if hardiness is not considered.

#### THE (WINTER-) HARDINESS ZONES

The introduction of the term and concept of *hardiness zones* in nursery and landscaping practice is a result of globalization. Back in the early 1900s, the activity of landscape architects was mainly restricted to their own country and plant material usually also came from the same region, from local nurseries. They produced plants for the local climatic zone so they performed well in the same country. Later (as a first sign of globalization) nurseries gradually moved further from the market, towards the climatic optimum (and also to cheaper land and cheaper labour). Finally, the nursery areas have developed not at a country but on a continental level. The elaboration and introduction of the hardiness zones is a result of these changes (Schmidt et al., 2008).

While initially started in the USA, this hardiness zone system has recently been extended to Europe. It was developed in close relationship with the step-by-step formation of a more united and more open Europe, with the removal of the borders to trade and movement of goods and the building out of a proper infrastructure (motorway system) for long-distance transport. These integration processes in Europe were also leading to the production and the application of woody plants not only on a country but also on a continental level. So here the term of hardiness zones should also be introduced not only in the landscape architecture practice but also in the respective teaching.

The first map of the hardiness-zone areas for woody plants in Europe was produced by W. Heinz and D. Schreiber (Mitteilungen der Deutschen Dendrologischen Gesellschaft Nr. 75) and published for the first time by A. Bärtels in his book 'Gartengehölze' (Eugen Ulmer, Stuttgart, 1981). The hardiness-zone areas are based on the average minimum air temperatures. Eleven areas with temperature fields of 5°C each are shown in Europe. In Central Europe only the areas from 5 to 8 exist. Their division in half-areas takes care of the relatively small climatic areas in Central Europe.

The map of hardiness-zone areas for woody plants in Europe is shown in Figure 2.4. Each area of the map shows the average hardiness of the hardiness-zone. However, in these areas many divergences based on the local climatic conditions

**Figure 2.4**
Winter-hardiness zones in Europe (after Krüssmann, 1989 and Schreiber 2007).

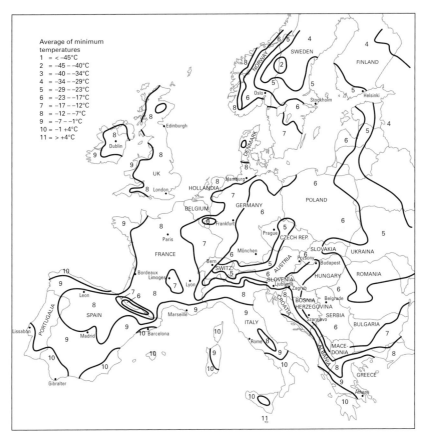

exist, such as the shelter of buildings especially in cities and also a position facing south giving a milder micro-climate, or the 'frost-pockets' like valleys, depressions and positions facing north giving a cooler micro-climate. This effect often causes the climatic conditions to improve or to decrease by one half-area (warmer or colder, respectively) (Kordes, 2009).

### Other points of selection

#### SELECTION BY THE FUNCTION OF LANDSCAPE TYPE

An important part of applied dendrology may be the selection by the function of landscape type which can also be considered as a *garden-by-garden planting guide*, for example. Each type and size of garden needs different species and cultivars. The plant material may strongly vary depending on whether it is designed to be used in public parks, streets and roadside plantings in urban or village areas, hotel gardens, gardens of schools and kindergartens, hospital gardens, churches, churchyards and cemeteries, etc. The respective recommendations, of course, contain many overlaps since most woody plants can be used for a range of functions (Bloom, 1976, 1994, 1996; Gorer, 1971; Lancaster, 1974).

## SELECTION ACCORDING TO ORNAMENTAL OR FUNCTIONAL VALUE OF TREES AND SHRUBS

In this respect, various groupings can be made according to specific features such as:

- *Flowers*: flowering trees, flowering shrubs, colour and time of flowering, fragrance etc.
- *Shape*: columnar, globular, weeping, corkscrew-like etc.
- *Foliage*: evergreen, deciduous, green, yellow, red, variegated, silvery.
- *Size*: large, medium-size and small trees, the same categories for shrubs.
- *Special use*: ground covers, hedges (pruned or non-pruned), climbing and trailing plants, roses as a separate group within woody plants (also in different size and colour); etc.

There are also many more points for grouping, but these usually depend on or are related to the other subjects taught in a given faculty or course. These are usually given in tabulated forms (often digitally, on a CD or DVD supplemented with search programmes and colour photos of the plants). All of them are useful aids in the process of planning, but *none of them can replace the real plant knowledge and the creativity of landscape students and architects.*

## *Practical dendrology and arboriculture (establishment and maintenance of plants)*

The work of landscape architects is not of course finished with the composition of landscape plans. An integral part of it is the consequent realization of the project and, later on, the maintenance of the completed green areas as economically and effectively as possible. Here once again applied dendrology, with the knowledge about the following operations, is important:

- *Basics of nursery practice*: the propagation, growing and marketing of woody plants. This is usually discussed from the customer's point of view: quality standards, size categories in the nursery catalogues, the best sizes for different purposes, interaction between production costs, prices, and availability.
- *Planting the area.* Principles of planting; planting of large trees, small trees, shrubs, conifers; training trees; hedging; methods of transplanting and establishment for different landscape conditions.
- *Maintenance operations*: soil cultivation, water- and nutrient supply, plant protection, weed control, pruning, cultivation; plant care.
- Tree care, *tree surgery*, other operations and practical skills. Tools for the task (see Figure 2.5).

## Opportunities for collaborative research

In practically all landscape architecture schools of Europe there is already a close collaboration between dendrologists and landscape architects. It is a natural symbiosis of specialists: the dendrologists develop the up-to-date knowledge of old and

# Trees: the living structure of the landscape

**Figure 2.5**
Pruning and other tree care activities are specialist work needing special equipment. (Photo Gabor Schmidt)

new woody plant species and cultivars, but it is the landscape architects who apply them in the parks, gardens or other green areas. Without actual planting and testing of species and cultivars, no information can be obtained as regards their hardiness, tolerance, sensitivity, possible pests and diseases (of course, this takes time). Equally, there are large resources in the existing dendrological collections (arboreta) to be made use of by landscape architects: these arboreta can be considered as results of long-term experiments. Those dendrological rarities which were planted long ago and which have survived certainly deserve more attention as plants which have already proved their adaptability to the extremes of local climate and soils. With this in mind, the main opportunities for collaboration are, in my view, as follows.

At a national level:

- Introduction and systematic trial of new woody species and cultivars in different landscape situations.
- Exploration of the resources of local arboreta and other dendrological collections for new species useful in landscape projects.
- Breeding of woody plants for local conditions – an evaluation by landscape architects with recommendations for further directions.
- Urbanization effects on woody plants – dangers and possibilities.
- Exotic woody plants inclined to escape into the natural landscape from cultivation.
- Climatic and local meteorological factors and their effects on the life cycle of woody plants.
- Colouration of flowers and leaves: the influence of genetic and climatic factors

At an international level, the LE:NOTRE network of landscape architecture schools provides an opportunity for joint international research, embracing the whole continent. Such are the questions which are important and studied more or less almost in every member country, but the methods are different or the results to date are only utilized locally. Examples are:

- Studies on the mechanism of stress-tolerance under different landscape conditions (stress enzymes, intensity of assimilation, chlorophyll content, etc.).
- The salt problem and research on salt tolerance of tree species in streets and along roadsides.
- Evaluation methods for trees in different landscapes and for street trees.
- Green space inventory in towns and cities
- The effects of woody plants on the environment.
- Vitality description of woody plants.
- Understanding soil conditions and tree growth.
- Approaches to planting trees in the landscape of urban areas.
- Management of landscape and urban trees.
- Legal frameworks of landscape and urban tree planting, protection and management.
- Professional organisations and their roles and structures; relationships with other professions.
- Elaboration of a joint nomenclature relating the teaching and use of dendrology and arboriculture (now under different names) in Europe. This could also include a Multilanguage Plant Vocabulary.
- A joint survey and assessment of plant production and nursery stock in the member countries.

## Concluding thoughts

Centuries ago, in the age of early baroque and landscape gardens, a handful of woody plant species were available (and sufficient) to create magnificent parks: *Quercus*, *Tilia*, *Malus*, *Ulmus* ssp. for majestic specimen trees and clumps, *Populus nigra* 'Italica' and *Salix alba* 'Tristis' to give accent and contrast, *Cedrus*, *Platanus* or *Corylus colurna* for exotic touches and *Buxus*, *Taxus*, *Carpinus* or *Tilia* for different sizes of pruned hedges and topiaries. The background of the park was the natural landscape and the framework the natural wood from which the park was literally 'carved out'.

The dendrological knowledge of the designers of those parks was limited too: they gave the outlines and the basic conception; for the rest – adding colours, fragrances, flowers etc – they relied upon gardeners. Later, as the world opened and masses of splendid new trees and shrubs arrived from East and West, landscape gardeners (who would eventually form a profession and call themselves landscape architects) started to use (and to learn about and from) an increasing number of woody plants and, about the same time, dendrology grew out of botany as an independent science and practice.

At present, a reasonable knowledge of dendrology is as important for every landscape student and practicing landscape architect as knowledge of currently available building materials and the ways of safe structural calculations for architects. Consequently, the teaching of dendrology (although at different levels) is an integral part of the curriculum of landscape architecture schools all over Europe.

My personal experiences and remarks in this respect are these: most landscape architect students take pleasure in learning dendrology. They like plants and it makes them feel superior to the other 'regular' architects. After their studies, the best landscape plans are made by those who simultaneously plan, realize and maintain the parks and gardens (or, at least, work in a company which does) and have a permanent feedback. I know many small companies consisting only of a married couple: the wife (a landscape architect) designs, while the husband (usually a landscape architect, a horticulturist or forester) plants and maintains the parks and gardens.

There are however many who tend unconsciously to forget the knowledge about trees and shrubs. It happens mostly if they only make plans and do not take part in the realization and the aftercare of the project. The worst case is if they start to make plans using unknown plants from foreign non-specialist books, colour plant catalogues or internet images. Unfortunately, plants cannot read and do not follow bad directions of the landscape architect. If applied incorrectly, they suffer and ultimately die. These poorly selected plants will sadly disappear from the garden within a few years and then the maintenance people will replace them with the proper (hardy) ones. Thus, the '*corpus delicti*' disappears and the green area heals itself. It will look beautiful again, but quite different from the original plan.

So my advice to every landscape student is this: stand with both feet on the ground and be realistic in your plant ideas when designing gardens, parks and other green places. Try not to forget but rather to refresh your knowledge of trees and shrubs and, in more complicated cases, do not hesitate to ask for the advice and collaboration of professional dendrologists. *It will be profitable for both sides!*

## Bibliography

Arnold, H. F. (1980) *Trees in Urban Design*. Van Nostrand Reinhold. New York, NY.
Bailey, L. H. (1933) *How Plants Get Their Names*. Macmillan Co. New York, NY.
Bailey, L. H. (1948) *The Cultivated Conifers in North America*. Macmillan Co. New York, NY.
Bailey, L. H. (1949) *Manual of Cultivated Plants*. Macmillan Co. New York, NY.
Bärtels, A. (1981) *Gartengehölze (Garden Shrubs)*. Eugen Ulmer Verlag, Stuttgart.
Bärtels, A. (1989) *Der Baumschulbetrieb (The Tree Nursery)*. Eugen Ulmer Verlag, Stuttgart.
Bärtels, A. (1989) *Gehölzvermehrung (Woody Plant Propogation)*. Ulmer Verlag, Stuttgart.
Bean, W. J. (1976–80) *Trees and Shrubs Hardy in the British Isles*. Vol. I-IV. (V, supplement). M. Bear and John Murray Ltd. London.
Bernatzky, A. (1978) *Tree Ecology and Preservation*. The Netherlands: Elsevier. Amsterdam.
Bloom, A. (1975) *Conifers for Your Garden*. American Garden Guild.
Bloom, A. (1976) *Plantsman's Progress*. I. Dalton. Lavenham, England.
Bloom, A. (1986) *Conifers and Heathers for a Year-Round Garden*. Aura Books and Floraprint Ltd. London.
Bloom, A. (1994) *Winter Garden Glory*. HarperCollins. London.
Bloom, A. (1996) *Summer Garden Glory*. HarperCollins. London.
Bloom, Alan and Adrian (1992) *Blooms of Bressingham Garden Plants*. HarperCollins. London.
Buczaki, S. (1986) *Ground Rules for Gardening*. Collins, London.
Chadbund, G. (1972) *Flowering Cherries*. Collins. London.

Coombes, A. J. (1994) *Dictionary of Plant Names.* Timber Press. Portland, OR.
Den Ouden, P. and Boom, B. K. (1978) *Manual of Cultivated Conifers.* Martinus Nijhoff, The Hague, Boston, London.
Dirr, M. A. (1978) *Photographic Manual of Woody Landscape Plants.* Stipes Publ. Company. Champaign, IL.
Dirr, M. A. (1984) *All About Evergreens.* Ortho Books. San Francisco, CA. Champaign, IL.
Dirr, M. A. (1990) *Manual of Woody Landscape Plants.* Stipes Publ. Company, Champaign, IL.
Dirr, M. A. (1997) *Dirr's Hardy Trees and Shrubs.* Timber Press. Portland, OR.
Dirr, M. A. (1997) *Photo-Library of Woody Landscape Plants on CD-ROM.* PlantAmerica. Locust Valley, NY. 7,600 images.
Dirr, M. A. and Charles W. Heuser, Jr. (1987) *The Reference Manual of Woody Plant Propagation.* Varsity Press, Inc. Athens, GA.
Dujesiefken, D. (1995) *Wundbehandlung (Wound Care).* Bernhard Thaleker, Braunschweig.
Elias, T. S. (1981) *Illustrated Guide to Street Trees.* The New York Botanical Garden. Bronx, NY.
Evelyn, J. (1979) *Silva* (facsimile copy of original work). Strobart and Son. London.
Fazio, J. R. (ed.) (1992) *Understanding Landscape Cultivars, Tree City USA Bulletin No. 26.* Arbor Day Foundation. 100 Arbor Avenue, Nebraska City, NE 68410, 8.
Fiala, J. L. (1988) *Lilacs.* Timber Press. Portland, OR.
Fiala, J. L. (1995) *Flowering Crabapples.* Timber Press. Portland, OR.
Fitschen, J. (2007) *Gehölzflora.* Quelle und Mayer. Heidelberg-Wiesbaden.
Forrest, M., Konijnendijk, C. C. and Randrup, T. B. (eds) (1999) *Research and Development in Urban Forestry in Europe: Report of COST Action E12* 'Urban Forests and Trees' on the State of the Art of Urban Forestry Research and Development in Europe. EUR 19108 EN. Commission of the European Communities, Luxembourg.
Gardiner, J. M. (1989) *Magnolias.* The Globe Pequot Press. Chester, CT.
Gencsi, L. and Vancsura, R. (1996) *Dendrológia – erdészeti növénytan (Forestry dendrology).* Mezôgazda Kiadó. Budapest.
Gorer, R. (1971) *Multi-season Trees and Shrubs.* Faber and Faber. London.
Gorer, R. (1976) *Trees and Shrubs: A Complete Guide.* David and Charles. North Pmfret, VT.
Griffiths, M. (1994) *Index of Garden Plants.* Macmillan Press Ltd. London.
Gross, W. (2002): *European Tree Worker – A Handbook of European Arboricultural Council.* Patzer Verlag. Berlin-Hannover.
Halász, T. (2004): *Mezövédö erdösávok, hófogó erdösávok (Windbrakes, Shelter Belts)* (in: Schmidt G. Varga G. ed.:Famutató, Hillebrand Kft, Sopron) 94–96.
Harlow, W. M. and Ellwood S. H. (1969) *Textbook of Dendrology*, 5th edn. McGraw-Hill. New York, NY.
Hardwicke, D., Toogood, A. R. and Huxley, A. J. (1973) *Evergreen Garden Trees and Shrubs.* Macmillan Co. New York, NY.
Hartmann, H. T., Kester, D. E. and Davies, F. T. (1990) *Plant Propagation: Principles and Practices.* 5th edn, Prentice-Hall. Englewood Cliffs, N.Y.
Hoffmann, M. H. A. (ed.) (2005) *List of Names of Woody Plants.* Boomteeltpraktijkonderzoek, Applied Research Institute for Nursery Stock. Boskoop, the Netherlands.
Iliev, I. (2001) *Genetika i szelekcija na dekorativnite rasztenija.* Sejani Ed. Szófia.
Jámor-Benczur, E. (2003) Ipari területek fásítása (Woody plants for industrial and devastated areas). In: Schmidt G. ed. (2003) *Növények a kertépítészetben* (Plant Use in Landscape Architecture), Mezögazda Kiadó. Budapest) 344–356.
Konijnendijk, C., Nilsson, K, Ranmdrup, T. B. and Schipperijn, J. (eds) (2005) Urban Forests and Trees. Springer Verlag, Berlin.
Kordes, G. (2009) *Die Kordes Jungpflanzen* Katalog. Bilsen. Germany.
Kramer P. J. and Kozlowski, T. T. (1979) *Physiology of Woody Plants.* Academic Press, New York.
Krüssmann, G. (1985) *Manual of Cultivated Conifers.* Timber Press, Portland, OR.
Krüssmann, G. (1989): *Manual of Cultivated Broad-leaved Trees and Shrubs* Vol. I-II-III. Timber Press. OR.

Krüssmann, G. (1990) *Manual of Woody Landscape Plants*. Stipes Publ. Company, Champaign, IL.
Krüssmann, G. (1997) *Die Baumschule (The Tree Nursery)*. Paul Parey, Berlin-Hamburg, 372–360.
Lancaster, R. (1974) *Trees for Your Garden*. Charles Scribner's Sons. New York, NY.
Lancester, R. (1993) A Brief History of Plant Hunting (in: Hillier ed. *The Hillier Manual of Trees and Shrubs*. Redwood Press Limited. Wiltshire, UK).
Lord, T. (ed.) (1997) *The RHS Plant Finder 1997–1998*. Dorling Kindersley. London.
MacDonald, B. (1989) *Practical Woody Plant Propagation for Nursery Growers*. B.T. Batsford Ltd., London, Timber Press, Portland, OR.
Mitchell, A. F. (1972) *Conifers in the British Isles*. Forestry Commission Handbook 33. HMSO London.
Mitchell, A. F. (1974) *A Field Guide to the Trees of Britain and Northern Europe*. Collins. London.
Mitchell, A. F. (1981) *The Gardener's Book of Trees*. J. M. Dent. London.
Mitchell, A. F. (1985) *The Complete Guide to Trees of Britain and Northern Europe*. Dragon's World Ltd. Surrey, England.
Moll, G. and Urban, J. (1989) 'Planting for Long-Term Tree Survival'. In *Shading Our Cities, a Resource Guide for Urban and Community Forests*. Ed. by G. Moll and S. Ebenreck. Island Press, Washington D.C.
Nilsson, K., Sangster, M., Gallis, C., Hartig, T., de Vries, S., Seeland, K. and Schipperijn, J. (eds) (2010) *Forests, Trees and Human Health*. Springer, Berlin.
Notcutts Nurseries Ltd. (1981) *Notcutts Book of Plants*. The Thelford Press Ltd. London and Thelford.
Ouden, P. Den and B. K. Boom (1965) *Manual of Cultivated Conifers*. M. Nijhoff. The Hague, Netherlands.
Probocskai, E. (1969) *Faiskola. (Nursery Practice)*. Mezögazdasági Könyvkiadó. Budapest.
Radó D. (2001) *A növényzet szerepe a környezetvédelemben (The Role of Plants in Environment)*. Levegö Munkacsoport. Budapest.
Randrup, T. B. and Konijnendijk, C. (eds) (1999) *Research and Development in Urban Forestry in Europe*. Office for Off. Publ. of European Communities, Luxemburg.
Rehder, A. (1990) *Manual of Cultivated Trees and Shrubs*. Dioscorides Press, OR.
Sandoz, J. L. (1999) 'Standing Tree Quality Assessments Using Ultrasound'. Acta Hort. 496: 269–277.
Schmidt, G. (2000) 'Ornamental plants in Hungary Part II. Open-ground Cultivation'. *International Journal of Horticultural Science* 6(1): 144–147.
Schmidt, G. (ed) (2003) *Növények a kertépítészetben (Plant Use in Landscape Architecture)*. Mezögazda Kiadó. Budapest.
Schmidt, G., Honfi, P. and Kohut, I. (2008) 'Integráció és szakosodás az amerikai és európai faiskolai termesztésben (Integration and Specialization in American and European Nursery Production)'. *Kertgazdaság* 40 (1): 87–92.
Schmidt, G. and Tóth I. (eds) (2006) *Kertészeti Dendrológia (Horticultural Dendrology)*. Mezögazda Kiadó. Budapest.
Schmidt, G., Tóth, I. and Sütöriné Diószegi, M. (2008): *Dendrológia, kertészmérnök és tájépítész BSc hallgatók részére (Dendrology for Students Studying Horticulture and Landscape Architecture)*. BCE Fac.Horticulture, Budapest.
Schreiber, D. (2007) Europe Hardiness Zone Map. Backyard Gardener. Available: http://www.backyard gardener.com/zone/europe1zone.html (visited January 2011).
Seneta, W. (1976) *Dendrologia. Panstwowe Widarnictwo Naukowe*. Warsaw.
Tar, T. (2003) *Arborétumok és botanikus kertek a közmüvelödésben (Arboreta and Botanical Gardens in Environmental Education)*. XI. ker Mıv.ház. Budapest.
Tóth, I. (1969): *Díszfák, díszcserjék. (Ornamental Trees and Shrubs)*. Mezögazdasági Könyviadó. Budapest.
Tóth, I. (ed.) (1999) *Magyarország arborétumai és botanikus kertjei (Arboreta and Botanical Gardens in Hungary)*. SO-MA-KER. Budapest.
Wandell, W. (1989) *Handbook of Landscape Tree Cultivars*. Gladstone, Illinois: East Prairie Publishing.

Chapter 3

# Space, place and perception

## The sociology of landscape

*Detlev Ipsen[1]*

**Introduction**

In the sociology of space landscape is an unusual theme. A volume of collected essays on the subject of landscape perception edited in 1990 by Gröning and Herlyn (Gröning and Herlyn 1996) contains works by a number of prominent sociologists, including Georg Simmel (1957), but the work of these scholars is marginal and moreover plays scarcely any role in sociology in general and in urban and regional sociology in particular. Landscape is a key concept one will find only in the work of Lucius Burckhardt – especially his short article about landscape as a transitoric space which has been rather influential in the German discussion (Burckhardt 1992). The only book about sociology of landscape I have been able to identify is written by Kaufmann (2005). The books and articles written in English concentrate more on specific issues like the change of land use, the ways of perceiving landscape or the development of tourism. Sometimes landscape is used as a metaphor for the historical meaning of specific geographical areas and time periods.

      This marginal position of the category of landscape in sociology is all the more astonishing, because other disciplines, especially geography and philosophy, deal quite specifically with questions of access to landscape, landscape perception and the significance of landscape for modern society. One reason for the reluctance of sociology to deal with the phenomenon might have to do with its origin. The rapid growth of cities, disruptions and conflicts associated with immigration and industrialization were the starting point, at least of urban and regional sociology. The first sociological institute was established in Chicago in order better to understand modern urban societies and to anticipate and regulate conflicts.

      The concentration of sociology on the city may have had the effect of obscuring landscape as an access point to space. The modern city understood itself as emancipated from nature and landscape, which were imagined as outside the city, at best as places for weekend excursions and summer holidays. Basically, however, most cities had left the countryside and thus the space where landscapes were located, behind them. Hard work in the fields and the forests, proximity to animals, lack of social space in the small cottages and villages were widespread experiences. Control of nature, the promise of comfort and security in the city were established components of urban culture. My impression is that sociology was unable to detach itself from them and so 'overlooked' the social aspects of landscape. Today we are in a different situation. Landscape, especially the future of landscape, is becoming a social problem in Europe and at the same time offers, as we shall later discuss, a fruitful approach to solving a series of imminent problems.

Four problem areas appear to me to be particularly relevant:

- It is becoming increasingly difficult to speak of town and country. In large parts of Europe, at least, new landscapes are developing between city and countryside. How are these new areas experienced? How is social life organized in them? Which tendencies should be pursued and which ones rather avoided?
- Landscape environment of a city is increasingly becoming a site factor, in as much as individual towns are scarcely any longer distinguishable from each other in terms of their social and technical infrastructure. This is definitely connected with a change of lifestyles and status symbols. Leisure activities in a landscape convey a sense of vital flair and social status.
- Demographic development in most European countries will in all probability lead to a clearly defined competition of individual areas for people and investment. If this in turn means peripheralization of individual regions, new concepts of landscape development are required. Will this lead to secondary reforestation, the emergence of extensively used savannas, new moor landscapes, the abandonment of villages and small towns? What possibilities and risks are to be expected for landscapes in these developments?
- Landscape is by no means identical with nature or the natural environment, but of course associated with it. Environmental problems themselves and their possible solutions are most closely associated with the relationship between nature and society. Landscape, or a particular understanding of landscape, can be a bridge to another relationship to nature and contribute to the formulation of strategies for sustainable regional development.

These are the aspects in particular that give rise to the discussion of landscape as a sociological access point to the development of spaces. First, the conceptual elements of landscape are discussed. On the one hand, it is important to point out the dual nature of the landscape concept, which refers both to the material nature of space and the construction of an image of space; on the other hand, by extension, landscape is developed as a necessarily interdisciplinary concept of spatial analysis.

## Landscape: an interdisciplinary concept

From the sociological point of view a contemporary definition of the term landscape will include its dual character: on the one hand, landscape is a material objective structure of space, and secondly a subjective culturally determined form of perception and evaluation of this material structure. The area of tension within the concept of landscape, caught between materiality and graphic quality, implies that landscape can only be grasped when different disciplines are related to one another. For one, the materiality of landscape relates to its natural features.

Geological formations, geomorphology, water balance, soil and local climate, wildlife – all have potential natural characteristics and interaction. In reality, landscapes are (almost) nowhere to be viewed as isolated from human influence. The treatment of nature and its cultivation affect the material flow and modify local characteristics. Consequently, the use of fertilizer modifies the fertility of the soil, the settlement density the local climate, the canalization of streams the water balance, etc. The treatment of nature is not only controlled technologically, but is subject to a variety of social rules. This is what we call the social structuring of landscape. Ownership rights or right of use restrictions through conservation laws are one example. All regulations relating to planning law are an important part of further social structuring of the landscape. The concept of culture is in the centre of the model. Culture is a system of interpretations and meanings; it contains codes which enable us to understand a landscape and the evaluations which are tied to landscapes. The image of landscape is probably the most important part of landscape culture. Therefore, regarding the relationship between landscape development and landscape image, the natural conditions and forms of utilization play an equally significant role for both.

At the same time, landscape development and landscape image influence each other. The representation of a subtropical river delta has characterized the image of the Pearl River Delta in China for a long time; tomorrow, as a model of the Megacity Pearl River, it could impact the future development. Figure 3.1 shows the elements of the interdisciplinary concept of landscape. The natural conditions, the land use and the social regulations form a triangle. The cultural forms of perception, the interpretations and images are located in the centre in order to hint at culture being viewed as an integrative element (Figure 3.1).

With this concept of landscape it becomes clear that landscape does not only relate to rural space; to a large degree, it also relates to urban spaces. Cities as well as the rural areas are linked to nature; both contain soil, water, air and wildlife.

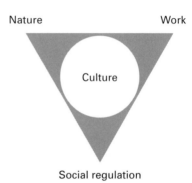

**Figure 3.1**
The concept of landscape.
(Source Detlev Ipsen)

**Space, place and perception**

Significantly dissimilar, however, is the use of nature, which differs in the density of people, pattern of functions, number of constructions, modes of the economy, but most of all in the extent to which work and nature relate to one another (Figure 3.2).

Not to be forgotten, the cultural interpretations of landscape are different for the city and the rural space as well. However, when an urban landscape develops

**Figures 3.2a and b**
Landscape as a frame. (Photos Detlev Ipsen)

as we can see it in different parts of Europe, containing the most diverging forms ranging from city to country, it becomes clear how useful the term 'urban landscape' is.

On the other hand this concept of landscape makes clear that, as a prerequisite of cooperation between sociology and landscape architecture, social scientists have to study the material aspects of landscape development: the fauna and flora, the soil and the water, the land use pattern and the aesthetics of landscape. Landscape architects on the other hand should try to understand the structure of social dynamics and define landscape more as a process than a stable constellation. If this general attitude is developed, the theory of modernization and landscape in transition will together find a way of understanding the interaction of society and environment much better.

## Landscape analysis

This concept of landscape should finally be subtly differentiated in order to show that we have not simply chosen a theoretical term to understand landscape, but that it is also an instrument for the analysis of developments. Behind the form of treatment and use of nature there are complex societal processes which can be classified into individual spheres of regulation. Thus, a pattern/diagram emerges which systematically names/identifies the components of the relationships within a landscape; one could also speak of a human ecological landscape system (see Table 3.1).

With the aid of Table 3.1 we can now formulate individual effects and interactions as questions or hypotheses: how does the infrastructure planning (systemic political regulation) affect the agrarian use of a landscape; will the intensity increase, decrease or will the areas in a landscape be distributed differently? How does that affect the water balance, and what consequences does this have on the expansion of individual plant families? Or: how do the changes within the value system shape the development of lifestyles, and what consequences does that entail for the leisure use of a landscape? Will that in turn affect the form of agriculture and forestry with the corresponding impact on flora and fauna? And conversely: what consequences will a modification of the water balance have on the supply of a densely settled area (e.g. qualitative problems with the potable water supply due to exceeded inflow of nitrate into the ground water), and how will politics react to this problem?

**Table 3.1** Levels of influence in a regional limited landscape

| Natural system | Land use | Social system |
| --- | --- | --- |
| Abiotic resources | Agrarian/forestry use | Systemic regulation |
| Geology | Extensive use | Economy |
| Climate | Leisure use | Cultural regulation |
| Water balance | Intensive | Values/norms |
| Soil | Diverse use | Interpretations |
| Biotic resources | Settlement/commercial use | Anthroposphere regulation |
| Plants | Spaces of lower density | Group building |
| Animals | Dense spaces | Intermediary organizations |

The urban landscape features a particular formation of the natural system, land use and social system. The vegetation of the city is characterized by diversity (Sukopp 1990), the land use by a tight network of spaces of flows (water mains, electrical wires, telephone networks, road networks) and the social structure by a high cultural heterogeneity and dense regulation networks. But even here, landscape is transitional (Burckhardt 1992): the dense mediaeval city differs from the widespread Fordist one in terms of landscape as well. As an example: in the mediaeval city, human and animal faeces were collected as manure and used in the gardens located near the city wall. In the Fordist city, human faeces are transported into creeks and rivers via sewers. If sewage treatment plants are available, the sludge usually has to be burned due to its being toxic. The urban landscape of the Megacity as it is developing in all parts of the world has yet a different quality: centres of intense development are characterized by industrial villages, agricultural plains, suburban settlements, islands of protected landscapes, industrial corridors alongside main traffic roads, etc. Heterogeneous land use creates differentiated landscapes and socially fragmented spaces.

## *The theory of modernization and the dynamics of landscape*

What brings sociology and landscape analysis together is the process of modernization. We will discuss the theory or, better, the theories of modernization in the next section. Now only the general idea has to be developed. In a famous formulation of Karl Marx and a very important book of Berman, modernization means that 'all that is solid melts into air'. Since the eighteenth century Western societies have found themselves in a kind of permanent revolution. To understand this process is the main subject of sociology. Very important starting points are

- the process of enlightenment and the growth of scientific understanding of human beings and the natural environment
- protestant ethics which can be understood as the spirit of capitalism
- primary accumulation and the development of internal and external colonization
- technological innovation and industrialization
- the development of capitalism: money as a general means of action
- the democratic revolution
- the demographic revolution and the ongoing process of urbanization.

Up to now the driving forces of this permanent modernization have been the following three complexes

- technological change (transport, information, biotechnology, energy)
- competition and market development (EU market, world market)
- social interests and conflicts (rich upper-class and poor underclass, ethnic and cultural conflicts, regional differences).

Landscape has become more and more transitional. As we know from the history of landscape there are a lot of natural factors which make landscape a

**Figure 3.3**
The panoramic view. (Photos Detlev Ipsen)

process. And it is a known fact that since the very beginning of human history there have been interventions in land use patterns and cultural ideas in the shape of landscape. But modernization changes the material structure and the understanding of landscape more intensively, more extensively and at high speed. To study the impact of modernization on landscape development and landscape design seems to me the key issue of cooperation between these two academic fields. To define this key issue we need a clear concept about the meaning of transitional landscapes (Figure 3.3).

## Transitional landscape

At the high point of the political ecologization of landscape the Swiss sociologist Lucius Burckhardt strikes a sober note. He writes: 'But landscape is clearly changing. We want to protect the landscape, but we do not know what we should hold on to . . . We have to consider two unstable phenomena: reality changes and, at the same time, the conceptual apparatus that is meant to determine that reality' (Burckhardt 2006: 65ff.). The theory of the transitional landscape excludes simple solutions. The concept, or one might say, the image of landscape changes both itself and reality as the use and materiality of a spatial sector change. Concept and reality are certainly related to each other, but are by no means identical. If we examine landscape development, we do well to keep them apart. The concept of landscape can change, but the materiality of landscape cannot. And the reality of landscape can change, while the image stays the same.

Let us use these considerations to do a small intellectual experiment. If concept and reality of landscape change in equal measure, one might describe this as a 'synchronous landscape concept'. With the construction of the first railway line to cross the Alps, from Vienna to Trieste via Semmering, the materiality of the mountain pass and its image changed. Being within easy reach turned a tiring and tedious journey into a pleasure trip. Hotels and cafes, walks and views sprang up. The Semmering became an ideal landscape in art and literature. The Viennese made the Alps their own (Kos). The change in the materiality of the landscape was assessed positively in aesthetic terms in order to produce a new landscape image. One sees already, however, that this synchronization can add something that is known in psychology as rationalization or dissonance reduction. Because one has decided in favour of a particular landscape, one assesses the real landscape that emerges as a result positively. Synchronicity does not have to go together with such forms of rationalization. The nature of synchronicity can easily be recognized by looking for the reasons for an aesthetic judgement that is an intellectually differentiated aesthetic.

The fewer the reasons, the more likely it is to be a dissonance reduction. One is much more likely to come across the other scenario: the reality of the landscape changes, but not the concept or the image of it. We decode the reality with signs and concepts which do not correspond to it. Many regard this as a critical position: by using old concepts of landscape to describe and assess new landscapes, the new is often felt to be a loss. Criticism feeds on this experience of loss. If one's concept of landscape is meadows, fields and forests, one regards a housing estate in the middle of meadows and fields an intrusion. There is no new concept for a new, urbanized landscape. I suggest that an assessment of new landscapes that uses concepts and images of a landscape that has gone should be described as 'regressive'. We reach back to old patterns and want to save the familiar by devaluing the changes instead of trying to understand them. We are familiar with this thought structure, by the way, from the criticism levelled at cities at the end of the nineteenth and beginning of the twentieth century.

Because it was no longer possible to find the village community in the city, the city was felt to be cold, dangerous and immoral. Similarly, even today the industrial landscape, the modern agricultural landscape and the urbanized landscape are evaluated critically because they do not correspond to the image of landscape formed by small farms (*bäuerliche Kulturlandschaft*).

When old concepts are applied to new material landscapes, arguments are often deployed that are not related to concepts, images or an aesthetic judgement, but identify the object itself as a disruption that is felt to be a loss. It is not that one deplores the loss of a landscape, but rather that one describes it as landscape destruction. Previously this sort of argumentation was common among ecologists. When one spoke of environmental pollution, the implication was that the Beautiful was also ecologically valuable. The ecologist Peter Finke wrote: 'There is often a considerable discrepancy between the usual evaluations of landscape as beautiful, charming . . . or damaged and their actual value from a landscape ecology perspective' (Finke 1992: 278). Also in the model of the regressive landscape concept

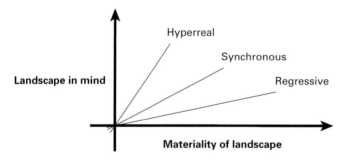

Figure 3.4
Types of landscape transition.
(Source Detlev Ipsen)

there are often psychological traps. It is not recognizable simply as a conservative concept, but is rationalized and disguised as critical and progressive.

In the third case, concepts and images of landscape are changed, whereas the reality is subject to relatively small changes or even none at all. This seems to be to have been the philosophy concealed in several large-scale landscape projects, such as the International Construction Exhibition Emscherpark (*IBA – Internationale Bauausstellung Emscherpark*). The dumps of long disused mines, the remains of steelworks, the river transformed into a sewage canal cannot be cleaned up. So, the landscape concept is changed: the Emscherzone became Emscherpark, the dump, by means of a tetrahedron; the rusting remains of a steelworks became an industrial park.

This can be an imaginary concept and, in the sense of Baudrillard (1978), an agony of the real. This, however, does not have to be the case. When the building of the railway made it possible for the first time to ascend the Alps without difficulty and enabled the masses (and not just English pioneer Alpinists) to discover landscape as beautiful wilderness, the real landscape changed little at first. The Alps remained the Alps. But soon hotels and holiday homes, walks and huts began to develop. Cow bells became souvenirs and the Zillertaler (name given to the people inhabiting the Zillertal) became a symbol of the original and good 'untamed'. In other words, the imaginary can be visionary and thus have incisive influences on the real landscape and thereby adjust reality to the imaginary. In this model of landscape construction there is not only the danger of remaining imaginary, but also of equating the imaginary or at least confusing it with reality. Image is literally not merely in the concept, but also in the matter itself (Figure 3.4).

## Areas of basic knowledge

With the relationship between the modernization of society and the landscape as process the basic connection between sociology and landscape planning/ landscape architecture unfolds. First, we will relate modernization theory in a more concrete way to landscape architecture. Later, three aspects will be addressed that might be particularly relevant for landscape architecture from a sociological point of view:

- The sociology of space and the changes made to space – time structures in their effects on landscape development.

- The emergent formation of classes, strata and lifestyle groups who develop claims on landscape use and landscape images.
- Mental landscapes in mind: the subjective meaning of landscape, the social conditions of aesthetic judgements and spatial images relating to landscape.

## *Modernization theory and landscape planning/landscape architecture*

Modernization theory will be clarified in four theoretical approaches to show the connection between sociological modernization theory and landscape planning/landscape architecture. Examples taken will be the theory of civilization (Norbert Elias), the theory of disembedding (Anthony Giddens), the theory of communicative action (Jürgen Habermas) and regulation theory (Michel Aglietta). Of course, these are only examples.

According to the thesis of Norbert Elias (1995), modernization has fundamentally changed the relationship of society to nature and, with it, the use and perception of landscape. In his principal work *Prozess der Zivilisation,* Elias developed the thesis that modernization is above all an increase in the control over external and internal nature. Whereas control of nature as environment was also recognized before Elias as a characteristic of modernization, the connection with control of the internal nature of man as a prerequisite for the control of external nature is a new perspective. The control of external nature presupposes foresight, for which spontaneous and emotional utterances of internal human nature are counterproductive. Elias demonstrates the refinement of eating culture and the development of hygiene as an important area of internal nature control. Gradually a distance from one's own nature has emerged, which supplements the distance that has emerged from the control of external nature. This double distance expresses itself in a marked inclination to avoid unplanned nature-determined events and utterances. Control of the waters is an example that characterizes landscape. At the same time it is clear in the case of water that the distance from nature is paralleled by nostalgia for nature. This is echoed in the expectations of landscape planning too. Images of naturalness, even wilderness that does not threaten, are sought in landscape. The dialectic between distance and nostalgia, alienation from nature and the need to assimilate to it constitute the socio-cultural and psychological area in which landscape architecture has to perform.

With his concept of disembedding, Anthony Giddens (1990) has attempted to formulate the social focus of the extremely complex process of modernization. The first laws to privatize the commons in England led to a situation where small farmers were no longer able to buy land, whereas capitalist landlords bought up the common property in order to graze sheep on them. As a raw material, wool led to a dynamic development of the textile industry. Numerous small farmers were no longer able to survive without the use of the commons and were compelled to migrate to the towns in order to find work. They left the village community, often with their families, and they also left the countryside that was familiar to them. This is what Giddens means by disembedding. It is often associated with changes in economic use and with changes of political regulation. Use and form of landscapes change within the framework of these processes. At the same time new groupings, interest groups and security systems develop in order to anticipate social disintegration. Landscape architecture finds itself in the middle of this area of disembedding and re-embedding. With large-scale schemes, such as the canalization of rivers, the building of dams, but also with the enclosure of protected areas that exclude groups of users, landscape planning can destroy places and thereby contribute to social disembedding. On the other hand, landscape design can create places that compensate for loss and contribute to new fixtures.

Habermas, in his theory of communicative action (1981), distinguishes three subsystems by which the modernization of society is determined: the economic system, whose means of action is money, the political system, whose means of action is power and 'lifeworld' (*Lebenswelt*), whose means of action is communication. Habermas develops the thesis that in the course of modernization the economy and politics impregnate 'lifeworld' with increasing strength. He calls this the colonization of 'lifeworld'. At the same time it is 'lifeworld' in which social innovations are able to develop. The colonization of 'lifeworld' by politics and the economy, in the medium term, undermines society's power of innovation and endangers the project of the Modern, which with all its contradictions has developed utopian energy for the 'lifeworld' of people: the emancipation from dependence on nature, facilitation of work, opening of cultural multiplicity . . . This approach requires thought in planning. Does a concrete approach of landscape design create room for manoeuvre for everyday communication or contribute to the colonization of 'lifeworld'? Many planners are commissioned to do their work by the government or the economy, but this fact alone by no means determines the content of a design. It is by no means a question of a schematic contrast between politics and the economy on the one hand and 'lifeworld' on the other, but rather a complex field of dependencies and chances of emancipation. In any case this approach places communication in the centre of planning and design. This does not of course mean simply addressing a naive participation approach, but rather aiming for a planning method that qualifies commonsense in matters of planning and aesthetic design. In landscape architecture one can count on openness among people. My own experience has shown that landscape has great value for most people (Ipsen *et al.* 2003) However, both the legal codification of use of landscape and the commercialization of landscapes have to be subjected to critical consideration if one, willingly or unwillingly, wishes to avoid the charge of contributing to the colonization of 'lifeworld'.

The examples cited so far all show that they can contribute to an assessment of the effects of planning and design in the modernization process. They serve to evaluate finished projects and can also, at an early stage, contribute to the development of planning ideas and design concepts. The regulation theory approach of Aglietta (2000) serves, on the other hand, rather to analyse problems and formulate planning questions.

We have to thank the Russian economist Kondratiev (1926) for the insight that capitalism not only moves in short growth and depression cycles but also exhibits long waves of growth and fundamental crises. Kondratiev made the connection between these long waves and key innovations. The invention of the smelting of iron with coal in the Severn Valley in Great Britain made not only the building of the first iron bridge possible, the Iron Bridge in Coalbrookdale, but also an entire age of railways, iron bridges and iron-glass architecture. The mobility of goods and people was raised decisively and the principle of speed in the circulation of goods developed with the building of the railways and, by means of new technologies, is still developing. The acceleration of travel stimulated a changed perception of landscape. Where Goethe was able in his sketches of his Italian journey to describe from his coach the close view of what was growing on the side of the road, the railway led to the development of the distant view. The panorama became the favourite landscape view, since the close view from the train was no longer possible because of the speed. It was soon followed by hills with a view, towers with a view and panorama scenic routes. Aglietta (2000) developed this fundamental approach into a theory of regulation. Certain periods of modernization can be described not only by predominant technical-scientific inventions and are characterized not only by a particular form of capital accumulation, but they combine with value structures in society, with organizational forms of the state and a specific space-time structure. Landscapes as a form of space-time structure have, according to the theory, forms of use that correspond to the regulation periods. The forms of perception, the aesthetic and evaluation of landscapes might also be determined by the form of regulation. One might suppose that the Western world, possibly even world society, is at the moment in a period of transition. The so-called Fordist regulation regime is crumbling and on all sides there is a need for reform. We will deal with the research topics associated with this with reference to sociology and landscape planning at the end of this article. A brief review of outgoing Fordist regulation will, I hope, clarify regulation theory at this point.

Regulation theory is based on the principle that social regulation of landscape is related to three regulation clusters which are connected to each other, but at the same time function independently of each other. The first area may be called, to use a term of Habermas, systemic regulation. This is a specific form of accumulation logic and the control of production and circulation and political control, which in principle are again both independent units.

Cultural regulation refers to values and time-specific perspectives, that is, concepts, images and models. Lifeworld regulation means lifestyles, consumption practices and social groupings of household forms developed and effective in a particular period in the form of groups of friends, associations, interest groups and political groups, if they are not completely parts of the political system. If these levels of regulation agree with each other and have a high degree of coherence, one can speak of a regulation regime. This does not exclude the possibility that temporal disparities, which are older and the germs of new regulation forms that may be effective in the future may co-exist. Coherence thus means a relatively fluid state characterized by tensions, contradictions and conflicts.

One of these regulation regimes is called Fordism. It characterized the economic and social conditions in the second half of the twentieth century. The elements of this regulation regime important for landscape development may be demonstrated by the example of a vintner. Anyone visiting the small village of

Ellerstadt in the Palatinate area of the Rhine in Germany finds it almost totally surrounded by vineyards. Seven full-time vintners work the largest section of the area. But if one had visited the village in the 1950s, one would have seen fields for grain, meadows for hay and pasture. In the stables and in the pastures one would have seen horses and cows, as well as pigs and hens. Of course, one would have found vineyards and been told that wine was sold in Mannheim and Heidelberg. In those days the village had a varied agriculture with varied agricultural activities. The fields looked a little as they now do in northern Alsace. Here a strip of vineyard, here a grain field, here a strip of orchard and meadow.

One vintner tells how the transition to vineyards was made. It all began when the feed for the horses started to require too much land. To keep pace with living standards, farmers wanted to and had to change to tractors, which could be bought for reasonable amounts of money. Now, wine-growing requires narrow tractors with short axles, unlike the broader more powerful ones needed for grains. Both would have been financially out of the question. The decision was also influenced by the market prices in favour of viticulture. Ellerstadt became a wine-growing village, surrounded by endless vineyards. Initially, all the relatives and friends came to harvest the grapes and a few outside workers were also employed. Their hands were often numb in the morning cold, and this increased the feeling of comfort when the pot of *Krummbeeren* (the name for potatoes in this region) and liver sausage was brought out to the vineyard. But fewer and fewer people wanted to do this job, and then the harvesting machines appeared. This was no cheaper, but at least one no longer had the trouble of having to get everyone together for the harvest. The harvesting machines needed large tracts of land, so the fields in the village of Ellerstadt were consolidated. Not only did the fields become larger, but they also became more suitable for machines and as geometrical as possible. Fordism is essentially a radical switch of efficiency standards and consumption patterns. The tractors were reasonably priced because they were mass-produced and no longer assembled by hand. Increased efficiency improved wages, and so almost everyone could afford to buy what was on offer. Even non-industrial areas, such as agricultural ones, were affected by the new standards. Efficiency here meant the highest possible use of machines and concentration on a few products. Farmers did not want to be left behind in consumption. In order to have the sort of urban home comforts, to own a motorbike or a car, the farmer had to work in a more rational way.

The use of artificial manure and herbicides guaranteed higher yields, but also caused considerable damage to the ground water. Drinking water supplies had to be centralized because the village pump no longer provided an adequate supply. An efficient community structure became necessary to replace the small village mayors and villages in which people had been largely self-sufficient. In the 1970s there was regional reform, and the community became larger, got a full-time mayor and administration. Land use planning became large scale. On the edge of some villages areas were set aside for housing development. Town-dwellers moved to the countryside, the car made commuting easier. A new motorway made travelling into the city easier and the villages became urbanized.

Fordism is also a form of regulation in which state control increases perceptibly. For agriculture, programmes of field consolidation were organized, minimum prices guaranteed, quality control introduced. Land use came to be regulated more strictly and more comprehensively. A system of development plans provided for land use in a functional structure to be agreed between the states, regions and communities and co-ordinated by the Federal State. Ecological compensation areas were assigned to overcrowded urban regions. Areas of intensive agricultural use were balanced by landscape and nature-preserve areas. In many places spatially functional order was reflected in the shape of the landscape. We can summarize the important elements in landscape changes during this period as follows:

- Monocultural areas increased substantially as a result of increasing specialization.
- The result of mechanization in agriculture is a geometrical field pattern, and at the same time small-scale topographical disparities have been levelled off.
- The abandonment of unproductive fields has led to the development of secondary wilderness.
- The growing importance of individual mobility has led to extensive housing areas, which are neither village nor urban.
- Increasing traffic has reduced the number of traffic-free roads and paths and led to what can be described as isolation of the landscape, in particular restricting the movement of small animals.

The factors determining this development have tended to contribute to a marked geometrization of the landscape. As a result, the landscape appears less differentiated and more homogeneous. The agricultural landscape has thus in the last fifty years changed considerably in its ecological structure and appearance.

## In-depth aspects of the relationship between sociology and landscape planning/landscape architecture

We wish to examine three aspects of sociological knowledge in landscape architecture in more detail. As noted earlier, it is a matter of the sociology of space, the role of social classes and lifestyle groups and the conditions under which landscape is perceived and experienced. Of course, there are many other points of contact, but they are mostly specific aspects of those mentioned above. Thus, demography, migration, the sociology of work etc. have an impact on landscape architecture. Ultimately these are questions of spatial sociology, or one can discuss them within the framework of group interests and the social structuring of landscape perception.

The in-depth aspects discussed here, space, social classes and the perception of landscape, are intimately connected with the modernization of society. Pre-modern concepts of space, such as the earth as a flat disk surrounded by an unnavigable sea behind which paradise is located, influenced the voyages of discovery and the beginning of modernization. Paradise is still a central concept of landscape. Colonialization and the plants brought back from the colonized regions decisively changed the landscape of the Mediterranean. As the working class emerged with modernization, it changed the architecture of urban parks, which were now intended to serve health needs and no longer simply the desire of citizens to stroll. In this sense we are still within the framework of modernization theory.

## Landscape as space

Let us turn first to the sociology of space. Landscape is an access point to social and physical space. In this sense, landscape, as a specific case of general approaches, is subject to the analysis of spaces. I myself always find it very useful to keep an eye on physical space even in the sociology of space. Volcanic landscapes develop a typical topography and basalt has an influence on vegetation forms. With respect to physical space, forms of social use, cultural narratives about space, specific special places etc. develop. Bearing this in mind, we can address ourselves to the sociology of space. Space and time are to be considered together not only in physics. They are constructions which undoubtedly have reference to matter, but they are also cultural constructions. Let us the take the concept of time-cycles. Day and night, the seasons as they run their course, the life-cycle between birth and death has been very differently assessed in their social significance, historically speaking. In antiquity and in the Middle Ages cyclical time was the element that gave structure to society. At the same time, time-cycles are always spatially related. The vegetation cycle is space determined. Use and the aesthetics of space are influenced by a time sequence. On the other hand, other concepts of time, such as the linear one, developed through spatial developments, such as the construction of the railway. A train leaves at 8 a.m. and reaches its destination at 6 p.m., time being a linear sequence. Linear time, which today is the predominant model and names an abstract sequence of event units, is already, by definition, spatial.

For a long time space was understood as something 'that was there', in which something then happened. The works of Von Thünen (1875) represent this understanding. In the Thünen circles, so-called after him, agricultural use changes around the town markets. Quickly perishable produce, such as vegetables, is to be found near the town, more storable produce, such as grains, outside. The space is there and is structured by transport technology, the perishability of the produce, transport costs and market prices. In this case one can speak above all of the agricultural aspect of space. Since the works of the French sociologist of space, Lefebvre (1991), the concept of space has changed: space itself is an activity and process. Interestingly, this view is related etymologically to the original meaning of space in the German-speaking world: the German word *Raum* (space) comes from *Roden* (clearing). Not everything is *Raum*, but comes from the activity of *Roden*. If one thinks of space as a process, it becomes clear once again that time and space have to be thought of together. Landscape, above all, is process space. Constant movements, short-term events, long-term transformations determine the naturally marked landscape. Social action is either stored in these processes – think of the routes taken by shepherds in Europe or Alpine pastures, which are oriented in accordance with heights or vegetation periods. This has long been the dominant form of the relationship between natural and social landscape processes in Europe. Or social processes are superimposed on and transform natural landscape processes – dams overlay the river and river landscape, urbanization covers the structure of the ground and models the topography. I would like just to mention that the relationship of natural and social landscapes to each other substantially determines the debate on sustainability. It is clear from urban ecology (Sukopp 1998) that, even with intensive superimpositions such as undoubtedly happen with urbanization, urban landscapes of this sort still demonstrate a varied ecology. Storage of social processes in natural ones and superimpositions are ideal types. With respect to urban vegetation, one might call it secondary vegetation that was rediscovered in the 1980s as quality in landscape architecture.

With social space processes a distinction is made between a space of places, which has been dominant during a long period in history, and a space of flows, which has been more and more important in the last decades. Planning and realization of space flows can be called 'spacing'. As a result functionally interconnected space flows emerge: traffic routes, water supply networks, waste channels, energy transport, commuters – and leisure activity movements of people, raw materials movements etc. These 'space flows', which have been known in principle since the aqueducts of Roman times, have become much denser and more rapid (Castells 2000, 2004). In addition to material flow spaces there are increasing numbers of flow quantities: speech flows, data flows, image flows, money flows. Space flows correspond to a clear division of labour, by means of which in principle an optimal allocation of economic resources is to be guaranteed. Up to the 1980s regulation of rivers, roads and high voltage wires not only disturbed many places, but even totally disrupted them. Squares disappeared, gardens were driven away and forests and meadows built over. This process is still visible in regions of the world with strong urban growth. On the other hand, places were created that are described in perception psychology as 'Gestalt'. A 'Gestalt' is more or less clearly distinguishable from a ground and this is especially true of places.

Places can be natural (crest of a hill, a river bank, a coastline) or artificial. This artificiality is described, in contrast to spacing – the creation of space flows – as placing. Space flows are oriented to functions and division of labour and are to be understood systematically. In this context economy and politics are important. Places are to be understood in the 'lifeworld' sense and this is also to be understood in a concrete way: One wants to live in this landscape, in this quarter, on this square. Now, most landscapes are not a 'place' in this sense, but are characterized by space flows.

Landscape planning and architecture are involved in the whole spectrum between space flows and creation of a place. Between these extremes there is not merely an important dialectic of landscape space, but also superimpositions. One knows this from 'railways': small gardens of the railway employees became places in the space flow of the railway, in that unused areas were put to use. Sewage plants and waste incinerators, logistics centres and motorways can become important jobs for landscape architecture and still are today. The economy of urbanization has created level areas, rows of the same or very similar living boxes. To create places here is at least a job for landscape architecture.

## Social classes, strata and lifestyles

The interest in landscape, its use and the perception of landscape are determined by the social position occupied by a person or group. In order to describe this position sociology distinguishes classes, strata and lifestyles.

Classes are distinguished by ownership or non-ownership of the means of production. For landscape architecture property ownership or the use of the land as leaseholder are decisive. This means not only farmers, but also companies for the mining of raw materials, owners of waters etc. Social strata are determined by income and education/knowledge. Property and education can be quite far apart from each other – there are groups with high income and low education and vice versa. This is known as status inconsistency and is often the reason for marked expressions of interests. Lifestyles are distinct patterns of behaviour on an everyday level: house, clothes, nutrition and cultural practices. The three concepts enable us to describe the differentiation of society and explain corresponding interests. A farmer belongs to the class of property owners, but if he has a low income can belong to the lower stratum, have a formally low educational status, but an enormous amount of practical knowledge. In many areas of South Germany it was this group that had distinct cultural lifestyles characterized by music and culture. Other farmers in a wine region, however, can belong to the upper middle stratum and have received a university education. Other classes, such as blue and white collar workers, have distinct views of landscape, according to stratum and education. In this way the lower strata value orderly landscapes, while the educated strata look for complex 'wild' places. Unlike classes and strata lifestyle groups develop demands on landscapes. Think of the thousands of kilometres of hiking, cycle and riding paths that have arisen in recent decades. The tendency is for the development of lifestyles to become even more important. Certain types of sport and cultural interests are developing into complexes distinguished by certain clothing, seasons and landscapes. These lifestyles are tied up with the interests of industry and trade, advertising and the media. The examination of the social differentiation seems to me important for two reasons. First, interest constellations and thus interest conflicts can be analysed at the design or planning stage. The building of a regional park has to know and take into account the interests of farmers, real estate, sports clubs etc. Second, trends and scenarios of future developments can be developed and thus the new challenges to landscape architecture identified.

## The mental landscape

As a corollary to the real material landscape there is a second landscape, the mental one.[2] It is with caution that we speak of a correlation – which may be low or high between the two and not an equation (*Entsprechung*). This means that there can and very often will be a gap between the mental image of a landscape and the objective counterpart. First, it has already been emphasized that landscape is a construct that emerges only because it is an abstraction from the variety of material elements that determine a delimitable geographical area. Our perception follows image-forming processes which reduce the complexity of the world. These processes can operate in very different ways individually or in certain historical periods. Thus, whole space types like the Alps or limited regions like Tuscany were seen and evaluated in quite a different way from today. And everyone knows that there are differing opinions about landscapes, because each evaluation is individual. Behind the differing evaluations there are factors that structure the relationship to material landscape. The understanding of how certain ideas and evaluations of landscapes arise is only of importance when one assumes that these ideas substantially determine the individual and social use and future development of a space.

I do not mean all types of development, but rather those in particular that have a direct effect on society's relationship to nature. Let us take an example: if the forest as landscape is not valued, if it is felt to be a threat, it might be difficult to enforce programmes against acid rain damage to forests. In Germany this damage has been described as the 'death of the forest' and thus come to be regarded as an existential question. This can happen only if society puts a high value on forests. A certain material property of a space is so structured in consciousness that a landscape and thus an image of landscape is perceived to be 'in danger'. This perception builds on something we call landscape consciousness.

What does landscape consciousness mean? If one assumes that landscape as an integrated concept addresses the natural space side of a space, its use and use history, the social structuring and its cultural meaning, then it is still not clear how this complex unit can affect action in a region. Our working concept for this is 'landscape consciousness'. The material and aesthetic, the economic and cultural aspects of a landscape become action-effective only if they are implicitly or explicitly present in the consciousness of the actors. Three dimensions of landscape consciousness can be distinguished analytically. Landscape consciousness has cognitive elements. There is knowledge about a landscape. This knowledge can be more or less differentiated. One can know that fertile soils are used intensively in agriculture in Europe. One can also know that these soils have been created by wind that, in the course of millenia, has blown fine silt or 'loess', which has been deposited in the areas known as '*Börde* soils' in Germany. The question of which social groups have or do not have knowledge about landscape is an empirical one. It is important for the effectiveness of knowledge to know whether this knowledge is professional, that is 'administered' by only a few people, or whether it is general. Secondly, there is an aesthetic dimension that embraces both – perception or non perception, the anaesthesia of landscape and the aesthetic evaluation. There are spaces that are present in landscape consciousness (the Lüneburg Heath, the Black Forest); for others there are no general names. Aesthetics condenses perception to a concept and assigns values to it. It is thus an important prerequisite for the question of whether communication can be made of a landscape, whether this communication can be made locally or generally. There are landscapes such as the Amazon rain forests which have produced a global aesthetics and can thereby enter the realm of global communication. Thirdly, there is an emotional dimension which manifests itself as place-relatedness or spatial identity. Home as a place of origin or a longer biographical personal history generally leads to an emotionally positive or negative tie. Figure 3.5 reflects this dimension and places it in relation to the communicability of landscape.

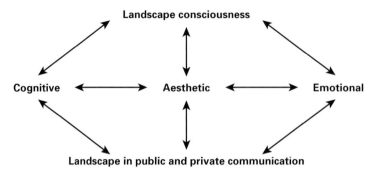

Figure 3.5 Landscape in private and public communication. (Source Detlev Ipsen)

In the further discussions of the dimensions of landscape consciousness these rather social-science dimensions can and have to be related to concrete aspects of landscape. Thus, knowledge of a landscape can be related to the natural data, to the use of a landscape, to the history of a cultural landscape or to cultural meanings. If one systematizes this association, the result is a table that determines the empirical field of landscape consciousness (see Table 3.2).

There is a dialectical relationship between landscape consciousness and the material landscape. Landscape consciousness always refers to a materiality of the environment from which consciousness is abstracted. By reducing complexity the 'endless' variety of the material world produces an image. These images in turn have an effect on the formation of the landscape, by exercising direct control of the use or indirect control of the political regulation of the use. We have already stated emphatically that the material space of the landscape and landscape consciousness are related to but do not correspond with each other. Landscape often changes, but the image remains the same in consciousness. The mountain rambler tries to exclude forest paths, high voltage wires, lifts, blocking of streams etc. from photographs, so that his image of the mountains is confirmed. Elements that disturb the image are omitted. Or he develops a critical view of the discrepancy between image and reality. An important reason for the tension between image and reality is the construction

**Table 3.2** The empirical field of landscape consciousness

|  | Cognitive relationship | Aesthetic relationship | Emotional relationship |
|---|---|---|---|
| Natural space | Biology, ecology, nature protection etc. | Nature aesthetics Nature observation | Love of nature |
| Use | Landscape history, place knowledge etc. | Perception of cultural space | Use ties Place significance |
| Social structuring | Ownership, legal regulations etc. | Particular places Personalities | Social networks Spatial milieus |
| Cultural significance | Stories, literature, painting | Symbolic significance of particular places | Dialect, home, identity |

of landscape consciousness that takes time and has been affected by certain innovations like the railway. At least for the twentieth century one can say about Europe that the asynchronicity of landscape development and landscape consciousness is significant and has given rise to important phenomena that have had an effect on landscape, such as nature and landscape protection.

## Research approaches

It must be emphasized at the beginning that there is such a small amount of theoretical and empirical research on landscape development and landscape architecture in sociology that one can easily compile an almost endless list of possible approaches. There is no basic research on landscape perception and use patterns. Also missing is the social scientific evaluation of important landscape projects, such as the restructuring of the Maas between The Netherlands and Belgium or about the effects of nature parks in Austria. Strikingly little is known about the different meanings of landscape in EU countries. Is the meaning of landscape in the consciousness of people in Mediterranean or in East Central Europe as high as we know it to be from countries like the UK or Germany? What are the mental landscape images? What is the relationship between land use and nature protection? Indeed, an examination of the sociology of European landscapes would not only be interesting, but could be very useful for the further development of European cultural landscapes (see Figure 3.6).

This is becoming all the more important as decisive shifts in landscape development manifest themselves. The main factors for these shifts are the emergence of urban regions with strong growth tendencies, the evacuation of certain

**Figure 3.6**
The relationship of rural and urban will be the main field in the sociology of landscape. (Photos Detlev Ipsen)

regions for demographic reasons, the shrinking of cities as residential and industrial areas are torn down and the emergence of new urban open spaces and the restructuring of agriculture. This area, which might be described as European Landscape Dynamics, is one where most of the research projects should be of an interdisciplinary nature. The following issues are the most urgent research questions:

- What should be the future development of urban landscapes in urban regions? There is already important initial research, such the network analysis of Baccini and Barabesi (2008), which however does not take account of sociology.
- What should be the future development of open spaces in shrinking cities? There is already an international project on Shrinking Cities that should serve as a starting point.
- Will new peripheral areas emerge as a result of demographic development and what strategies for landscape planning should be developed?
- What will be the impact of the change in agriculture to the production of sustainable raw materials on the landscape?

Sociology can bring into this research complex the questions of use and interests of various social groups, the question of landscape consciousness, aspects of participation and the mediation of conflicts of interests.

## Concluding remarks

The relationship between sociology and landscape architecture can be very productive and innovative, but can also present problems. In sociology, and also in urban and regional sociology, the subject of landscape architecture is known only to a few. Mostly, the research is devoted to the use of urban open space. The least research has been done on the development of landscapes which are not used for tourist purposes. There are also important works about the Alps or sea coasts written either by sociologists or sociologically oriented geographers. In order to take stock of the state of research it would be necessary to examine, in a study related to this article, the research on rural sociology and the sociology of tourism, the studies in the field of ethnology and anthropology for possible landscape associations.

For further work I suggest the module of a course 'Sociology of Landscape Architecture' be developed in a workshop of sociologists, social geographers, anthropologists and landscape architects. Part of this course would certainly be devoted to the examination of important arguments on the modernization of society. There would also be part devoted to the methods of sociology, such as interviews, observations, group discussions etc. with reference to landscape architecture. On the whole, however, it is not necessary in my opinion to start from the social sciences, but from the questions and problems which arise in landscape architecture. What can sociology say about the development of leisure landscapes? What contribution can sociology make to the ecology of urban landscapes? What could a park that corresponds to an individualized society look like? How should the design of the open space look in a multicultural city? Only then is it likely that landscape architects would take an interest in sociology and the cooperation between the two disciplines lead to creative results.

## Notes

1 Since submission, Detlev Ipsen has sadly passed away.
2 'Cultural landscape in mind' – The name of a research module of the Austrian programme for cultural landscape research (cf. Strohmeier 1997).

## Bibliography

Aglietta, M. (2000) *Ein neues Akkumulationsregime. Die Regulationstheorie auf dem Prüfstand*. VSA-Verlag. Hamburg.
Baccini, A. and Barabesi, L. (2008) *Interlocking Editorship. A Network Analysis of the Links Between Economic Journals*. Department of Economics. University of Siena 532.
Baudrillard, J. (1978) *La précession des simulacres*. Traverses 10. Paris.
Bell, S. (1999) *Landscape. Pattern, Perception and Process*. Spon Press. London.
Benson, J. F. and Roe, M. H. (eds) (2000) *Landscape and Sustainability*. Spon Press. London, New York.
Berman, M. (1988) *All That is Solid Melts into Air. The Experience of Modernity*. Penguin Books. Canada.
Bölling, L. and Sieverts, T. (2004) *Mitten am Rand*. Verlag Müller + Busmann KG. Wuppertal.
Bruns, D., Ipsen, D. and Bohnet, I. (2000) 'Landscape dynamics in Germany'. *Landscape and Urban Planning* 47, 143–158.
Burckhardt, L. (1992) 'Ästhetik der Landschaft'. In Kos, W. *Die Eroberung der Landschaft*. Wien.
Burckhardt, L. (2006) *Warum ist Landschaft schön?* Martin Schmitz Verlag. Berlin.
Castells, M. (2000) *The Rise of the Network Society*. Blackwell Publishing. Oxford.
Castells, M. (2004) 'An introduction to the information age'. In Webster, F. (ed.) *The Information Society Reader*. Routledge. London and New York, 138–149.
Corner, J. (1999) *Recovering Landscape: Essays in Contemporary Landscape Architecture*. Princeton Architectural Press. New York.
Cosgrove, D. E. (1984) *Social Formation and Symbolic Landscape*. Croom Helm. London & Sidney.
Elias, N. (1995) *Über den Prozess der Zivilisation. Soziogenetische und psychogenetische Untersuchungen. Bd. 2: Wandlungen der Gesellschaft. Entwurf zu einer Theorie der Gesellschaft*. Suhrkamp. Frankfurt/M.
Finke, P. (1992) *Die Eroberung der Landschaft* (ed. Ausstellungskatalog), Wien.
Giddens, A. (1990) *Consequences of Modernity*. Stanford University Press. Stanford.
Gröning, G. and Herlyn, U. (1996) (ed.) *Landschaftswahrnehmung und Landschaftserfahrung*. Lit-Verlag. Münster.
Habermas, J. (1981) *Theorie des kommunikativen Handelns*. Suhrkamp. Frankfurt
Hoskins, W. G. (1955) *The Making of the English Landscape*. Hodder and Stoughton Ltd. London.
Ipsen, D. (2006) *Ort und Landschaft*. VS Verlag für Sozialwissenschaften. Wiesbaden.
Ipsen, D., Reichhardt, U., Schuster, S., Wehrle, A. and Weichler, H. (2003) *Zukunft Landschaft. Bürgerszenarien zur Landschaftsentwicklung. Arbeitsberichte des Fachbereichs Architektur, Stadtplanung, Landschaftsplanung*, Heft 153, Universität Kassel. Kassel.
Jackson, J. B. (1984) *Discovering the Vernacular Landscape*. Yale University Press. New Haven.
Kaufmann, S. (2005) *Soziologie der Landschaft. Reihe Stadt, Raum und Gesellschaft*. VS Verlag für Sozialwissenschaften/GWV Fachverlage GmbH. Wiesbaden.
Klijn, J. and Vos, W. (eds) (2000) *From Landscape Ecology to Landscape Science*. Dordrecht, Boston, London. Pp. 73–80.
Kondratiev, N. D. (1926) 'Die langen Wellen der Konjunktur'. In: *Archiv für Sozialwissenschaft und Sozialpolitik*. 56, 573–609.
Kos, W. (ed.) (1992) *Die Eroberung der Landschaft*: Semmering, Rax, Schneeberg. Ausstellungskatalog. Falter Verlag. Wien
Kost, S. (2009) *The Making of Nature. Eine Untersuchung zur Mentalität der Machbarkeit, ihre Auswirkung auf die Planungskultur und die Zukunft europäischer Kulturlandschaften. Am Beispiel Niederlande*. Metropolis. Marburg.
Lefebvre, H. (1991) *La production de l'espace*. Anthropos. Paris.

McNamara, K. M. (1995) *The Urban Landscape*. Selected Readings. University Press of America. Lanham, Maryland.

Penning-Rowsell, E. C. (1986) *Landscape Meanings and Values*. Unwin Hyman. London.

Sieferle, P. (2003) 'Die Totale Landschaft'. In Oswald, F. and Schüller, N. (eds) *Neue Urbanität – das verschmelzen von stadt und landschaft*. GTA Verlag.Zürich.

Simmel, G. (1957, 1913) 'Philosophie der Landschaft'. In Simmel, G. (ed.) *Brücke und Tür*, F.K. Koehler. Stuttgart, 141–152.

Strohmeier, G. (1997) 'Kulturlandschaft im Kopf'. In Grossmann, R. (ed.) *Wie wird Wissen wirksam*. iff texte, Wien/New York.

Sukopp, H. and Wittig, R. (eds) (1998) *Stadtökologie*. Fischer. Stuttgart.

Von Thünen, J-H (1875) *Der Isolierte Staat in Beziehung auf Landwirtschaft und Nationalökonomie*. Wiegand, Hempel, & Parey. Berlin.

Wylie, J (2007) *Landscape*. Routledge. London.

Chapter 4

# A prospect of time

Interactions between landscape architecture and archaeology

*Graham Fairclough*

## Introduction

The vision of a widely inter-disciplinary and integrated field of landscape studies has in recent years taken a much wider hold on the scientific and academic imagination of all those working in the many disciplines concerned with landscape research, understanding and practice. Inter-disciplinarity is now a commonplace aspiration in the field and even though the difficulties of attaining such a goal remain manifold, it might be said that a post-disciplinary position is already becoming the next horizon to reach. The European Landscape Convention has also provoked a coming-together of various landscape-focused disciplines and has become a widely used framework for teaching, study and practice.

Archaeology is an inherently multi-disciplinary field of research and practice, standing on both sides of the humanities and sciences divide in the subjects it covers, the theories it follows and the methods and techniques it uses. It is therefore not surprising that the landscape-facing parts of archaeology have been expanding for some years and have been seeking greater connections with other landscape disciplines. This applies not only in management and planning related practice but in research as well. Like landscape architecture, landscape archaeology is well placed to respond to the new challenges and opportunities thrown up by the new paradigms emerging in landscape studies (Fairclough and van Londen 2010, Fairclough 2011).

## A landscape of archaeologies
### Archaeology beyond landscape: the parent discipline

Archaeology as a whole operates at many scales, and many practitioners regard 'landscape' as one of them. It has been said that 'Landscape is central to archaeological practice, and has been for some time' (Schofield 2007). Archaeology's initial, one might say its traditional, focus has been sites, monuments and artefacts large and small, but among its origins – and from the sixteenth century onwards – is a concern with topographic descriptions and the study of the land and its human face. In more recent decades historic buildings, another of the discipline's starting points, from the eighteenth century, have re-emerged as a major topic for archaeological study, creating in the process another 'handle' on landscape and its inherited character. Archaeology's scope extends to much smaller scales as well, to human and animal remains, to artefacts (such as flint blade or potsherd), to individual deposits and soil layers ('contexts' in the jargon of archaeological excavation), and far beyond into the microscopic scale of, for example, pollen grains, the microscopic structure of soils and most recently DNA. Archaeological work at these scales may not appear very directly relevant to landscape architecture, but the knowledge it produces can be synthesised and 'filtered' through landscape archaeology – rescaled by generalisation – so that it does become relevant. The practice of garden archaeology is a demonstration of this in a context familiar to landscape architects.

Archaeological methods are immensely varied. Excavation – 'digging' – is merely one of the most popularly recognised; the full spectrum also includes field observation and survey, artefact analysis and conservation, site and garden restoration, many 'hard' scientific analytical techniques and, quite simply, thinking about the impact of the past's material traces on the present. The discipline's theoretical base is also diverse. All 'archaeologies' share a core of interests, however, notably a concern with the passage of time and its effects, and with the changes and continuities that result from it, and above all else, the study of the human actions and lives that stand behind the material remains from the past that survive in the present. All these are also of course fundamental aspects of landscape character as well (Fairclough 1999, 2003).

Archaeology in almost all of its fields is becoming increasingly landscape oriented or at least landscape aware. The past few years alone have seen a plethora of books (as well as many journal articles and even whole journals, such as *Landscapes*) published in many countries (see the references section of this chapter). Landscape is a common theme of numerous national European conferences (several sessions at each of the annual conferences of the European Association of Archaeologists, for example).

The word landscape may indeed be one of the most used (or over-used?) term in recent archaeological literature. Taking a random example to show the increasing importance of landscape concepts in all types of archaeological research, in the *c.*200 index entries of a book on New World archaeology reviewing recent work on 'The Archaeology of Communities' 'landscape' has 32 page-references, a total exceeded only by words that are much more central to the main theme of the book, namely boundaries, villages, settlements, households, houses, society, family and identity (and, ceramics, the archetypical archaeological obsession).

One explanation for this interest of archaeologists in landscape is that archaeology is ultimately a way of telling stories about the past, of drawing up biographies of people and places, or narratives of travel to different times and places, describing journeys through the past. 'Landscape' defined as a perception of the world is one of the best ways that exists to explain people's place in the world and their origins, identity and roots.

Landscape study within archaeology has a long history. Archaeology was already partly about landscape from its earliest origins, as the earliest, often travel-based, antiquarian accounts of (for example, in England) John Leland (1502–52) or John Aubrey (1626–97), and their equivalents in other countries, show, with their topographic and antiquarian attempts to describe and understand the land and identity. In the modern period, however, during the (re)-invention of archaeology for the new European nation states of the mid-nineteenth century, landscape for a while became less central to archaeology than monuments and buildings (cathedrals, castles, and megalithic monuments, for example). It was 'rediscovered' in the early and mid twentieth century because of the emergence of a variety of new methods or ideas.

In the UK, as an example, these new methods include the emergence of aerial photography as a new way of seeing landscape – notably O.G.S Crawford (Crawford and Keiller 1928), regional studies (e.g. Cyril Fox 1938), the development of local history (notably W.G. Hoskins, 1955), and the refinement of increasingly sophisticated techniques of large scale ground survey (e.g. Crawford 1953, Taylor 1974, Aston and Rowley 1974, Aston 1984, Bowden 1999). More recently, new techniques such as systematic field walking (Bintliff *et al.* 2000), often involving environmental archaeology (Castro *et al.* 2002) or geophysics, and often being fully integrated 'landscape surveys', have provided much wider views of the past and drawn other archaeologists into the debate about landscape. Even more recently, very large scale rural rescue excavation in the last part of the twentieth century, in many parts of Europe (e.g. lignite opencast in Germany, massive area excavation in the Dutch sandy regions, gravel extraction sites in the Thames valley), or excavations before major infrastructure, such as the recent Heathrow airport expansion, single large linear developments such as the cross-Channel rail links and TGV lines in England and France, or the response to the construction of whole networks of new motorways as in Ireland or Hungary, have given new very detailed insights into the time-depth and extensive patterning of large territories, thus enabling 'landscape evolution' to be studied. Since the 1960s, archaeological post-processual and later phenomenological theory (e.g. Bender 1993, Tilley 2006) has also contributed to the current diversity and health of landscape archaeology.

## What is landscape archaeology?

Several distinct branches of archaeology are referred to as landscape archaeology, but they are unified by a number of attitudes that stem from their parent discipline. These include the study of material culture to understand human history and the present day environment, therefore aligning much of landscape archaeology with a material view of landscape. They also share recognition of the need to see beyond the 'sites' or 'monuments' that had come to dominate archaeological practice to the point of being seen as real things rather than as constructs designed to make the complex more simple. This desire to work at 'landscape scale' – to study the whole not just some of the parts and to look at the supposedly empty spaces between sites – is one of the origins of modern landscape archaeology (Figure 4.1). The different branches of landscape archaeology also share an interest in ideas and theory

**Figure 4.1**
'... to study the whole not just some of the parts...': A 'traditional' – 'cultural' – landscape at the village of Blindcrake, at the edge of the Lakes National Park: small scale strip fields reflecting mediaeval land divisions in a distinctive settlement region of NW England (UK). (NMR 20146/21 [2004]). ©English Heritage. NMR

as well as methodology. Most of all, they share a concern for understanding the long story of how people have made our landscape over time, and thus for understanding our own present day landscape, which also allows landscape archaeology to engage with the perceptual, ideas-based view of landscape as mentality (e.g. Johnson 2007).

Archaeology is primarily a research-based historical discipline. But the knowledge that it creates about landscape contributes to understanding of the present as well as of the past. Its data is the surviving physical and material remains of the past, and it therefore always operates within a present-day context. In other words, archaeology studies the past within and as part of landscape through time, which particularly in the context of the European Landscape Convention has implications for the balance to be sought between change and conservation, and thus design (Fairclough 2007a). All types of landscape archaeology increasingly consider understanding as being a first step towards managing change in the environment and towards shaping future landscape. No longer (if they ever did) do most archaeologists seek knowledge of the past only for its own sake; instead they seek to find ways to influence the shape of future landscape. Landscape archaeology is thus developing an identity as an 'action' discipline as well as a research discipline, one which wishes to use information about the past in a practical way to shape future landscape. Because landscape architecture is principally defined as a design-based discipline, but one which is strongest when design is informed by knowledge and theory, this chapter will argue among other things that the two disciplines can be highly complementary.

## Diversity of approaches

Many different archaeological approaches are described by their practitioners as landscape archaeology. Recognising this is perhaps the first requirement for examining how archaeology and landscape architecture can collaborate.

For some archaeologists (as, for example, for some ecologists), landscape is mainly a matter of scale, extent and inter-relationship. Simply to act over large extensive areas (e.g. through field survey or excavation) is a form of 'landscape archaeology', although it is not always explicit whether landscape is the object or method of study. For others, landscape offers a special temporal rather than spatial scale that allows past environments (usually for a particular wide period of (pre)history, as in for example the Swedish 'Ystad Project' (Larsson *et al.* 1992) or the Bronze Age (as in Fleming 1988)) to be understood as landscape. Such approaches frequently focus on the mediaeval period, particularly on settlement patterns, and landscape archaeology has become a key method of mediaeval archaeology (e.g. Austen 1999, Rippon 2001, Roberts and Wrathmell 2002, Laszlovszky and Szabo 2003, Williamson 2003, Turner 2006b, Martin and Satchell 2008 and many others).

Foremost among the latter are researchers, including historical geographers as well as landscape archaeologists, who study settlement and its patterns at community, regional or national scale (e.g. Roberts and Wrathmell 2000). Garden archaeology is also a sub-branch of this sub-discipline, dedicated to understanding a particular type of past landscape. For prehistoric research at this scale, the approach is heavily environmental; it uses techniques such as pollen analysis, environmental indicators such as invertebrate remains or soil morphology, to reconstruct on paper or virtually the world in which our predecessors lived and which they changed and shaped as stages on the way to the landscape we have today.

Other landscape archaeologists, however, recognise that the word landscape contains a concept beyond that of simple size or scale and beyond a particular period of the past. One relatively new branch of landscape archaeology is closely linked to heritage management and spatial planning. This studies present day landscape – landscape as perception, as in the ELC, our own time's concept of landscape, that is, 'landscape character', a material artefact derived from a long and deep past that can be studied archaeologically in the same way as can any other aspect of material culture (e.g. Macinnes and Wickham-Jones1992, Herring 1998, Fairclough *et al.* 1999, Fairclough 2003, Macinnes 2004, Rippon 2004, Fairclough and Wigley 2006, Turner 2006a, Turner and Fairclough 2007, Austin *et al.* 2007, Nord 2009). Related to this work is the work of archaeologists who try to 'excavate' past perceptions of landscape from the landscape-scale material culture, often prehistoric, in which they are embedded. They try to imagine the landscapes that people in the past might have conceptualised.

Landscape archaeology is geographically diverse as well, differing across national borders just as much as landscape architecture does (see e.g. Fairclough and Møller 2008). Practice across Europe is varied for all manner of reasons, from historical to cultural to geographical, but not least because of the very different character of national landscapes and their archaeological remains.

Differences are found for example in countries where the study of the history of landscape is led by geographers, historians or archaeologists: these groups have different source materials, but also, just as importantly, different theory and starting assumptions. Some countries in central Europe (Kuna and Dreslerová 2007), for example, only relatively recently gained widespread access to modern aerial photography, one of the discipline's key techniques; others, such as the Netherlands, have a long tradition of very large-scale excavation which coloured approaches to landscape study and has created new viewpoints focused on the idea of landscape biography (Hidding et al. 2001). In Britain landscape archaeology was led along two quite different but complementary paths by, on the one hand, its long tradition of analytical field survey of earthwork remains from the late nineteenth century, and on the other, its very strong connections with local history. In other countries there is a strong tradition of map regression; in some countries, landscape archaeologists see designed landscapes as their major focus, others focus more on (agri)'cultural' landscapes (e.g. Pungetti and Kruse 2010); in some countries environmental-determinist interpretations about landscape have dominated, elsewhere it is settlement theory, whilst others approach landscape through a framework of cultural and social theory. Such examples of theoretical, methodical and technical diversity can be multiplied, and are being uncovered through annual conferences of the European Association of Archaeologists (EAA) (e.g. Darvill and Godja 2001, Fairclough and Rippon 2002, Meier 2006), and can be illustrated from the results of many European landscape partnership projects and networks that are all to a greater or lesser extent already inter-disciplinary (e.g. Clark et al. 2003, del Arbo and Orejas 2005, Bartels et al. 2008, Compatangelo-Soussignan et al. 2008, Fairclough and Møller 2008, Orejas et al. 2009). On a global scale similar ideas are growing through the products of many debates and workshops at the meetings of the World Archaeological Congress (such as Ucko and Layton 1999 or Hicks et al. 2007).

## Cognate disciplines studying the history of landscape

Landscape archaeology has its own neighbouring disciplines, some very closely related and others more distantly linked. They include landscape history, historical geography and the branches of landscape architecture that are concerned with the history of parks and gardens, with monumental heritage (e.g. Doukellis and Mendoni 2004) or (in some countries, e.g. Norway or Sweden) the interdisciplinary field of cultural landscape studies (Jones and Olwig 2008). All these disciplines, to one extent or another, share landscape as a subject of study in a geographical sense, but they operate in different time frames and more importantly with different attitudes to how we know about the past. Landscape historians are constrained by reliance on documentary evidence, for example, but within those bounds enjoy a clarity and relative certainty of fact and knowledge. Landscape archaeologists do not need to recognise any chronological boundaries and can instead be concerned with landscape of every date from earliest prehistory to last year. They can scan a deeper time than documentary evidence allows, but from a different epistemological starting point. For landscape archaeologists, however much they use historical documents or maps where available, their primary source is the material world itself, which through the idea of landscape can be made to tell us about people's lives both past and present.

The question of whether research relies on the evidence of historic documents and maps or on the material evidence of the landscape itself is the most fundamental difference between the various disciplines that study the origins and history of landscape, affecting not just methods but also results (although what follows is a relatively personal view, not shared by all landscape archaeologists or historians). Usually, of course, both documents and the landscape's materiality are taken into account, but in different measure. Archaeologists do not necessarily accept that an interpretation drawn from documents always has priority over the interpretation of material remains. There can be conflicting stories about the past that may be equally valid in different contexts, and this is perhaps particularly so in the case of landscape.

Different approaches can however lead to different questions being asked, and the character of the questions influences the answers obtained. Archaeologists tend to assume that landscape has a very long history even if they have not yet found the evidence. They look for the evidence or build models of the past which can then be tested; they are comfortable with the ideas of uncertainty and that the past cannot be definitively known but that interpretations and hypothesis, sometimes multiple and conflicting, are always possible. Historians can sometimes forget that their documents do not tell the whole story: that there were times before documents, and that there have been past actions and therefore consequences (in terms of how landscape looks today) that were not always or accurately and fully (even in recent centuries) described in contemporary documents.

Conversely of course, historians can study things that were described in documents or maps but which may have vanished long ago, and things and events (for example battles) that no longer have material form. Landscape archaeologists deal with the 'stuff' that still remains in the world. Whatever date or age we attribute to these, whether 10, 1,000 or 10,000 years, it is important to archaeologists that these landscape components are simultaneously part of the present landscape as well as from the past. Whether we can see them on the surface of the land, or only on air photographs as crop- or soil-marks, through geophysical remote sensing or only after digging them out of the ground, as for example with previous layouts of designed gardens, it is only their survival at some level into the present that allows us to study them. Whereas historians deal with the past itself, although inescapably interpreted as a reflection of the present, archaeologists (as a gross over-generalisation) deal with the past in the present, although it is a present imbued with a time depth that gives it its distinctive character. Landscape archaeologists more than historians consider that landscape exists 'in the present'; time depth in landscape is not just about origins, but also about the density and richness of the past.

It comes naturally to archaeologists therefore to take the next step beyond knowledge creation of using research to support landscape protection and management. There is for example far more of a discipline of applied archaeology than there is of applied history. In part, this is because the discipline seeks to protect its own resource: landscape, and all the archaeological remains that make up the environment, are in effect the discipline's raw material, its archive. There is a wider, social, motivation, however, which is relevant to this review of landscape archaeology as a neighbouring discipline of landscape architecture: archaeology's focus on the landscape as it exists in the here-and-now, and its desire to use knowledge about the past to shape future emerging landscapes by management and design provide a sound basis for interdisciplinary collaboration with landscape architecture.

## Heritage management and landscape

As mentioned earlier, a great deal of landscape archaeology is now carried out in the framework of heritage management (for example, in general text books on heritage management such as Graham and Howard 2008 or Fairclough *et al.* 2008). The aim is to understand landscape character and landscape history in order to influence decisions about change and to shape the modification and creation of future landscapes. This is partly because landscape subsumes many parts of the archaeological resource (the historic environment), and partly because landscape in its own right is a subject for both archaeological research and management.

At the same time, it appears that growing numbers of landscape architects are concerned with managing (planning) the inherited landscape as well as with creating (designing) new parts of the landscape. In this sense landscape architecture is becoming a research as well as a design practice. Both activities benefit from understanding the origins and history of landscape and the complex of historic and social processes that have created it. Landscape archaeology can help to provide this information. Like landscape architecture it can be forward looking in its objectives, and it can help to influence design by understanding the historic context of any new landscapes. The two disciplines are more complementary than is generally recognised.

## Landscape as a meeting place

Landscape is the domain of no single discipline, but is rightly shared by many disciplines. Landscape is always the sum, or more than the sum, of its many parts. The European Landscape Convention is quite clear on the need for inter-disciplinary collaboration, and many disciplines are already pursuing this goal (e.g. Palang and Fry 2003). Archaeology is habituated to being inter-disciplinary: it employs a wide range of practical and methodological techniques from most of the natural sciences as well as from humanities-based disciplines; there is also a well-known and very strong tendency for archaeology to make theoretical borrowings from other disciplines, most notably in recent decades from sociology. Landscape archaeology is inherently even more multi- and inter-disciplinary.

The generic definition of landscape architecture that was adopted for the LE:NOTRE network, in which context this chapter was written, is careful not to lay claim to all aspects of landscape work. It defined landscape architecture as being concerned with 'mankind's (sic) conscious shaping of his external environment' (Figure 4.2). This definition is both closely targeted but also incomplete. 'Targeted', because 'shaping' indicates that landscape architect is primarily an 'action' discipline concerned with designing and making, hence the strong links to horticulture and connection with historic designed landscapes like parks and gardens. 'Incomplete', because 'conscious' can exclude important aspects of landscape and indeed assumes that people only act outside, rather than within, nature. Use of the concept of consciousness apparently places large areas of landscape beyond the scope of landscape architecture (beyond the pale or the ha-ha), such as the vast bulk of landscape that was created by human agriculture, industry or other actions, the areas that the WHS criteria isolate as 'organic', or that others sometimes call (tautologically) 'cultural landscapes'. This is the part of landscape on which landscape archaeology can be most informative, and another reason why landscape architecture and archaeology can be complementary disciplines.

A design/research distinction is sometimes drawn between the two disciplines, landscape architecture being first and foremost a designing, making, acting discipline ('create, maintain, protect and enhance places' in the words of the definition), landscape archaeology essentially a discipline of research, understanding and knowledge. This is not entirely justified. Landscape architecture is also an

**Figure 4.2**
'. . . the conscious shaping of his external environment . . .': ornamental landscaping at Downton, Herefordshire (UK), an 18th century industrialist's 'castle' and improved 'nature'. (NMR 23392/08 SO4474/19 [2004]). ©English Heritage. NMR

academic research discipline as well as a field of practice, and its theoretical base grows yearly more sophisticated; landscape archaeology comfortably contains both research and scientific strands and a strong strand (within all types of archaeology) of resource management. This is partly because the speed at which landscapes have been seen to be changing, and archaeological sites to have been lost, over the past 75 years has rightly or wrongly been a major stimulus to landscape archaeology, alerting archaeologists to the erosion of their data and resources, and encouraging them to try to influence future landscape as well as to record and study its past. The threat of losing something also provoked a desire to understand more, and the enormous amounts of new knowledge gained by large scale 'rescue' excavations in recent decades in advance of modern development or agriculture, because of the scale of new discoveries and knowledge, has more or less revolutionised the way that archaeologists think about landscape. Archaeologists can no longer easily hold to the belief that sites are infrequent and widely spaced – instead, everywhere holds archaeological potential and more still remains to be discovered or uncovered than has so far been found.

The apparently ever-growing conservation and heritage management movement (in particular, 'archaeological resource management') is finding that landscape is one of the more useful ways of approaching the issue of major change and threat. It goes beyond simply looking after the settings of monuments to become a wider connection to environmental integration (from sustainable development to the European Landscape Convention). It can be relevant to issues such as identity and local distinctiveness. This is another area for closer collaboration between landscape architecture and archaeology: both disciplines sit on the cusp of past and future, and at the interface of people with places. Particularly in the fields of cultural landscape, landscape management and theory there seems to be a very strong case for closer collaboration at many levels.

## What should landscape architects know about landscape archaeology?

The question of what type and level of understanding about landscape archaeology would help students of landscape architecture is approached in three ways in the following pages. The more basic concepts of landscape archaeology that might inform undergraduate landscape architecture training are discussed first, followed by a description of three main branches of the discipline and, finally, suggestions for which methods and techniques of landscape archaeology might be useful for landscape architecture.

### *Basic concepts*
### An archaeological definition of landscape?

There are many possible definitions of landscape. Since the 1960s (mainly), archaeology has been a very theory-conscious discipline and all its definitions and concepts tend to be continually reviewed and re-worked; that for landscape is often particularly contested. Landscape archaeologists regard landscape – often all at the same time – as something to be *studied*, something to be (intellectually) *constructed* and re-constructed (not physically but virtually, unearthed from the past), something to be *used* to frame other archaeological debates (e.g about the nature of past societies), to be a bridge to other disciplines (ie landscape as unifying concept, as meeting place), to be a metaphor for narratives about the past and the present, and (increasingly), to be protected, managed and planned for a sustainable future.

## Subjectivity versus scientism

Some approaches in landscape archaeology treat landscape as essentially material. There are schools of landscape archaeology which are very objective and positivist, and which treat the landscape as a material thing, a physical rather than perceptual construction, little more than a synonym for 'environment'. Such approaches tend towards environmental theory, and also teach of successive 'layers' of landscape, each separate and ultimately retrievable and knowable, rather than seeking for evidence of their continued cultural presence and impact on present day landscape, instead of recognising the flow and fluency of change and continuity, re-use and renewal.

Many landscape archaeological approaches, however, recognise landscape not as a material thing *per se*, but rather as primarily an idea, and definitely not as synonymous with environment, although they insist that it is usually conceptualised and 'constructed' from material remains inherited from the past. Because landscape is seen as an idea not a thing, ('as perceived by people', as the Convention reminds us), qualitative and subjective measures and interpretations are important. Aspects (components) of landscape can be measured to some extent, but because landscape's essence is perceptual, ideational and qualitative, many parts of landscape archaeology avoid overly quantitative and scientific approaches to theory and interpretation even though some of them rely on highly scientific and technical processes to obtain data and analysis.

## Archaeological theory, the character of 'archaeological knowledge'

The importance of reflective practices and critical thinking as part of landscape architecture, and the recognition of more than one epistemology in any discipline, cannot be easily overstated. Knowledge is not fixed ('cut and dried') but is a living, diverse network of ideas. It will be difficult to achieve the democratising aims of the European Landscape convention (or the Faro Convention on the value of cultural heritage to society) without recognition of the validity of multiple strands of thought and 'knowledge'. Expert knowledge, itself not monolithic, is only one form of these. Landscape archaeology similarly recognises the importance of theory and reflexivity, with their implications for how people perceive landscape today. This is perhaps one of the strongest bases for collaboration between the two disciplines.

Theory within archaeology is not approached in the same way in all European countries. As a generalisation, post-1950 theoretical approaches were historically most common in the UK, Scandinavia and north-west European schools, but new theoretical positions have been developed elsewhere, notably in France and Germany, while Central and Eastern European countries tend to have distinctively different theoretical positions, often towards a stronger positivism. The existence of

European *fora*, of which the European Association of Archaeologists is the most important in this context, is gradually revealing the patterns and diversity of theory across the continent.

## The politics of landscape – contested landscape

As a consequence of the focus on theory, landscape archaeologists less often take landscape as a concept that is given and absolute but as one that is fluid and contingent on all sorts of political and other attitudes, as is any other part of culture. This can take the form, for example, of landscape being an expression of cultural values written backwards onto the past, or of appropriation by one community (of place or of interest) or another. The practice of landscape architecture, as landscape itself, can also be regarded as being entwined with politics and power negotiation, just as landscape itself embodies conflict; such things are naturally at the core of archaeological thinking, and the ways in which, and the extent to which, landscape is politics, and is contested, would be fruitful and necessary ground for collaboration between disciplines.

## The present past

A commonly-held public view of archaeology (held too by some archaeologists) is that it is a discipline concerned only with the past: intent on understanding what happened in the past, occasionally straying into the management debate but only to try to preserve its data sources or to discourage change or even recreate what has been lost. Some of these assumptions can be particularly common in terms of landscape, but the picture they paint is not representative.

Alternative views exist, particularly in the more reflexive and theoretically aware schools of archaeology and increasingly in heritage management circles. These views note that archaeology is above all about the study of change in history. This is an area of study that encompasses both very recent and current change, and also (through archaeological resource management) fosters an interest in future change, design and planning. This is in other words a sort of 'socially-embedded' archaeology, its practitioners being actors in the construction of landscape as well as witnesses to change.

Archaeology's sources – all inherited material culture – exist in the present whatever their date of origin, and whether we can see it or have to dig it up first. This is most clearly true of landscape. In perceiving (creating) landscape, the remains, traces or influence of the past do not always need to be visible, but our knowledge of their hidden existence (or even that they once existed, and survive through its effect on what came after, the cause and effect of the 'long chain' that leads to and binds the present) is an important aspect of perception (Figure 4.3). The concept of the past being in the present, of its 'present-ness', topicality and relevance to the modern world, can be particularly important at landscape scale, where landscape is the best example of the time-layered palimpsest (the historic environment) that makes up the world.

# A prospect of time

**Figure 4.3**
'... the past being in the present...': the military airbase at Scampton, Lincolnshire (UK), used since 1916, its runway extended for Cold War nuclear bombers diverting the Roman road from its 2000-year-old path; historic landscapes are not only ancient. (NMR 12854/12 [1996]). © Crown copyright. NMR

## The character of landscape

An understanding of the fundamentally historic and cultural character of landscape is what landscape archaeology can most uniquely contribute to the full interdisciplinary study and management of landscape. Its contribution can fall under several headings, all helping to understand the patterns, relationships, interactions and overall character that we can see in today's landscape. Some of the more important ones can be summarised here:

- Time in landscape (that is, the extent to which today's landscape is a patchwork, and a distillation, of all previous landscape and environments; landscape is a product of time as much as being a spatial phenomenon)
- Human agency (that is, human decision making, cultural processes, and the side-effects of human behaviour, actions, inactions and decisions through time: the over-riding of environmental determinants, a concern with people and society not nature)
- Cause and effect (that is, how one 'layer' in the landscape through time is a response to previous layers of landscape and human actions, and the frame for later landscape change; also how changes to process and behaviour affects landscape more widely)
- Historical processes, that is, all the collective social, political, economic, land-use, religious processes, drivers and fashions that have shaped landscape (and still do), above and beyond individual design decisions, which themselves invariably are shaped by wider social pressures because landscape management and conservation are themselves historically-contingent social processes.

These themes are viewed differently by landscape archaeologists compared with landscape architects and other disciplines.

## Scale and spatial patterns

The scales at which landscape archaeology operates may not always be the same as those used by landscape architects, or for that matter, landscape ecologists. Scale varies from national or regional patterning to sub-regional (as in Historic Landscape Characterisation) or parish and township scale; some landscape archaeology of the traditional detailed kind operates at very small, almost site scale (see, e.g., Lock and Molyneaux 2006). All scales of course have in common an interest in space in various forms: the use of space by people in the past, the way people have created space by enclosure of land, the way landscape is created by the inter-relationship of sites and actions. Landscape architecture and archaeology are complementary in the way they relate to temporal scale and temporal patterning, archaeology can help to provide a temporal frame for landscapes. Equally, spatial issues strongly unify the two disciplines; space provides common ground for both.

## *Branches of landscape archaeology*

There are many ways to classify different schools of landscape archaeology. Muir (1999) for example identified schools based on 'landscape history' (largely in terms of field techniques, and to an extent seeing landscape as being the same as 'environment'), on 'scenery and structure' (visual appraisal – similar to the English usage of the concept 'countryside'), and on landscapes of the mind, of politics, and of aesthetics. Ashmore and Knapp (1999) offered the idea of distinctions between 'constructed', 'conceptualized' and 'ideational' (loosely, emotional plus imaginative) landscapes, each having different theory and method as well being ostensibly different facets of any area of land. UNESCO proposes a threefold division of landscape into 'Designed', 'Organic' and 'Associative' landscapes as a (problematic) way to simplify landscape for the practical purposes of selecting candidates for WH status.

There is another way to classify types of landscape archaeology that has proved useful in trying to think through with other archaeologists how to unify the disparate approaches within the overall discipline, in particular in the context of using landscape knowledge gained archaeologically to support the protection, management and planning (to use for convenience the three tools identified by the European Landscape Convention). It is probably therefore also helpful in trying to find natural bridges between landscape archaeology and landscape architecture, the first primarily a knowledge discipline, the second primarily an action discipline, but both with aspects of the other approach.

This perspective is based on a classification of approach and method instead of trying to classify different types of landscape. Its scheme is modelled on the intended outcomes of research, and may be useful for conveying something of the diversity of approaches in landscape archaeology. Any attempt to produce elaborate and definitive classifications of landscape is a fairly sterile exercise, however, because scientific approaches do not sit perfectly comfortably with landscape as perception. It might be wise to be cautious about reducing the complexity, diversity and richness of the whole landscape to simple classifications, whether theme based, area-specific, or topic-based.

The three (complementary) branches of landscape archaeology loosely defined below effectively map out much of the field covered by landscape archaeology. They all share common methods and some theories; and they speak, largely, a common language. They do, however, have different objectives and their connection to landscape architecture is potentially very different. These are, briefly:

1	archaeology through landscape (at 'landscape scale')
2	the archaeology of present day landscape character (i.e., perception)
3	the archaeology of past perceptions of landscape.

## Archaeology at landscape scale

This is the most traditional form of 'landscape archaeology'. As a practice (or suite of practices), it is several decades old (from Crawford 1953 and before, through Taylor 1974 to Bowden 1999), and it is the branch with the closest links to landscape history and landscape geography. It is also the most positivist branch, as befits one that evolved hand in hand with official state cartography, concerned straightforwardly with creating factual knowledge about the past. This is often simply the study of past environments, a major concern being 'settlement' and land cover succession, rather than perceived landscape as defined by the Convention. It is sometimes difficult to distinguish its results from regional history (or historical geography). It does however possess great inter-disciplinary strength as it builds on documentary and cartographic knowledge, and a very wide range of archaeological techniques.

It tends to have a focus on the past, as in attempting to understand, for instance, the mediaeval or Bronze Age landscape (or rather the environment of those periods, as this approach also tends to treat landscape as a synonym for environment, and less often engages with questions of perception). It has little clarity about scale, but accepts as landscape almost anything that involves more than one 'site' and the space between. (One senior British archaeologist joked that some archaeologists call their research 'landscape' if the excavation trench is bigger than 1 hectare.) In other words, it uses a concept of landscape as a framework for intra-site work, yet it retains many of the theories and paradigms that have been developed for conventional monument or site archaeological study. For these reasons, when practitioners of this branch of landscape archaeology engage with conservation and planning, or with archaeological resource management, they sometimes adopt a preservationist stance aimed at keeping individual components of the landscape in their original fabric.

This branch's concept of landscape is not really very close to what landscape architects (or the Convention) means by the term 'landscape', although slightly paradoxically, some of its results ('hard facts' about past landscapes and landscape evolution) are among some of the archaeological products most used by landscape architects. The results of this branch of landscape archaeology can be fed into perception once they are 'known' as 'facts' by observers (although it has to be said that the results of archaeological research are not always as accessible to landscape architects or the public as they ought to be). This is an important approach, and perhaps one that is most widespread across Europe. It is, however, predominantly about sites *in* landscape, not about landscape *per se*. This is also landscape archaeology predominantly as a knowledge discipline.

## The archaeology of landscape

This approach uses the concept of 'landscape' in the perceptual sense that the Convention proposes, that is, landscape as something that is the product of action and interaction between humans and nature as perceived by people. It thus places landscape in the present, no matter how historical it is. This type of landscape is not only implicitly cultural, but it is doubly cultural – it is cultural because it is perceived by people (and so is a human construct of the mind or emotions), and it is cultural because its material manifestation is very largely the product of human agency and actions over centuries and millennia; in other words. Landscape uses cultural heritage as its raw material. Archaeology makes a central contribution to this concept because it studies time and human actions at landscape scale and does so in the light of present day material remains.

Landscape archaeology of this type can make a special and largely unique contribution to this holistic construction of landscape, usually through generalising and interpretative methods such as historic landscape characterisation. It can focus on explaining how time, human agency (culture) and successive layers of change are present in the contemporary landscape, forming a large and influential part of its character. This type of landscape archaeology explains and illustrates the past in the present, and creates what might be called 'concertina'd time scales' in order to understand the depth of time that is part of landscape character. It works with ideas such as *palimpsest*, time as an aspect of space, time-depth, and with concepts such as the existence of the past in the present, and also past 'landscape' reflecting the transience and historical-specificity of our perceptions (e.g. Fairclough 2003). This is necessarily an inter-disciplinary field, because it makes no claim to be able to know everything about landscape but to contribute to a broader more holistic concept. Landscape architecture could benefit from more familiarity with this branch of landscape archaeology, but it is not widespread across all Europe.

This branch also promotes understanding of how a recognition of the extent to which both environment and landscape is humanly made allows a constructive approach to future change and landscape management. It is the field of landscape archaeology that is most active in the fields of conservation, spatial planning, archaeological and heritage resource management, and landscape management. This is the realm in which landscape archaeology comes closest to being a designing and shaping (as well as an understanding) discipline. Here, for the same reason (and because it deals with the past in the present landscape, not with the past itself) landscape archaeology has the greatest potential to develop a closer practical relationship to landscape architecture. It can do this by sharing methods (cf. the common language and shared assumptions, despite different scales and theory, implicit not just in Historic Landscape Characterisation (HLC) methods carried out usually by archaeologists but also in Landscape Character Assessment (LCA) methodologies usually used by landscape architects or geographers). Because of the way that its scale, resolution and technical aspects are established, HLC can act

The three (complementary) branches of landscape archaeology loosely defined below effectively map out much of the field covered by landscape archaeology. They all share common methods and some theories; and they speak, largely, a common language. They do, however, have different objectives and their connection to landscape architecture is potentially very different. These are, briefly:

1   archaeology through landscape (at 'landscape scale')
2   the archaeology of present day landscape character (i.e., perception)
3   the archaeology of past perceptions of landscape.

## Archaeology at landscape scale

This is the most traditional form of 'landscape archaeology'. As a practice (or suite of practices), it is several decades old (from Crawford 1953 and before, through Taylor 1974 to Bowden 1999), and it is the branch with the closest links to landscape history and landscape geography. It is also the most positivist branch, as befits one that evolved hand in hand with official state cartography, concerned straightforwardly with creating factual knowledge about the past. This is often simply the study of past environments, a major concern being 'settlement' and land cover succession, rather than perceived landscape as defined by the Convention. It is sometimes difficult to distinguish its results from regional history (or historical geography). It does however possess great inter-disciplinary strength as it builds on documentary and cartographic knowledge, and a very wide range of archaeological techniques.

It tends to have a focus on the past, as in attempting to understand, for instance, the mediaeval or Bronze Age landscape (or rather the environment of those periods, as this approach also tends to treat landscape as a synonym for environment, and less often engages with questions of perception). It has little clarity about scale, but accepts as landscape almost anything that involves more than one 'site' and the space between. (One senior British archaeologist joked that some archaeologists call their research 'landscape' if the excavation trench is bigger than 1 hectare.) In other words, it uses a concept of landscape as a framework for intra-site work, yet it retains many of the theories and paradigms that have been developed for conventional monument or site archaeological study. For these reasons, when practitioners of this branch of landscape archaeology engage with conservation and planning, or with archaeological resource management, they sometimes adopt a preservationist stance aimed at keeping individual components of the landscape in their original fabric.

This branch's concept of landscape is not really very close to what landscape architects (or the Convention) means by the term 'landscape', although slightly paradoxically, some of its results ('hard facts' about past landscapes and landscape evolution) are among some of the archaeological products most used by landscape architects. The results of this branch of landscape archaeology can be fed into perception once they are 'known' as 'facts' by observers (although it has to be said that the results of archaeological research are not always as accessible to landscape architects or the public as they ought to be). This is an important approach, and perhaps one that is most widespread across Europe. It is, however, predominantly about sites *in* landscape, not about landscape *per se*. This is also landscape archaeology predominantly as a knowledge discipline.

## The archaeology of landscape

This approach uses the concept of 'landscape' in the perceptual sense that the Convention proposes, that is, landscape as something that is the product of action and interaction between humans and nature as perceived by people. It thus places landscape in the present, no matter how historical it is. This type of landscape is not only implicitly cultural, but it is doubly cultural – it is cultural because it is perceived by people (and so is a human construct of the mind or emotions), and it is cultural because its material manifestation is very largely the product of human agency and actions over centuries and millennia; in other words. Landscape uses cultural heritage as its raw material. Archaeology makes a central contribution to this concept because it studies time and human actions at landscape scale and does so in the light of present day material remains.

Landscape archaeology of this type can make a special and largely unique contribution to this holistic construction of landscape, usually through generalising and interpretative methods such as historic landscape characterisation. It can focus on explaining how time, human agency (culture) and successive layers of change are present in the contemporary landscape, forming a large and influential part of its character. This type of landscape archaeology explains and illustrates the past in the present, and creates what might be called 'concertina'd time scales' in order to understand the depth of time that is part of landscape character. It works with ideas such as *palimpsest*, time as an aspect of space, time-depth, and with concepts such as the existence of the past in the present, and also past 'landscape' reflecting the transience and historical-specificity of our perceptions (e.g. Fairclough 2003). This is necessarily an inter-disciplinary field, because it makes no claim to be able to know everything about landscape but to contribute to a broader more holistic concept. Landscape architecture could benefit from more familiarity with this branch of landscape archaeology, but it is not widespread across all Europe.

This branch also promotes understanding of how a recognition of the extent to which both environment and landscape is humanly made allows a constructive approach to future change and landscape management. It is the field of landscape archaeology that is most active in the fields of conservation, spatial planning, archaeological and heritage resource management, and landscape management. This is the realm in which landscape archaeology comes closest to being a designing and shaping (as well as an understanding) discipline. Here, for the same reason (and because it deals with the past in the present landscape, not with the past itself) landscape archaeology has the greatest potential to develop a closer practical relationship to landscape architecture. It can do this by sharing methods (cf. the common language and shared assumptions, despite different scales and theory, implicit not just in Historic Landscape Characterisation (HLC) methods carried out usually by archaeologists but also in Landscape Character Assessment (LCA) methodologies usually used by landscape architects or geographers). Because of the way that its scale, resolution and technical aspects are established, HLC can act

## A prospect of time

as a bridge between LCA and more scientific (or statistical?) ecological approaches. The arena for this coming-together is offered by the operationalisation of the ELC: landscape understanding, spatial planning and public participation.

### The archaeology of past landscapes

This field of research (in essence, although it has some much earlier origins) arises mainly from new theoretical concerns of the 1970s and 1980s, that is the use of archaeological techniques to explore whether and how our predecessors created perceived landscape, or simply modified their environment. It thus touches on our views of how (if at all) we are different from our prehistoric predecessors, raising issues of consciousness, and of how early people saw themselves in relation to the world (Figure 4.4). One definition of its aims is that it studies social meaning in ancient landscapes, and regards 'ancient landscapes' as products of perception just like our own landscape is today.

This work involves using material remains from the past at landscape scale to understand past perceptions of landscape: whether for example the prehistoric laying out of the land into tribal or community territories reflected contemporary cosmographic models, hunting patterns and resource exploitation, ritual and resources, or how notions of landownership and politics 'wrote' themselves into the landscape and thus endorsed their own power. Some archaeologists talk of 'sacred landscapes', 'ideological landscapes', 'landscapes of settlement and subsistence' – these all have in common being past landscapes, landscape constituted at a particular

**Figure 4.4**
'. . . products of perception . . .': the Scillies (UK) – St Martins Island and Tresco beyond, looking into the west; a flooded land, today's islands are the tips of hills that were farmed and settled until late prehistory; lost field walls emerge at low tide; an historic seascape as well as an historic landscape. (NMR 23893/12 SV9217/1 [2005]). ©English Heritage. NMR

period of the past by a particular group of people. They are themselves intangible, though usually (being archaeologists' constructs) they are anchored to one extent or another in material objects that survive physically in at least one sense, even if perhaps hidden below ground or resident only in patterns between visible remains, material culture. This is not only a matter of prehistoric times: the creation of 'landscapes' for purposes of social consensus and control, for example by land enclosure of communal fields from the sixteenth century onwards, can be seen as an aspect of the invention of capitalism, and there is also the example of the archaeological models of eighteenth century landscapes of social control.

The relevance of this branch of landscape archaeology to landscape architecture, and to the understanding, managing and designing of the current landscape, is twofold. It is another dimension to understanding the stages by which our environment, and therefore our perceived landscape, came to be the way it is, to look like it does. It is particularly useful because it does not rely wholly on economic explanations, and escapes a little from an assumption that the landscape was only ever a place to make food, to eke out existence while responding to environmental constraints. It helps to put human choices, resulting from past perceptions, into the centre of the stage in the long centuries and millennia before 'conscious' landscape 'design' can be recognised. Secondly, it provides a long prehistory for our modern, quite recent (only a few centuries) Euro-centric notion of landscape, and it allows us to remind ourselves that approaches and concepts that we might think are natural givens such as aesthetic rules or the scientific quantification of biodiversity are as historically specific as, say, sky-based perceptions of landscape were to people in the Neolithic.

## *Methods and techniques*

The main techniques and sources used by landscape archaeologists offer much to landscape architecture in terms of analysis and understanding of the historic dimension as a basis for planning, design and management. Typical examples of methods and techniques are:

- Aerial photography (which apart from revealing invisible archaeology also importantly offers (rather than the horizontal view of traditional landscape appreciation) a new, distant and vertical perspective that emphasises time-depth and palimpsest as well as a different scale)
- Historic Landscape Characterisation (HLC), a way of generalising and synthesising what archaeologists know about the landscape into a format best suited for amalgamation with other forms of landscape understanding
- Regional studies
- Long-term change and historic process, revealing why landscape looks as it does, and can be perceived in the way it is
- Use of historic maps (and map regression) to understand the time layers and processes of change
- Use of the results of large-scale excavation, analytical survey of earthworks, science-based environmental study (e.g. micro-morphology of soils to wood

remains to pollen and snails), including understanding of the limits and opportunities of detailed archaeological data and knowledge
- Theory, including time-based and spatial experiential and phenomenological aspects; also the concept of archaeology as performance (engaging with landscape, not simply studying it); all concepts that merge with issues of popular and local perception.

The use of landscape archaeology to explain the historic processes, many extinct or modified, that have created the environment and which in turn underlie perception is a particularly valuable contribution the discipline can make to landscape architecture. This should involve all three branches defined above, and could be amalgamated with an understanding of a basic narrative of landscape and environmental development since the Neolithic (or earlier), tied into present day survival and impact on current landscape character.

This would help to build another bridge between the understanding that landscape archaeology provides and the need to use it to support management, action and new design. An understanding of the processes that created landscape in the first place is essential to managing landscape, whether by maintaining the processes, by finding a different, modern, proxy process which would create similar landscape effects, or by accepting that landscape must change because new processes are unavoidable. Here, landscape archaeology moves close to landscape ecology, in its concern with the mechanics rather than the appearance of landscape. Archaeology's emphasis on the impacts and relevance of successive and continual past change is an advantage here. The role of change in the past (or, more accurately, the visible, legible or in some other way perceivable evidence for past change) sits at the core of landscape character: not as a process which might destroy landscape, but as one of its fundamental attributes, the results of the historic processes that both created and continue to modify landscape, and still ensure that it continues to evolve.

Finally, the role of landscape archaeology in creating stories that explain the present can be valuable for landscape architectural practice. The basic products of landscape archaeology are narratives about the past and (therefore) about the identity of the present. There are many ways of telling stories. These are generally qualitative not quantitative, subjective not objective, multiple and inclusive not singular, exclusive and 'right'. Such complex narratives, always trying to capture the contested political character of landscape and the identity that are drawn from it – that is, narratives and stories, metaphors, biography – are all essentially about people. It is people who stand at the centre of landscape as actors and agents in its physical creation, as those who create it in perception, and as those who change it and enjoy or use it. Landscape archaeology's stories about people and their lives throughout history (waiting to be read in the landscape) complement ecology's description and classification of natural systems. Its questions of 'how' and 'why' provide a platform for answering the 'what next?' asked by spatial planners. Landscape archaeology tells stories; landscape architecture paints pictures – a natural complementarity.

## Opportunities for collaborative research

Both landscape architecture and landscape archaeology are disciplines with a great diversity of research traditions, interests and methods, with variations across Europe that reflect national cultures, although not always in the same ways in the same country. A starting point for collaborative research programmes and projects might therefore be comparative and reflexive research to investigate at various scales and in different countries the parallel histories and character of the two disciplines and where the potential for stronger or closer inter-relationship most lies.

The new few pages offer a first attempt to identify areas for collaboration at research level between landscape architecture and landscape archaeology. First is a discussion of how landscape architects and landscape archaeologists might share their differing cultures, then suggestions for specific landscape archaeology approaches or methods that landscape architect researchers might find most useful in their tool-kit, and third a number of more concrete – if very general and broad – opportunities for research and areas within which research might be fruitful for both disciplines.

### *Cultures in landscape archaeology*

Inter-disciplinary collaboration begins with an understanding of the different culture that underpins each discipline. For landscape archaeology (indeed for archaeology in general), this culture includes:

- a very strong tendency towards, and an experience of being multi-disciplinary, both within its own broad scope (e.g. between hard science, 'soft' science, humanity-based approaches, even artists and performers) and between itself and other disciplines (traditionally history, geography and geology and more recently palaeo-ecology, and in the form of theoretical borrowings, anthropology or sociology)
- a strong interest in theoretical approaches, and in particular the value of having a diversity of possible approaches in use, in order to produce multiple strands of understanding and if necessary conflicting interpretation
- archaeologists study 'landscape' for more than one reason, studying it as a context for individual sites as well as an object of study and analysis in its own right, and using it as the vehicle for experimental and theoretical engagements with both past and present perceptions
- the discipline remains interested in the past for its own sake, with understanding as a goal in its own right, and one rationale for conservation archaeology is to protect the discipline's resource, its 'archive', i.e. the historic environment
- in looking at landscape in its own right, archaeology treats it first and foremost as an expression of material culture (in which social, cultural and human meaning is embedded), which speaks for itself through the application of archaeological theory rather than necessarily through historic documents or imposed aesthetics
- the growing strength of the applied practice of archaeology, centred on management and conservation, so that as well as being a research discipline archaeology is increasingly socially embedded in action (although not usually yet a design discipline in the way that landscape architecture is)

- change in past and present is often seen as interesting and worth studying for itself, without necessarily being something to mitigate, avoid or encourage. This study throws objectives and aims of landscape management into high relief.

## Approaches and methods

Landscape researchers might experiment with adapting some landscape archaeology methods to the objectives of landscape architecture, for example through the practice of landscape character assessment. Examples of appropriate methods (some were discussed above as well) could be:

*Aerial photographs*: offer different ways of seeing things, notably a vertical rather than horizontal view of landscape, which provides the distance that reminds us paradoxically that landscape is enacted and embedded, which draws attention to larger patterning and which emphasises time depth as well as present appearance. The aerial view also focuses on form and function more than fabric and condition, and makes the nuances of human influence and choices in the past more visible above the crude, self-evident determinacy of topography. Moreover, the existence in many countries of long series of photographs from different dates shows landscape change and thus emphasises the time scale element of landscape.

*Historic Landscape Characterisation (HLC)*: a difference of scale (neither the site scale of landscape design/garden design, nor the Landscape Character Assessment (LCA) scale of district or sub-region, nor even the scale of ecological survey); a scale that reflects human activity especially at community or township level, a method that recognises the grain of human land uses through time, cutting across not following the topographic grain; also a different way of using GIS. It is relevant as a generalising method with specific spatial and temporal values and perception scales (Aldred and Fairclough 2003).

*Regional study*: landscape archaeology is also interested in regional patterning, and has been for a long time since (in the UK) Fox (1938), and onwards through Rackham (1986) to Roberts and Wrathmell (2000). Does landscape architecture operate at similar levels? Can we move jointly towards European wide maps that show cultural and historical variation across the continent rather than the quantified land cover maps with which we are familiar that in effect largely map present day climate and soils: a map to show cultural patterns, and the human regional identities of Europe? A series of national projects to develop methods and approaches would be helpful.

*Long-term change*: the study of this is the essence of archaeology, but is sometimes missing from landscape architecture (and exploring why might itself be a good self-reflexive research topic). Research into how landscape architecture (both in managing existing landscape and in designing new ones) could take better account of past change, and follow existing trajectories of change, would be valuable. How can knowledge of the past in the landscape be used for future design? Is it the bridge between aesthetics and biodiversity (between landscape architecture and landscape ecology?).

*Historic maps (and map regression)*: probably a technique already shared, but worth further exploration, not least to explore how maps are used in landscape architecture (the need for source criticism, for example) or to explore how far historic maps can be used not merely to tell us what was there at the date when the map was drawn, but to use old maps as a way of understanding (through 'archaeological' techniques of analysis) what had already been changed in previous periods (i.e. reading historic maps as a *palimpsest* that contains clues and evidence for previous landscape and for change, as opposed rather than just as a depiction of what was there at a particular point in time (and one not usually that distant).

*Excavation*: a basic, almost defining, tool of archaeology, and some parts of Europe now have such very extensive areas excavated that enable large areas of land to be interpreted at a landscape scale. This is the basis for 'landscape biography' as used in the Netherlands, and for landscape histories as used in the UK. Admittedly this is an issue of understanding past environments more than modern landscape, but it could be a research topic based on exploring how far such knowledge (that is derived after all from the complete removal of previous landscape, usually by mining, extraction or infrastructure) can be used in the landscape re-design process (spatial planning as well as landscape architecture) of future landscapes. Similar use might be made of the results of large scale analytical field survey.

*Analytical survey*: a technique that has been used extensively in Britain for many decades but which is perhaps less common elsewhere in Europe (and one might ask why that is). It comprises the detailed analysis and survey of past environments ('relict landscape' in UNESCO parlance) that still survive visibly on the surface of the land as earthworks. These are archaeological components of the present day landscape, and their more detailed understanding can readily influence landscape perception, whether they are large sites (e.g. hill-forts) or expansive areas of prehistoric abandoned field systems (which of course tell us much about long term agri-environmental transformations).

*Science-based environmental study*: this brings hard science into the discipline, and (to some extent at least) also an environmental approach rather than a cultural or social one. It concerns, among other things, the extent to which soils are now humanly modified, the sequences of vegetation cover through time and the effect of people on natural successions, and thus starts to indicate how far biodiversity is itself a cultural construct. These are basic archaeological questions and techniques that have connections with landscape ecology as well as architecture (e.g. the ancient grassland flora on prehistoric burial mounds, such as Bjäre in Sweden) and more broadly with landscape character. There should be scope for research to identify ways in which this detailed knowledge can influence landscape management and design.

*Process*: any collaborative research would be useful that explores how to increase landscape architects' awareness of both economic and material landscape processes above and beyond aesthetics (e.g. prehistoric, medieval and modern farming patterns) and of symbolic or associative aspects of both past and present.

*Theory*: a major area for research opportunities. One area for theoretical collaboration on theory might be the politics of landscape (contested landscapes) as already discussed above.

*Popular and local perception*: capturing public perceptions of landscape, particularly at local level, is a major unaccomplished task for almost all landscape archaeology work to date, although it is one that many practitioners recognise as desperately needed. It is perhaps an area where landscape archaeology can learn from landscape architects, while at the same providing another field for exploring new theoretical stances. How to do it? Do we use existing tools (e.g. HLC) as a frame for capturing popular perception and stories? Is LCA too small a scale? Do we start with popular views, and use professional tools simply to systematise it? How should experts and politicians respond if public opinion privileges 'lifestyle' over 'sustainability'?

## *Opportunities for fruitful research*

This section offers a few outline suggestions for more specific collaborative projects, covering both knowledge research and action research. Its main aim is to initiate a debate on the sort of work that might be possible and useful. Topics have been suggested on the basis of those that would not only gain new knowledge and understanding of the story of any given area of landscape, but that would also create a broader generic understanding of the ways in which the two disciplines might strengthen each other by sharing theory and perceptions. Partners for this research could be sought not only in the university sector, but within national and regional heritage management authorities, with EAA (the European Association of Archaeologists) and EAC (the European Archaeological Council), and within existing and future European funded networks such as the Culture, Interreg or COST programmes, and in future the implementation of the recommendations of the ESF/COST Science Policy Briefing on landscape research, with for example its proposed high level research themes: landscape as universal commons, the roots and routes of landscape (mobility and lifestyle), reactions and resilience in landscape (responding to transformations), and 'road maps', inherited landscape as the baseline for future changes. Inter-disciplinary research into all of these will have important 'past-time' (historical and archaeological) dimensions.

## Knowledge research – better understanding of the past

The scope for research using landscape archaeology to understand past environments at landscape scale, and to understand the historic dimension of the present landscape, is more or less unlimited. Amongst the large range of possible topics, identifying those most suitable for collaborative research between the landscape archaeology and landscape architecture disciplines is difficult. The following list concentrates on 'big questions' that would contribute to the landscape architect's feeling for landscape's time depth and familiarity with archaeological theories, techniques and concepts. These are first examples offered for discussion purposes. The aim of all would be not merely to study the thing itself, but to explore the extent to which their remains are still physically or mentally part of modern people's landscape or could be with greater familiarity. These first examples are chosen as far as possible as common European-wide phenomena; the study of their extent, comparability, differential survival and regional diversity might illuminate trans-continental similarities and help to define discrete individual (national or regional) personalities of landscapes across Europe.

Extant/retrievable prehistoric and early origins of rural landscapes: for example:

- Neolithic monuments and 'sacred landscape' (the first time we can see people 'making' landscape?)
- The spread and character of settlement and land-use in the second millennium BC (the origin of Europe's agricultural settled landscape?)
- Relationship in the landscape of humans to nature/sea level change/land reclamation (the real interface of nature with culture?)
- Upland/lowland links and transhumance, all periods (with reference to the different dates and impact of disappearance as well as to survival, since even long-extinct transhumance can very often affect the current landscape)
- Enclosure and appropriation of land (the origins of capitalism written in landscape?)
- Mediaeval townfields and villages (the creation of 'modern' structures?)

Townscape is a second major area: urbanisation, its effects on pre-urban and urban fringe, but also the landscape of towns themselves. Landscape character is rarely purely rural in twenty-first-century Europe, and along with spatial planning it is a key driver in creating new urban landscape (Fairclough 2008, Hall 2006). Townscape often still seems to be an undeveloped field in landscape architecture, most urban landscape architecture restricting itself to open space within towns (those small parts of town which retain some slight vestigial resemblance to countryside, and are still green or at least not built-over). Townscape is much more than this, however, encompassing the whole of the built area, hard and soft space, both public and private realms; as the ELC says, landscape can be urban as well as rural. Landscape archaeology, like historical geography, has a lot to say about townscape and its evolution (Figure 4.5).

Major transitions and change in landscape is a third area. Much study of landscape emphasises continuity and survival landscape, thus creating a powerful if mythical narrative of 'timelessness'. Archaeology focuses on change (in some ways archaeology is the study of change) and the following examples are suggested as major areas where long-duration change in the past can be studied. Such research would secondarily facilitate reflection on the aims of landscape management and design: keeping the remains of the past or designing within the spirit of previous change; maintaining existing landscape character (the 'top layer') or creating new character that reflects (and embeds ) past landscape character and thus maintains local distinctiveness. Such topics might be:

- Industrialisation (all types)
- Comparisons between the twentieth century experience across Europe, from

# A prospect of time

**Figure 4.5**
'. . . transitions and change . . . continuity and survival . . .': in the middle of England, through 18th century fields, mediaeval ridge-and-furrow and buried out of sight but not out of perception prehistoric fields, a communications corridor for thousands of years, Roman road, modern roads, railway, canal all pass south to north within a kilometre or two of 'DIRFT' (Daventry International Rail-Freight Terminal), a landscape marking 21st-century lifestyle of mass consumerism. (NMR 23286/20 [2003]). ©English Heritage NMR

capitalist-led industrial farming in the west to the state-led collective structures of the east
- Later twentieth-century landscape – all bad?, wholly destructive or also creative? or something to study, to emulate or improve (i.e. an archaeological approach to the recent and contemporary past) (Fairclough 2007b, Penrose *et al*. 2007, Harrison and Schofield 2010)
- The balance through time between continuity and change, re-use and replacement
- The effect of changing/lost management processes.

The methods to be adopted will need to be considered. Truly interdisciplinary collaborations would share perspectives as well as methods to create a new broader understanding as follows.

- *Communication*: Comparison/Synthesis of different modes of communication in the two disciplines: narrative / biography '*versus*' images and maps.
- *New technology*: Shared use of new technology: GIS for data/ideas capture and interpretation; virtual visioning.
- *Sources*: shared sources, exchanges of ways to use them.

Graham Fairclough

## Action research – shaping future landscapes in the light of understanding past landscape

Landscape architects do not want to become archaeologists! Their main interests are not research into the past but the making of new landscapes, big or small; in many countries landscape architects are taught alongside regional planners. But nevertheless they could benefit from using some archaeological techniques, for example:

- the mechanisms by which the past is read in the present landscape (Scazzosi 2002, 2004, Fairclough 2003, 2007a, Rippon 2004), what this adds in value, significance, context etc, and how to ensure that design continues to allow the future's past to be legible
- how future landscape can be designed to fit into and with inherited landscape
- management and conservation; future landscapes, the 'so whats?' of landscape characterisation and research – how to put knowledge to practical use.

This area of collaboration requires some place- and research-based case studies; much research could be done by collecting examples and drawing out lessons, finding out how the two disciplines each regard the different applicability, results and impact on both past and future landscape. A convenient and robust framework would be provided by the three tools set out by the European Landscape Convention, that is 'landscape protection', 'landscape management' and 'landscape planning'.

All these ELC instruments revolve around the 'so whats?' of landscape research. We might understand more about past landscapes, landscape history and current landscape character, but so what? What are our objectives for (or with) landscape (Fairclough 2006)? Should the remains of the past be kept or allowed to disappear when development is proposed? Do we seek always to preserve fabric, or is sometimes legibility of more subtle traces enough? A landscape architecture practice better able to understand the impact of the past (or its absence) on contemporary landscape would be able to support decisions about when to keep things from the past (protection), when to use them differently (management), or when to change them or their surroundings (planning). How we wish to treat landscape (our objectives) is closely allied to how we perceive landscape, and in this the two disciplines presumably differ, but constructively.

The management of existing landscape character – landscape planning rather than landscape design – is beginning to be as much a part of the practice of landscape architecture as is the design of new landscape. Landscape architects involved in managing existing landscape (so-called 'cultural landscapes') may well be those for whom the European Landscape Convention offers most opportunities. This wing of the profession acts as a research as well as an action profession, and in the UK and a few other countries is already closely allied to spatial planning, nature conservation and historic management conservation.

## Concluding thoughts

This chapter briefly outlined some reasons for landscape archaeology and landscape architecture to find natural and constructive common ground. Landscape archaeology may be one of landscape architecture's most relevant neighbouring disciplines simply because the landscape that exists now is almost wholly an inheritance and the product of past generations. All landscape has been 'built' and 'designed', as a form of human architecture, even when modern ideas defined by the idea of conscious design might regard them as 'merely' vernacular, organic or cultural. And, evidently, the work of landscape architects practising today will also become part of archaeology at some stage in the future.

There appears to be one major distinction between the two disciplines, however, in that landscape archaeology is primarily a research discipline, where landscape architecture is an action/design practice. In fact, this creates useful complementarities. Landscape archaeology is becoming ever more interested in using its understanding of the landscape's development to shape sustainable future landscapes, while the place of knowledge and research, and associated theory, in landscape architecture is also expanding. There is thus an overlap and a common interest which offers further opportunities.

The European Landscape Convention is an extremely important catalyst for inter-disciplinary work on landscape. Its ideas and recommendations are already being influential. Its definition of landscape is very wide, with human action and decision-making (design, conscious or not) at its centre, and it escapes from the exclusiveness of special designation, insisting that all landscape everywhere is part of Europe's common heritage; it insists on the need for inter-disciplinary and trans-frontier collaboration; it defines three types of instrument (protection, management and planning) which unite the concerns of all aspects of landscape architecture as well as heritage archaeological resource management; it emphasises the democratic aspects of landscape and the importance of education. Most of all it is forward looking, seeking to use wider and deeper understanding of landscape to inform its future shape, an aim of both landscape architects and landscape archaeologists.

Landscape archaeology principally studies the effects of change in the past. The information and understanding it provides concerns the present-day landscape and can be influential on future change. One of the many reasons for understanding the history of landscape change is to help monitor landscape change, and therefore to manage it more completely. Another is to recognise that change itself is actually a part of landscape's character. It might be said that there is no landscape without people (because landscape is perception) but equally (and for that reason) there can be no landscape, past or present, without humanly induced change. Everything that we appreciate or dislike about an area of landscape today is the product of past changes, increasingly humanly induced or guided, over the past few thousand years. Often indeed it is the effect of change that people most appreciate in landscape that they admire or cherish; multi-period landscape, *palimpsests*, are the areas that most attract tourists for example, and a frequent popular criticism of recent landscape change, at least in the UK, is that it wipes away all earlier traces and is too bland and one-dimensional. Not to acknowledge all this can lead to an untenable resistance to future change, the defence of form without function, and the loss of authentic time depth in favour of a manufactured 'timelessness'.

These issues are at the heart of what archaeology contributes to landscape, which can be summarised as understanding the effects of the passage of time allied to human agency and actions over long periods. Other history-related landscape disciplines offer some of this, but archaeology offers a much longer time-depth and its focus on materiality connects it to human action and society. Archaeology focuses on the historic processes that created landscape, and on the material effects of those processes that still survive, so that its research is rooted in the present as well as the past.

Landscape archaeology offers more connections with landscape architecture than sometimes seems to be recognised. There are clear connections between landscape history and so-called 'cultural landscapes'. Landscape architecture education and research would benefit from greater historical awareness and from awareness of the future historical contingency of our own times. Archaeology's theoretical and methodological basis and its use of 'new' technologies of GIS and remote sensing connect it to many strands in current developments within landscape architecture. Landscape archaeology in many guises and in several countries is a heavily theorised discipline. Because many of its practitioners are comfortable with self-reflective critiques of knowledge, contested analysis and cultural and social theory, it relates readily to the issues of complexity, conflict and contestation that sit near the heart of landscape. Its access to deep time and to a variety of historical contexts encourages a detached view of, for example, the universality of current aesthetic theory.

In particular, there is a lot of scope for synergetic inter-disciplinary work between the two disciplines in the field of urban landscape. Landscape archaeology studies townscape as much as rural landscape, and of course in particular the transition between the two at various times in history, a very relevant body of knowledge in the light of current transformations, whether urban expansion, the modification of urban fringe areas, the peri-urbanisation of rural areas or the spread of ex-urbanisation. Many landscape architects work mainly only with open space in urban areas, but a future integrated landscape architecture really needs to deal with cities and towns as well as with open space within urbanised areas. The whole urban fabric past and future is a subject for the attention of landscape architects. Landscape archaeology, alongside urban historical geography and spatial planning, would be a natural partner in this enterprise.

Finally, there is the question of the emergence of European-wide networks, communities of practice, and of an interdisciplinary field of landscape research and of landscape studies. In this large context, networks like LE:NOTRE have been crucially important. There is no exact equivalent in the archaeological world, although the European Association of Archaeologists (EAA) brings together archaeologists from all European countries and from both the academic, commercial and heritage management sectors, and in the process (notably through its annual conferences which always include several sessions on landscape) is encouraging the evolution of common attitudes, aims and methods. National professional organisations in several countries such as the UK's Institute for Archaeologists (IfA) are starting to rationalise and guide professional practice. The European Archaeological Council (a standing meeting of the national heads of archaeological heritage across Europe) compares and exchanges best practice across Europe and in 2002 published its strategy for landscape study and management, to which a policy of landscape, archaeology and agricultural reform will soon be added. These established networks within landscape archaeology as within landscape architecture (and now much more widely promoted by the Landscape Science Policy Briefing, ESF/COST 2010, Fairclough 2010) create the common ground for the co-operations that were presaged by LE:NOTRE's initiative in carrying out its review of neighbouring disciplines.

## Bibliography

Aldred, O. and Fairclough, G. 2003. *Historic Landscape Characterisation: Taking Stock of the Method*, Taunton: English Heritage and Somerset County Council, available: http://www.english-heritage.org.uk/characterisation (last consulted 7 July 2006).

del Arbo, M-R. and A. Orejas (eds) 2005: *Landscapes as Cultural Heritage in the European Research*, Proceedings of COST A27 Workshop, Madrid 2004.

Ashmore, W. and Knapp, B. A. 1999: *Archaeologies of Landscape*, London, Blackwell.

Aston, M. 1985: *Interpreting the Landscape: Landscape Archaeology and Local Studies,* London: Batsford.

Aston, M. and Rowley, T. 1974: *Landscape Archaeology – An Introduction to Fieldwork Techniques*, Newton Abbott, David and Charles.

Austen, D. 1999: *The Medieval Countryside*, London.

Austen, D., Rippon, S. and Stamper, P. 2007: 'Historic Landscape Characterisation, Themed Issue', *Landscapes* 8(2), Windgather Press (Oxbow).

Barker, K. and T. Darvill (eds) 1997: *Making English Landscapes* (Oxford: Oxbow Monograph 93), 70–91.

Bartels, C., Ruiz del Arbo, M., van Londen, H. and Orejas, A. (eds) 2008: *Landmarks: Profiling Europe's Historic Landscapes*. COST A27 LANDMARKS, Bochum.

Bender, B. 1993: *Landscape: Politics and Perspectives*, Oxford, Berg.

Berlan-Darque, M., Terrasson, D., and Luginbuhl, Y. (eds) 2007: *Paysage: de la connaissance à l'action* (*Landscapes: From Knowledge to Action),* Editions Quae, Paris.

Bintliff, J, Kuna, M., Venclová, N. (eds) 2000: *The Future of Surface Artefact Survey in Europe*, Sheffield: Sheffield University Press

Bloemers, J. H. F and Wijnen, M-H. 2002: *Protecting and Developing the Dutch Archaeological-Historical Landscape*, Brochure NWO.

Bloemers, J. H. F., Kars, H. and van der Valk, A. (eds) 2010: *The Cultural Landscape and Heritage Paradox: Protection and Development of the Dutch Archaeological-historical Landscape and Its European Dimension*. Amsterdam: Amsterdam University Press.

Blur, H./B. and Santori-Frizell, B. (eds) 2009: *Via Tiburtina*, Swedish Institute in Rome, Rome.

Bowden, M. 1999: *Unravelling the Landscape, an Inquisitive Approach to Archaeology*, Stroud, Tempus.

Castro P. V., Chapman, R. W., Escoriza, T., Gili, S., Lull, V., Micó, R., Herrada, C. R., Risch, R., Sanahuja Yll, M. E. and Verhagen, P. 2002: 'Archaeology in the South-east of the Iberian Peninsula: A Bridge between Past and Future Social Spaces', in Fairclough and Rippon (eds) 2002: 133–142.

Clark, J., Darlington, J. and Fairclough, G. J. (eds) 2003: *Pathways to Europe's Landscape*, Heide, EPCL/EU.

Clark, J., Darlington, J. and Fairclough, G. 2004. *Using Historic Landscape Characterisation*, London: English Heritage/Lancashire County Council.

Compatangelo-Soussignan, R., Bertrand, J-R., Chapman, J. and Laffont, P-Y. (eds) 2006: *Marqueurs des Paysages et Systèmes socio-économiques,* Actes du Colloque COST de Le Mans (7-9 December), Presses Universitaire de Rennes.

Crawford, O., 1953: *Archaeology in the Field,* London: Phoenix House.
Crawford, O. and Keiller, A. 1928: *Wessex from the Air,* Oxford: Clarendon.
Council of Europe 2000: *European Landscape Convention,* European Treaty Series 176. Available: http://www.coe.int/t/dg4/cultureheritage/heritage/Landscape/default_en.asp (accessed 21/12/09).
Darvill, T. and Godja, M. (eds) 2001: *One Land, Many Landscapes,* Oxford, British Archaeological Reports (S987).
Doukellis, P. and Mendoni, L. G (eds) 2004: *Perception and Evaluation of Cultural Landscapes,* National Hellenic Research Foundation, Meïλethmata 38, Athens.
ESF/COST 2010: 'Landscape in a Changing World – Bridging Divides, Integrating Disciplines, Serving Society', *Science Policy Briefing* 41, Strasbourg / Brussels.
Everson, P. and Williamson, T. (ed). 1998: *The Archaeology of Landscape,* Manchester.
Fairclough, G. J. 1999: 'Protecting Time and Space: Understanding Historic Landscape for Conservation in England', in Ucko and Layton 1999: 119–134.
Fairclough, G. J. 2003: 'The Long Chain: Archaeology, Historical Landscape Characterization and Time Depth in the Landscape', in Palang, H. and Fry G. (eds) 2003: *Landscape Interfaces: Cultural Heritage in Changing Landscapes,* Landscape Series 1, Kluwer, Dordrecht, 295–317 (reprinted with slight revisions in Fairclough *et al.* 2008: 408–424).
Fairclough, G. J. 2006: 'Our place in the Landscape? An Archaeologist's Ideology of Landscape Perception and Management', in Meier, T. (ed). *Landscape Ideologies,* Archaeolingua, Budapest, 177–197.
Fairclough, G. J. 2007a: 'L'histoire et le temps: gérer paysage et ses perceptions (History, Time and Change: Managing Landscape and Perception)', in Berlan-Darque *et al.* 2007: 149–162.
Fairclough, G. J. (2007b) 'The Contemporary and Future Landscape: Change & Creation in the Later 20th Century', in McAtackney, L., Palus M. and Piccini, A. (eds) *Contemporary and Historical Archaeology in Theory, Paper for the 2003 and 2004 CHAT Conferences* (Studies in Contemporary and Historical Archaeology 4, BAR International Series 1677), Oxford, Archaeopress, 83–88.
Fairclough, G. J. 2008: '*Rus infra urbe:* Pre-industrial Landscape Below Modern Cities', in Compatangelo-Soussignan *et al.* 2008: 229–241.
Fairclough, G. J. 2009: 'New Heritage Frontiers', in Council of Europe 2009: *Heritage and Beyond,* Strasbourg: Council of Europe Publishing. Available: http://www.coe.int/t/dg4/cultureheritage/heritage/identities/beyond_en.asp (accessed 19/12/09), 29–41.
Fairclough, G. (2011): 'Look The Other Way – From a Branch of Archaeology to a Root of Landscape Studies', in Kluiving, S. J. and Guttmann-Bond, E. (eds) *Proceedings of LAC2010.* Amsterdam University Press, Amsterdam
Fairclough, G. J. and van Londen, H. 2010: 'Changing Landscapes of Archaeology and Heritage', in Bloemers *et al.*
Fairclough, G. J. and Wigley, A. 2006: 'Historic Landscape Characterisation: An English Approach to Landscape Understanding and the Management of Change', in del Arbo and Orejas 2005: 87–106.
Fairclough, G., Lambrick, G. and McNab, A. (eds) 1999: *Yesterday's World, Tomorrow's Landscape: The English Heritage Landscape Project 1992–94,* London, English Heritage.
Fairclough, G. J. and Møller, P. G. (2008): *Landscape as Heritage – The Management and Protection of Landscape in Europe, A Summary by the Action COST A27 "LANDMARKS", Geographica Bernensia G79,* 269–291.
Fairclough, G. J. and Rippon, S. J. (eds) 2002: *Europe's Cultural Landscape: Archaeologists and the Management of Change,* EAC Occasional Paper no 2, Brussels and London: Europae Archaeologiae Consilium and English Heritage.
Fairclough, G. J., Harrison, R., Jameson Jnr., J. and Schofield, J. (eds) 2008: *A Heritage Reader,* London, Routledge.
Fleming, A. 1988: *The Dartmoor Reaves: Investigating Prehistoric Land Divisions,* London, Batsford.
Fox, C. 1938: *The Personality of Britain,* Cardiff, National Museum of Wales.
Graham, B. and Howard, P. (eds) 2008: *The Ashgate Research Companion to Heritage and Identity,* Aldershot, Ashgate.

Hall, M., 2006: 'Identity, Memory and Counter-memory: The Archaeology of an Urban Landscape', *Journal of Material Culture* 11(1/2), 189 *A Heritage Reader* 209.

Harrison, R. and Schofield, J. 2010: *After Modernity, Archaeological Approaches to the Contemporary Past*, Oxford, Oxford University Press.

Herring, P. 1998: *Cornwall's Historic Landscape: Presenting a Method of Historic Landscape Character Assessment*, Truro, Cornwall Archaeological Unit.

Hicks, D., McAtackney, L. and Fairclough, G. J. (eds) 2007: 'Envisioning Landscape Archaeology', *One World Archaeology* 52, Walnut Creek, CA: Left Coast Press.

Hidding, M., Kolen, J. and Spek, T. 2001: 'De biografie van het landschap. Ontwerp voor een inter- en multidisciplinaire benadering can de landschapsgeshiedenis en het culturele erfgoed', in J. H. F. Bloemers *et al.* (eds) *Bodemarchief in Behoud en Ontwikkeling. De Conceptuele Grondslagen*. Den Haag: NOW, 7–109.

Hooke, D. 2000: *Landscape: The Richest Historical Record*, SLS Supplementary Series 1, Amesbury, Society for Landscape Studies.

Hoskins, W. G. 1955 and 1977: *The Making of the English Landscape*, London: Hodder and Stoughton, original edition 1955, reprinted 1977 with new introduction.

Jones, M. and Olwig, K. (eds) 2008: *Nordic Landscapes*, Minneapolis, Minnesota University Press.

Johnson, M. 2007: *Ideas of Landscape*, Oxford, Blackwell.

Kelm, R (ed.) 2005: *Frühe Kulturlandschaften in Europa: Forschung, Erhaltung und Nutzung*, Albersdorfer Forschungen zur Archäologie und Umweltgeschichte, Band 3, Heide, ÄOZA and Boyens.

Kolen, J. and Lemaire, T. (eds) 1999: *Landschap in meervoud: perspectieven op het. Nederlandse landschap in de 20ste/21ste eeuw*. Jan van Arkel, Utrecht.

Kuna, M. and Dreslerová, D. 2007: 'Landscape Archaeology and "Community Areas" in the Archaeology of Central Europe', in Hicks *et al. Envisioning Landscape Archaeology*, 146–171.

Larsson, L., Callmer, J. and Stjernquist, B. 1992: *The Archaeology of a Cultural Landscape*, Acta Archaeologica Lundensia 4(19), Lunds Universitet.

Laszlovszky, J. and Szabo, P. (eds) 2003: *People and Nature in Historical Perspective*, Central European University and Archaeololingua, Budapest.

Lock, G. and Molyneaux, B. (eds) 2006: *Confronting Scale in Archaeology: Issues of Theory and Practice*, Springer, New York.

Londen, H. van, 2006: 'Cultural Biography and the Power of Image', in W. van der Knaap and A. van der Valk (eds) *Multiple Landscape. Merging Past and Present*, Wageningen, NWO/Wageningen University, 171–181.

Macinnes, L., 2004: 'Historic Landscape Characterization', in K. Bishop and A. Phillips (eds) *Countryside Planning: New Approaches to Management and Conservation*. London, Earthscan, 155–169.

Macinnes, L. and Wickham-Jones, C. R., 1992: 'Time-depth in the Countryside: Archaeology and the Environment', in L. Macinnes and C. R. Wickham-Jones (eds) *All Natural Things: Archaeology and the Green Debate*. Oxbow Monograph 21, Oxford, 1–13.

Martin, E. and Satchel, M. 2008: 'Wheare Most Inclosures Be', East Anglian Fields: History, Morphology and Management, *East Anglian Archaeology* 124, Suffolk County Council Fields.

Meier, T. (ed.) 2006: *Landscape Ideologies*, Archaeolingua Series Minor 22, Budapest, Archaeolingua.

Muir, R. 1999: *Approaches to Landscape*, London, Macmillan.

Nord, J. M. 2009: 'Changing Landscapes and Persistent Places', *Acta Archaeologica Lundensia*, Series in Prima Quarto, No 29, Lund, Lund University.

Orejas, A., Mattingley, D. and Clavel-Lévêque, M. 2009, *COST A27 – From Present to Past through Landscape*, Madrid, CSIC.

Palang, H. and Fry, G. (eds) 2003: *Landscape Interfaces: Cultural Heritage in Changing Landscapes*, Landscape Series 1, Kluwer Academic Publishers, Dordrecht.

Penrose, S. with contributors 2007: *Images of Change – An Archaeology of England's Contemporary Landscape*, London: English Heritage.

Pungetti, G. and Kruse, A. (eds) 2010: *European Culture Expressed in Agricultural Landscapes – Perspectives from the Eucaland Project*, Rome, Palombi Editori.

Rackham, O. 1986: *The History of the Countryside*, London, Dent.

Rippon, S. 2001. *The Transformation of Coastal Wetlands: Exploitation and Management of Marshland Landscapes in North West Europe during the Roman and Medieval Periods*, Oxford, OUP.

Rippon, S. 2004. *Historic Landscape Analysis: Deciphering the Countryside*, York, CBA.

Roberts, B. K. and Wrathmell, S. 2000: *An Atlas of Rural Settlement in England*, London, English Heritage.

Roberts, B. K. and Wrathmell, S. 2002: *Region and Place, A Study of English Rural Settlement*, London, English Heritage.

Scazzos, L. (ed.) 2002: *Leggere il paesaggio. Confronti internazionali (Reading the Landscape: International Comparisons)*, Rome, Gengemi.

Scazzos, L. 2004: 'Reading and Assessing the Landscape as Cultural and Historical Heritage', *Landscape Research* 29(4), 335–355.

Schofield, J. 2007: 'The New English Landscape: Everyday Archaeology and the Angel of History', *Landscapes* 8:2, 106–125.

Taylor, C. C. 1974: *Fieldwork in Medieval Archaeology*, London, Batsford.

Tilley, C., 2006: 'Introduction: Identity, Place, Landscape and Heritage', *Journal of Material Culture* 11(1/2), 7–32.

Turner, S. 2006a: 'Historic Landscape Characterisation: A Landscape Archaeology for Research, Management and Planning', *Landscape Research* 31, 385–398.

Turner, S. 2006b: *Making a Christian Landscape*, Exeter: University of Exeter Press.

Turner, S., 2007: *Ancient Country: the Historic Character of Rural Devon,* Devon Archaeological Society Occasional Paper 20, Exeter.

Turner, S. and Fairclough, G. J. 2007: 'Common Culture: Time Depth and Landscape Character in European Archaeology', in Hicks *et al.*, 120–145.

Ucko, P. J. and Layton, R. (eds), 1999: 'The Archaeology and Anthropology of Landscape: Shaping Your Landscape' *One World Archaeology* 30, London, Routledge.

Williamson, T. 2003: *Shaping Medieval Landscapes: Settlement, Society and Environment*, London, Windgather Press.

# Part 2

# Art, landscape ecology, historical geography and forestry

Chapter 5

# Space, place and the gaze

## Landscape architecture and contemporary visual art

*Knut Åsdam*

### Introduction

There are many similarities and important crossovers between my field of visual art and that of landscape architecture. Writing from within the art world, and from my last fifteen years of extensive international exhibition practice as an artist, filmmaker, writer and researcher within the contemporary arts, my relationship to landscape architecture is not so broad, but I have in several of my installations engaged landscape architects either for advice or in cooperation on how to solve a particular project. Furthermore, the idea of landscape and, in particular, that of the garden, has been central to many of my works (Figure 5.1). My paper here, then, is not based on comprehensive knowledge of landscape architecture but has at its core my thoughts about the tangents and relations between visual art and landscape architecture and it is in particular trying to relate experiences within visual arts that can be thought to be useful in the practice of landscape architecture.

From my position as a visual artist, although multidisciplinary, there are immediately two different directions with the theme of landscape architecture and art that come to mind. One is to elaborate the differences between the two and how they can work together as different fields, the other direction would be to look at their intersections and how they do not need to be such separate categories. Seeing that landscape architecture already feeds ideas to art, it is interesting for me to see what is useful in visual art that can be transferred or used within landscape architecture. Perhaps this could be thought to be different from stylistics and more a case of approach and of asking the right questions in relation to a site. Stylistics, in this case, would mean to emphasise the style or look of certain artworks, aesthetic tendencies, and transfer them to another context. The problem is that the aesthetic properties at play in an artwork within the visual arts are (at best) reacting to conventions within the field, and play at visual politics and discussions there. Although

Figure 5.1
Knut Åsdam
The Care of the Self Finally Edit, 1999–2007
– DETAIL
Architectural installation with film: Finally (2006). Dimensions variable. 'Night time' park consisting of trees, bushes, grass and undergrowth with paths leading to different areas. The film Finally (2006) projected within one of the 'squares' of the park. © Knut Åsdam 2007. Courtesy of Gasser & Grunert, Inc. NYC and Galeria Juan Prats, Barcelona

many of the visual referents for a work of art relate to the world outside art institutions, the logic and effect of the aesthetic play can become lost by a mere transfer of a 'look' from one category to another if there is no similar play of referents within that other category of work. This is why it is often better to look at transferring strategies and approaches rather than a set of aesthetics. (You often see this within the visual arts too, e.g. in the relationship between art and political activism, where some artists make activist claims on works that merely puts activist aesthetics within the gallery space. In my opinion, better works are done when the strategies and modes of working have been transmitted from activism to art or from art to activism.).

One particular discussion that has involved a lot of energy and maintained an interest within the fine art context over the last 20 years is the analysis of space and place. With space, in this context, I am thinking first and foremost of a spatial formal category and of volume. With place, I am thinking of the situation when space becomes socially specific e.g. by usage and history. Place is situated space that is marked by the everyday, by economy, laws, deviations, politics, struggle, change and pleasure. To understand place, you have to understand the society in which space is allocated. This is not only evident where people would expect to find this, i.e. in installation art, but also within film/video art and audio art. It seems to me that this is something that is akin to the concerns of landscape architecture, since I believe that landscape architecture deals with the meaning of these terms 'space' and 'place' and what they mean in society. This is possibly a resource from an art discourse that can contribute to other fields concerned with space and place. In the visual arts the focus has not been on the aesthetic *per se*, but on a sociological and historical understanding of place together with a keen understanding of the body and the gazes of the viewers that are all at play in the experience of a work. If we deviate slightly from Foucault's argument that in the twentieth century the primary obsession was with space, I would perhaps say that we are now moving towards an analysis and work that is mostly concerned with *place*, i.e. a space that is actual, political and temporal but fully inscribed and re-inscribed with culture (Foucault, 1986).

## *Land art*

There are obvious developments within art that are relevant to landscape architecture, one of course being land art. I will not elaborate much on that here since I believe that this relationship is too obvious and not the most expansive of observations on art and landscape architecture. However, I think it is a necessary part of any study of landscape architecture to consider land art. Land art itself is both in response to the idea of nature, and the experience of unplanned landscapes, as much as it is also a response to culture and the histories of gardens and cultured landscape and cultivated space. However, land art is also a response to art history and positions itself as a counterweight to much art that privileges the city as the site of experience (although that is of course where most of the landscape art is mediated – in the galleries in the cities and towns). In a more interesting way art-historically, land art also emphasises process and impermanence as some of its main ingredients. It is also very concerned with 'use' and a 'light' materiality – for example in the works of Hamish Fulton, the art-works are really only skeletal documentations of walks in the form of a photograph with a text or only a wall text relating to the experience of the walk (Fulton *et al.*, 2002; McKibben *et al.*, 2002). Here the link to conceptual art is strong. It is not land art as an attention to an object that is important, but the idea or concept of an action which brings a special relationship between viewer/participant and an idea of 'landscape'. So perhaps, to generalise a little, one can say that there are three different – but sometimes overlapping – tendencies within land art.

The first tendency is to deal to some degree with landscape as a medium itself: one such example could be Robert Smithson's Spiral Jetty, where a form is drawn into the landscape and the landscape is treated as a cultural medium (Tsai, 2004). The scale of these works is often monumental, even though the incision in the landscape is small. This might have something to do with both the sheer scale of these works, but also with the idea of a whole landscape as a framework, which as such is extremely monumental and is within the context of art soaked in traditional ideas of the sublime or grotesque as an interface to that which is outside language. This could also be said of the monumental line walks of Richard Long, where he creates a line through a vast landscape by turning rocks or pressing down straw. An example is his white line through a landscape that some have compared to imperialist tendencies because of their placement in what from a Western point of view can be considered an exotic landscape and the problem of transforming a landscape left to its own devices by adding an articulation of culture and thus changing its value.

Secondly we have the task of making an impermanent object in a landscape – here the scale is usually much smaller and the object returns back to nature rather quickly. A typical example would be Anthony Goldsworthy with his forms of leaves left on a river or temporary sculptures made from sticks, leaves, straws or fruit, and which only exist for the accidental passerby, or in the imagination of the viewer encountering the documentation in an art context (Goldsworthy and Friedman, 1993). In this field, context is preeminent. It has been noted by many observers that in Grizedale Forest in Cumbria, UK, people now see natural objects and mistake them for art because within the context of this forest they have become so accustomed to the practice of impermanent art objects, which have been placed there for over 40 years (Grant and Harris, 1991).

Thirdly we have the conceptual approach. Here there is often little done to the landscape, but it is more the idea of an action, observance or history of the landscape that puts it into a discursive cultural context. Often the artwork consists of a photograph of the landscape with an accompanying text, again as in Hamish Fulton's work (Figure 5.2). The idea of landscape here is first and foremost tied to its role as a signifier: what is signified by the landscape? The beauty that often accompanies a distant overview (e.g. how large photographs of collapsing urban areas or war can have an uncanny beauty even when you know that what is depicted is horror or destruction, due to the formal play, sense of remoteness and lack of tangible detail that is in a distant view) is juxtaposed with what exactly is at play: the landscape is also a space of ownership, historical and economical processes, or perhaps a colonial axis. Even a landscape that is left alone is, today, necessarily deliberately left alone and sometimes perhaps a simulacrum of nature.

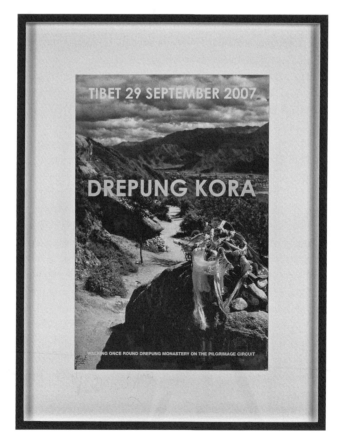

**Figure 5.2**
Hamish Fulton: Drepung Kora, (2007). Iris print, 45 × 30 cm. (Photo courtesy Hamish Fulton/ Galleri Riis)

# Space, place and the gaze

## Landscape and interpretation

However, another way of thinking about the 'meaning' of a landscape is no less political, and concerns the psychological aspects of the viewers' response. This is where the previously noted societal and historical aspects are internalised in someone's own personal and emotional narrative. One responds to it from one's own history and experiences, which were also experienced in the way they were because of what landscape represented to oneself in that society at that time. It has to do – unconsciously – with landscape interpretation, in a dynamic sense, according to one's position and needs at a certain time. Interpretation is an important aspect of how someone relates to landscape in any of the forms I have mentioned above, but also in how society relates to it. If one thinks about the societal, economic and communal role of landscape, one enters into the ideas of how landscape is used and therefore interpreted and how it is given meaning. The Los Angeles-based group Center for Land Use Interpretation (CLUI) (http://www.clui.org) straddles the art world and other discursive worlds in an interesting way. The members do not see themselves as an artist group as such, but they do present exhibitions and are often discussed and appear in the art context. CLUI's approach is both to celebrate and to analyse land use and to use an active way of interpreting landscape and land use – through their own or another's ideas – and to reveal the narratives of changing interpretations of land sites through recent history (Simons, 2008; Coolidge and Simons, 2006).

One example can be a landscape that has been subject to heavy industrial or military exploitation for decades and which is then abandoned and left as a man-made wasteland, emptied of people and undergoing a slow process of re-appropriation by nature (Figure 5.3). These shifts do not only follow economic or political priorities, but are also, in a dynamic way, dependent on what the land, the site, means for people in each instance. To put it more simply, one could imagine

**Figure 5.3**
Navy Target 103A, Imperial County, California. (Credit: CLUI Archive photo)

that a greater focus on pollution and health might give an old, polluting mining complex less clout at a municipal or national level, and place an increased focus on the availability of landscape to the public for recreation which might also increase the pressure on local authorities to make space available to people.

     Economic change is perhaps more determinative though, and usually space is only made available for public uses like recreation after a commercial use has failed or collapsed. CLUI seem interested in revealing and engaging in the dynamic of interpretation, of what land has meant and what it means both now and locally. This approach is surely conceptual but it is not concerned in representing this in the form of an art-work, but rather through discussions, workshops, site visits, lectures *and* exhibitions. Some of the workshops occur on poignant sites and involve art-making or other articulations or formal organisations at the site. CLUI's work therefore easily interfaces with different interpretative worlds, landscape architecture included.

## Art – landscape architecture interactions

As opposed to merely looking at the art forms that have materially or physically most in common with landscape architecture, I will attempt to articulate here what I think is perhaps more interesting for the discourse and interaction between the two fields. This is to look at concerns or approaches within art that deal particularly with context and experience that are useful in art and that can overlap with concerns within landscape architecture in terms of creating an approach rather than a formal method. In order to play with the idea of a renewed direction of landscape architecture as art we should perhaps look at what is particular or characteristic today in the art discussion rather than what merely appears to resemble landscape architecture. There are several main areas that I think will be vital for students of landscape architecture to engage in from within the discussion of contemporary visual arts, and as a route if they wish to relate to the discussion of place and space and also to a renewed interest in narrative and the cinematic.

### *Space/place – a place in encounter with the viewers as historical and political subjects*

Perhaps the most poignant development within the contemporary arts in recent decades is the adoption and investigation of sociological and historical models of space and place. This has been done not only within the obviously three-dimensional arts like sculpture and installation art, but it has also been a dominant development within photography, video, sound art and also within social art (art based on social dynamics). In the 1970s the investigation was, in my opinion, quite marked by the fact that the whole engagement with installation and 'room' was a new field and many of the works pioneered an engagement with the idea of space as if opening a new language. One can think of artists like the theoretically orientated Donald Judd and even to the psychologically orientated Bruce Nauman in this respect (Serota *et al.*, 2004; Kraynak, 2003). Simultaneously, what also developed in art within the 1970s and further in the 1980s was an understanding of the political or social context – through the work of feminist, gay, lesbian and 'queer' art or activism. At the same time, philosophers who had engaged in our historical and social understanding of space and place also came to the forefront, such as Foucault's *Of Other Spaces* (1986), Bachelard's *The Poetics of Space* (1958) and Deleuze and Guattari's *A Thousand Plateaux* (1980).

I also believe that the social struggles of the 1960s through to the 1980s, with their questions about what constitutes the public and hence the meaning of public space (i.e. the civil rights movements, student activism in the streets, establishment of new newspapers etc), also influenced the understanding of what space, place and subjectivity could mean across the field of cultural production. It became clear to many artists that just to deal with the formal exploration of space or merely with the psychological-spatial relationship in a reduced way is not enough because these experiences of spatial relations relate to an everyday world full of details, contradictions or struggles of usage, temporary changes etc. and as a consequence there has been, in the late 1980s and through the 1990s, a renewed understanding of space and a new explosion of installation arts, even to the extent of incorporating narrative. We are here dealing with an idea of space that is not about being able to make a light space or a dark space, a small space or a large space, or filling it with colour or material, but we are dealing with a space in which people struggle, eat, make love and contest with one another; it is a space marked by the body and which in return marks the body too. This is a result of a deepened understanding of the viewer's body and subjectivity as experiencing and acting and also an understanding of the rich temporality and changing nature of experience itself. It is based on the 'stuff' of everyday life. These artists have become especially interested in the histories, usages and practices of space – in *place* – a space that is specific and has specific historical, social and personal readings or experiences.

What is amusing in the context of which I write this chapter, is of course that what is opened up within the arts through a look at the everyday social and historic spaces of society is a look at prisons, brothels, graveyards, schools – and not least – to city parks and gardens – the realm of landscape architecture. The historical garden, with its representation of a micro-cosmos is paid attention, as is the contemporary city park with its outlet for desires in both daytime and night time – that of the fantasy of nature or that of the cultivated symbiosis, and that of the possibility of illicit desire – i.e. drugs or sex – during the night time.

Artists like Dan Graham and I have perhaps been particular examples in this regard. With Graham there are elaborations of the ideological and historical framing of garden design, while in my own work there is a focus on the usage of the garden or city park, through making narrative spaces in which the viewer finds him/herself in a relationship or narrative and fantasy in relation to other viewers (Graham, 1993). I should be precise here that when I write about use or usage, it is perhaps in a different meaning from the functionalistic use criteria often given by a client who wants a park or a garden to be built, instead it is rather a look at how people use spaces as much by error as by design, as much by deviance as by compliance.

What is perhaps most important here is not that art can find landscape architecture as its subject, but that the analysis and openness leads to an understanding of what 'makes a place tick' – what are the dynamics, readings, narratives, usages and histories of a place, and how this can stimulate or inspire the language of design that will be used. This is perhaps where contemporary art and landscape architecture can cross each other's fields of interest. Not that these relations should protect some approach that has to illustrate these points but that a sensibility and a keen analysis of place informs the projects from the beginning, not just the spatial-architectural concerns.

Outside the pure artistic approaches that do not have the same kind of restrictions as the practical tasks of landscape architecture, what seems to be a shared horizon in the understanding of place, or at least an area that potentially has some degree of transference or overlap, has to do with an analysis of what makes the experience of a place specific. This is where understanding place is more important than understanding space. We need to look at the details, histories and usages of a place, including its temporary changes and repeated temporary processes. With temporary changes and repeated temporary processes, I am thinking about several aspects. In that a place is as much defined by its use as by its interpretation, it follows that a place that is used by different user groups at different times of the day or night might change its meaning and cultural sense of place several times a day. An example of this might be a park that in the daytime is a playground for children and people taking a stroll can be known by families with children as a haven for them. The same site might be known to some ambitious runners as having the best environment for running. At lunch time, it might be known to some only as the place that carries the soul of the city through its decades-old roller-disco environment. In the night time, this park could be the local scene of sexual cruising and be known as a cruising site to another group of the public. A section of the park, known to only some as an important part of their world, could be a place to sleep. Then in the morning, the street–cleaners move in, it becomes their workplace and then the cycle starts again.

In this example, the park is a particular site holding different 'meanings' to different groups and individuals. Furthermore, it can only maintain that identity to those groups or in the popular imagination of the town or city because it is repeated – every day or every week. So this requires both temporal variation (a site is not the same all the time at all hours) and repetition (the significance of a place at a certain time is only kept up by repetition by its protagonists: people, animals, plants etc.). So usage is as definitive as design.

## *Viewer/user – the history of the viewer/user*

The notion of the 'viewer' – or later 'user' and 'participant' – in visual arts has gone through a history that probably has some parallels with the development of 'the subject' that the gardens or parks are for. To be brief, when the arts gained relative autonomy from the church, the viewer went from being either the public that came to church or the wealthy powerful person – the patron, to the realm and society of the bourgeoisie. In fact the development of the 'free' arts is intrinsically linked to the development of the Western bourgeois subject. This still follows the art-object like a disease in my opinion!

As a subject rather than an experience, the art-object is still to a large degree a question of connoisseurship and defining taste, although many relations have changed. There are specifically two aspects that lead to this: one is that the art-world is surprisingly reliant on the art-work as an art-object. This is not only seen with the typical examples of painting, or sculpture – works clearly identified as objects, but also in works that seek to expand the formal notions of the art-work. Within moving image works, film, for example, it is often seen as problematic if the work does not have an object-logic, a circular logic that delineates it and makes it complete as an entity. Works that are long and sequential have difficulties within the art economy of distribution and also with critics, because, I think, it is difficult for many to grasp it as an identifiable discursive object because the borders are harder to grasp. Performance, for all its other radical potential (e.g. the physical, temporal relation between the performer and the viewer, the avoidance of a material object) is allowed a role within the art world as long as it forms a temporary aesthetic object and as a less temporal discursive object, and invokes notions of authorship.

With the lack of objective validators (which is not necessarily desirable), without a real reliance on art criticism (but rather art promotion and distribution), with strong reliance on categories and historical conventions of art theory and closely bound to an economy of networking, a large role for the economy of the private collector, and the fragile economy of the artists, the art field is highly dependent on the opinions and taste of the distributors of the work and of collectors, curators and even critics. One would think that we would have come further after Duchamp or Rainer, but there isn't really an easy way out of this and the plurality of artistic work methods has been easily absorbed by the bourgeoisie and its role of defining and digesting good taste through culture. However, one has to remember that in the temporary meeting between the viewer and the art work or art articulation, other relationships occur, and transgressions of the relation to the economy of the art work happen.

In the twentieth century the concept of the viewer was expanded to include women, and there was also a development and problem formulation of the viewer in an attempt to establish a different idea of the viewer as the user and participant as well as a new development of the social within art, as well as a different idea of the artist being in a constitutive relationship to the art-work. This ranges through the work of artists like Joseph Beuys or Adrian Piper (Borer, 1996; Altschuler, 1997). More democratic, open and egalitarian ideas have since developed and it is now quite common to collaborate, or to split authorship and to make works that engage a wider audience in their experience and dissemination into the wider culture. To a large extent this has taken place through the investigations of the viewer through feminism and 'queer' theory. Here we are presented with a subject that has a body as an active part of the relationship to any art articulation – whether through desire or through vulnerability to space or social context. If the body of the viewer is understood in this way, the physical and bodily relationships that are set up are perhaps even more important than the way a landscape composition looks.

The aim here could be to steer away from the idea of landscape architecture as a picture, but understand it as a physical design for subjects that feel, move, possess weight and size, rest, move at different speeds, and also look at each other and at the space.

## The gaze

If we understand the body of the viewer as a force of understanding subjectivity in a place, we then should look at the use of the gaze in the work too. Some would talk about the gaze as a desiring look, a look that includes its own body and that of others into an understanding of a place. But for Lacan it was not a look defined by the subject, but rather an unsuspected space or a 'hole' in the interpreted 'scopic' field – in the field of vision – that opens up and threatens the stability of the subject (Lacan, 1977). It, which could be the pupil of the eye of a person momentarily appearing or an incomprehensible juxtaposition of objects or the appearance of an animal which momentarily breaks down the symbolic order of the world of the subject, exposes it to possibility of assimilation and the grotesque beyond language and exposes one to limits of desire. In cinema it is the uncontrollability of the next frame coming that makes it such a welcome example for psychoanalytical readings of the gaze. In installations and in landscape architecture, one can imagine that sight-lines and their limits and possibilities of accident and surprise, if you like, are important in order to involve the moving and experiencing subject in a dynamic relationship to the text or subtext of the art-work or landscape design. It would amount to an involvement intimately connected to desire through the projections of fantasy caused by the surprises involved that would occur in the processes of distance or assimilation for the subjects involved in a work or a landscape. This can then be made specific from the frame in which this can occur: the cultural and material aspects of the work (Figure 5.4).

## Narrative – cinematic

The idea of a desiring body and that of the gaze in landscape is also where an idea of the performative comes into play. In a play between the different manifestations

**Figure 5.4**
CLUI Field Session at the Desert Research Station. (Credit: CLUI Archive photo)

of the body of the viewer – where the visual vista is only one among many possibilities, the performative, in the sense of the temporal dynamics between different viewers in the place, becomes important. The architectural space becomes a framework for this, and here one of my pet-readings of space and place can come into play: the idea of narrative place.

The narrative place is one that generates a narrative relationship between participants and the place itself. It is something that we can think of in relation to a place, not in terms of illustrated clues of a story – like an open book, a half-empty coffee cup – but in how a place creates a sense of fantasy in its relationship between one viewer and the place or between many viewers engaged in that place. Typical examples of this are situations where one's presence is dramatised, as in a labyrinth, or dissolved and overwritten by the gaze of the other, as happens in a nightclub or other dark space, where you have a sense of a look but not necessarily of what or who is looking: it is rather a subjectivity of a place rather than a person, a subjectivity that you might assimilate. It is important here to really understand an environment not simply as a kind of enclosure, but something that is active, even discursive, in relation to your experience and of how you experience others in that space and in how others will experience you.

In a more experimental or even abstract reading, it would be very interesting to see an idea of the cinematic within landscape architecture: a set of sequenced narrative possibilities, with surprises or 'blank spots' like possible scenes laid out for the participant to experience and connect with. The element of time involved here would be the time defined by the movement of the viewer or user, who then seamlessly interprets and articulates his or her experience. This evokes a dynamic of interpretive distance and assimilation that happens in the viewing process of films and, as mentioned, also in particularised architecture – like the club – but surely this is also part of the history of relating to landscape. To experience something like landscape involves a multitude of relations: e.g. that of the changing movement, that of thinking clearly about that experience – or about something else, that of daydreaming or of being specific, of feeling tired or invigorated, or stressed or lost – one after the other, at the same time (Figure 5.5). If before this dynamic had romantic, national or sublime overtones, with today's multicultural and pluralistic society, we are given other possibilities for what that dynamic of experience or assimilation with landscape and nature could mean.

## *Cross-disciplinarity and cross-media approach*

An important part of visual art's ability to orientate itself and to keep its cultural picture relevant is its ability to include the largest possible disciplinary and media approaches. Perhaps this is something that could be thought of in relation to landscape architecture too. Why is it that we think as a default of landscape architecture as wholly consisting of design with plants? Perhaps more radical approaches are possible, including new technology, new materials as visible components (not just under the ground), or even landscape architecture as facilitating events and performances – temporary formations. An urban landscape is not a new thought, and perhaps it is possible to embrace materials and situations that are seemingly contradictory – new

Knut Åsdam

**Figure 5.5**
View of the CLUI Programme:
A Tour of the Monuments of the Great American Void.
(Credit: CLUI, photo by Steve Rowell)

materials, components, temporary formations with the idea of landscape, just as this very idea of landscape occurs in unplanned ways by the city itself in its everyday life. Alternatively, at the other extreme, let the vegetation take over more – to become anarchistic.

This raises the question: how can landscape architecture be allowed a more radical and active position in society? From my limited exposure to the work processes of landscape architects I realise that they are often given a task defined by others in terms of politically motivated architectural plans, and have to make architectural spaces defined by the architects' work of buildings. I think landscape architecture needs more autonomy as a practice – not only in parks and open areas, but within cities and dense areas. We also know how important are the spaces between or intersected by buildings. The actual areas we move through every day – while we seldom have access to the buildings – on our way to work, school, the shop or a friend – need to be considered more crucially in terms of environments by landscape architects independent of the concerns of the owners of the buildings, be they public or private.

## Research into the goal and context of the art-work

Within contemporary art a goal is often set as to what kind of work one wants to develop – this is akin to many other practices, of course. But one generally has an idea of what kind of art one wants to make. After that it is too general to say anything specific about working methods of artists or what pertains to art. Perhaps the only thing would be that there is a sense of the context of the work. If you are a painter you understand the context of painting and the problems of painting in order to develop your work and if you work with film or installation you also have an understanding of their role in art that attracts you to working with them in this case. Apart from that it is too unspecific to say anything about research approaches of contemporary art *per se*. However, one can look at the tradition of conceptual art or of installation art and one will find that artists use research methods in the development of their work. It often starts with what I have elaborated on here, an analysis of context – be it the discursive, the spatial, the social or some other.

From there one starts to think about possibilities, and about both formal and critical ideas of trying to resolve the work, or arrive at a good idea. I think that in general an art practice borrows and uses strategies from many other fields in terms of research and also for finding ideas. One ends in a symbiosis between art discussion, news, social or political processes or movements within culture – in film, music, mass culture and so on. Art in this case also shows itself as an interpretative and productive field. It digests and reproduces life's little or big movements as they are mediated by any of the means we know. If there are any things that students of landscape architecture should learn from the field of art – in terms that are not about processes that are already intrinsic to other research disciplines – then it is the appetite for other fields of work or culture.

## Concluding thoughts

I think there is more to the relationship between contemporary visual art and landscape architecture than a mere collaborative situation or a sense of mimicry of style – which are probably the most usual ones. Besides, given autonomy, landscape design is an art form in its own right. The visual arts digest everything they come in contact with, and constantly incorporate new discourses and working methods and media into their own work and discourse. However, it is not just digestive, it also produces relationships and interlinks areas that are otherwise seen as separate. If I were to speculate on a potential relationship between the disciplines, it will only be that they need not be seen as so separate. Perhaps what needs to change is the way they look at each other, especially the way landscape architecture looks at art. It would be my desire to see a gaze from the discipline of landscape architecture that is not so concerned with looking to art for style, but for ways of understanding place and people's interaction with it, the physical, psychological and social space.

## Bibliography

Altschuler, B. (1997) 'Adrian Piper: ideas into art,' *Art Journal 56*, 4, 100–101
Ault, J. and Beck, M. (2003) *Critical Condition: Selected Texts in Dialogue*. Zeitgenössische Kunst und Kritik. Zollverein
Bachelard, G. (1958) *The Poetics of Space* (trans. M. Jolas). Orion Press, New York.
Borer, A. (1996) *The Essential Joseph Beuys*. Thames and Hudson, London
Coolidge, M and Simons, S (2006) *Overlook: Exploring the Internal Fringes of America with the Center for Land Use Interpretation*. Metropolis Books, New York.
Deleuze, G. and Guattari, F. (1980). *A Thousand Plateaus* (trans. B. Massumi). Continuum, London and New York.
Flam, J. (1996) *Robert Smithson: The Collected Writings* (2nd rev. edn). University of California Press, Berkeley.
Foucault, M. (1986) 'Of Other Spaces,' *Diacritics* 16, 22–27.
Fulton, H., Vettese, A. and Bartlett, P. (2002) *Hamish Fulton Walking Artist*. Richter Verlag, Düsseldorf.
Goldsworthy, A. and Friedman, T. (1993) *Hand to Earth: Andy Goldsworthy Sculpture, 1976–1990*. H. N. Abrams, New York.

Graham, D. (1993) *Rock My Religion: Writings and Projects 1965–1990* (ed. B. Wallis). MIT Press. Cambridge, MA.

Grant, W. and Harris, P. (eds) (1991). *The Grizedale Experience: Sculpture, Arts & Theatre in a Lakeland Forest*. Canongate Press, Edinburgh.

Grosz, E. (1995) *Space, Time and Perversion: Essays on the Politics of Bodies*. Routledge, New York and London.

Kraynak, J. (ed.) (2003). *Please Pay Attention Please: Bruce Nauman's Words: Writings and Interviews*. MIT Press, Cambridge, MA.

Lacan, J. (1977) *The Seminar XI, The Four Fundamental Concepts of Psychoanalysis* (ed. by J.-A. Miller, transl. A. Sheridan). W.W. Norton & Co., New York.

Lambert-Beatty, C. (2008) *Being Watched: Yvonne Rainer and the 1960s*. MIT Press, Cambridge, MA.

McKibben, W., Scott, D. and Wilson, A. (2002) *Hamish Fulton*. Tate Publishing, London.

Piper, A. (1997) *Out of Order, Out of Sight: Selected Writings in Meta-art, 1968–92* vol. 1 (Writing Art) MIT Press, Cambridge, MA.

Serota, N. (ed.) (2004) *Donald Judd*. Tate Modern and D.A.P. London and New York.

Simons, S. (ed.) (2008) *Up River: Man-Made Sites of Interest on the Hudson from the Battery to Troy (With Topographical Map)*. (Center for Land Use Interpretation/American Regional Landscape) Blast Books, New York.

Tsai, E. (2004) *Robert Smithson* (2nd edn). University of California Press, Berkeley.

Chapter 6

# A shared perspective? On the relationship between landscape ecology and landscape architecture

*Bob Bunce*

## Introduction

The purpose of this chapter is to compare the disciplines of landscape ecology and landscape architecture in order to indicate the ways in which they can be mutually beneficial. Landscape ecology is a relatively young discipline in comparison with landscape architecture. Although the term was originally proposed in the 1930s, it was not until the 1980s that the first full major international meeting was held in Veldhoven, the Netherlands, in 1981, although the Slovak Academy of Sciences had previously organised international landscape ecology meetings on a three-year cycle. In these early meetings, discussions were mainly concerned with theoretical concepts and descriptive work. Landscape ecology has now developed into a distinct discipline with a strong science base involving hypotheses and concepts that can be tested in the field and to which statistical procedures can be applied. Underlying the science is an overriding desire to express landscape in a holistic way and to apply ecological concepts at the landscape level. Here is where landscape ecology differs from mainstream ecology, which largely studies aspects which are not concerned with interactions at the landscape level or scale. In particular, many detailed process and modelling studies do not involve the recognition of the complexity of landscapes and indeed, in many cases, specifically avoid such issues. In addition, many mainstream ecological research study sites are selected in isolation from the matrix of patches which comprise the surrounding area.

Landscape ecology is usually defined as the holistic understanding of the relationships between ecological components of landscapes and their interactions. In this context the word landscape is used as a term to include the complete mosaic of habitats and biota that occupy a given area of land. An underlying motivation of many landscape ecologists is that such developments should lead to practical applications for social benefit. Amongst the practitioners of landscape ecology there is a strong recognition of the role of man in the functioning and development of landscapes and how the understanding of these relationships can promote beneficial landscape management, at both strategic and local scales. These last three points especially converge with many of the principal features of landscape architecture.

The main principles of landscape ecology may be summarised conveniently as follows (all are interrelated and should not be considered to be independent):

- *Landscape structure and pattern*. This involves the study of fragmentation, metapopulations and the associated roles of corridors in connectivity and isolation of habitat units.
- *Landscape description*. This involves the definition of the complexity of landscapes and the role of the component elements.
- *Biodiversity at the landscape level*. This involves the principle that the assessment of biodiversity must be determined and integrated at different spatial scales.
- *Landscape function*. This involves the study of ways in which landscape elements interact in order to eventually understand how landscape can be managed in an integrated way.
- *Monitoring and the assessment of change*. This involves the detection of change within patterns, biodiversity and function, accepting the complexity of landscapes as an essential support for forward planning.

The study of landscape ecology is supported and encouraged through the International Association for Landscape Ecology (IALE), which currently has around 1800 members worldwide (a further 700 if the separate Dutch landscape ecological organisation is included). A world congress is held every four years and regional chapters are now present in all continents except Africa. Most European countries have independent chapters with variable frequencies of meetings but with a joint congress every four years. Activities are reported in a bulletin, which together with details of publications can be found at http://www.landscape-ecology.org. The symposium volumes from the World Congress in 2007 and the European Congress in 2009 provide valuable collections of papers representing the state of the art. There is also an IALE directory of addresses of all members which can be obtained via the website.

IALE could provide the structure for future collaboration with the landscape architects (perhaps by linking with IFLA, the International Federation of Landscape Architects) through:

- providing information on future meetings
- holding joint meetings
- holding meetings to discuss individual issues
- promoting joint publications.

A further option would be to invite individual landscape ecologists to participate in teaching programmes in university departments of landscape architecture (unless they do so already in some institutions).

There is therefore a high degree of overlap between landscape ecology and many of the sub-disciplines of landscape architecture such as landscape planning and landscape assessment. The principal difference is that landscape ecology is science-based (Forman and Godron, 1986) and, whilst both disciplines involve the role of man, landscape architecture specifically involves design and the enhancement of places to be functional and beautiful. The latter objective is only indirectly involved in landscape ecology (although there is a school of 'ecological aesthetics' concerned with the aesthetic content of natural environments and processes), while the functional and sustainable aspects of landscape architecture are in direct agreement.

Recently in landscape ecology conferences and scientific outputs there has been an increasing emphasis on implementation of the results of the science in planning at both strategic and local scales (e.g. Fry and Särlov-Herlin, 1997; Dramstad et al., 1995). These activities converge with landscape architecture and emphasise that there should be increasing collaboration – as indeed is being promoted by many in landscape architecture who consider that landscape ecology offers important tools for landscape analysis (e.g. Bell and Apostol, 2008). Such collaboration would be beneficial to both disciplines.

However, it is also necessary to point out that many practising landscape architects do not utilise these concepts. Certainly, many recent development projects in the UK show a disregard for anything but a 'gardening' approach and in my view show a complete lack of any environmental appreciation whether of landscape ecological or mainstream ecological concepts. Whilst it would be necessary to quantify this comment, the management of motorway verges in Britain seems to be set in a style appropriate to the 1960s and epitomises the lack of environmental considerations in practical management. Raising the awareness of landscape architecture practitioners with landscape ecological principles would therefore be a most useful activity and be mutually beneficial.

The participation of landscape architects in many of the landscape ecology meetings shows that there is a recognition of shared objectives. Indeed, the discussion below demonstrates the significant extent of existing collaboration and the degree of overlap between the two disciplines. In practice, many recent presentations at landscape ecological conferences show that the application of landscape ecological principles to design may often not add to the costs of a given project, and may contribute much towards ecological sustainability. In reality, many recent planning regulations in EU countries involve landscape ecology principles e.g. in the Netherlands and Slovakia, although they are often not specified as such e.g. badger tunnels, deer crossing points and 'ecobridges'.

## Overview of the main subject areas of landscape ecology
### Areas of basic knowledge in landscape ecology

First, it is important to note that all landscape ecological topics are relevant to landscape architecture but not vice versa e.g. construction materials are only necessary for the latter. The structure below was designed for comparison with landscape architecture and other groupings have been made for different purposes – e.g. in the World Congress in 2007, topics were divided into subjects such as urban ecology and climate change and wetlands.

Over the decades there have been several inspirational figures in landscape ecology such as Zev Naveh, Richard Forman and Jacques Baudry. These scientists have helped to formulate the main concepts of the discipline such as corridors and meta-populations. These are summarised in the following section.

## Landscape structure and pattern

This aspect covers the core activities of landscape ecology and could be defined by the observation that a given landscape is more than the sum of its component elements. Thus, many organisms often live in separate landscape elements and require a combination of these in order to survive: for example, a wood pigeon nests in trees or forests, but feeds in arable fields or grasslands. Whilst this is widely understood by people who live in the countryside, the science in landscape ecology has been concerned with formalising these relationships (Bellamy et al., 1996). Furthermore, many species have more subtle relationships e.g. the use of hedgerows by insects and birds to cross otherwise hostile crop fields (Hinsley et al., 1995; Fry, 1995). In this respect the scale is an important issue with different organisms living at contrasting landscape scales e.g. a vulture may have a home range of several kilometres but a robin only a quarter of a kilometre.

Whilst island biogeography theory, in its many guises, is often quoted as the stimulus for much landscape ecological work on fragmentation and meta-population studies (Olff and Ritchie, 2002), this was more concerned with direct observations of species and area, rather than the underlying processes involved. The realisation that not only agricultural intensification but other landscape processes such as land abandonment were leading to isolation of elements in the landscape with implications for extinction has provided a major stimulus for much of the recent work e.g. on birds (Hinsley et al., 1996). Some of this work is not included in the primarily landscape ecological literature, but in journals where the fragmentation of heathlands in southern England is described (Webb, 1990). As discussed below, much of this research has direct relevance to landscape policy, planning and design (Figure 6.1).

In the 1980s, the recognition of the significance of isolation at the landscape level led to the development of the hypotheses of meta-populations (Arnaud, 2003) where the possibility for movement of individuals between patches was restricted and could lead to extinction or a failure to colonise a new or existing patch. There has been much recent work on this subject and an extensive literature has been built up that has demonstrated the principal features of the movement of birds between patches in the landscape. There are also now many papers for other faunal groups (Dramstad and Fry, 1995) and, whilst each taxon has its own distinctive scale and patterns to which it responds, there is no doubt that connectivity is of vital importance. A common feature however, is the importance of the arrangement of the landscape patches within the landscape e.g. the distance between them combined with the dispersal ability of particular species. Whilst much of this work has been carried out in intensively farmed European landscapes, the contrasting scale and destruction of forests in Australia has also led to important conclusions to be drawn of the effects of isolation with important implications concerning the minimal area required for survival by viable populations of certain species.

### A shared perspective?

**Figure 6.1**
Landscape structure: An example of a fragmented landscape in Languedoc, in southern France. Small, separate patches of tall scrub mixed with semi-natural grassland, deciduous forest (poplar plantation and oak), pine plantation and pure grassland, with a road running across the centre of the picture. Grass strips connect the patches. (Photo Freda Bunce)

One of the effects of isolation is that dispersal becomes a critical factor in maintaining viable populations in contrasting landscapes. It has long been known that some woodland plant species are unable to colonise new areas of woodland, hence the recognition of these ancient woodland indicators (Peterken and Game, 1984). However, the limitation of colonisation by other groups e.g. fungi were not so well understood but there are descriptions of dispersal processes from a range of taxa that show that colonisation by some groups is likely to be limited in new situations. This is important when considering the current high potential for change in European landscapes and the necessity for taking positive management measures to alleviate potential population declines, or even extinctions.

The study of connectivity between landscape elements and the role of different types of corridors has become an important part of the way dispersal can be understood (Bunce and Howard, 1990). Although a review in the 1990s of the role of corridors concluded that there was little direct evidence, this is not now generally accepted and more recent studies conclusively show their importance. Such positive evidence supports previous observational studies, e.g. of bat movement along hedgerows, and circumstantial evidence, e.g. the spread of scurvy grass along major roads in Britain between 1957 and 2000. Whilst the role of some linear elements has been overstated, e.g. the potential of hedgerows for the dispersal of plant species, many linear elements are undoubtedly important for movement of propagules or individuals (Vermeulen and Opdam, 1995), but also as habitats in their own right – modifications in planting and design schemes to reflect this knowledge can therefore achieve significant benefits. The role of linear features as reservoirs of biodiversity is especially important in intensively managed agricultural landscapes.

The above studies have been carried out through the application of a range of qualitative and statistical methods. These are essential to produce results whose significance is not in doubt and which show that real differences can be isolated from background noise. Geographical Information Systems (GIS) have been especially widely used in studies of landscape pattern. There are now statistical packages for all the main analyses that are required e.g. FRAGSTATS. The problem is that inappropriate techniques are used if there is inadequate tuition and support.

The results from this area of work have been widely applied in the policy arena and design at the landscape scale (Fry and Särlov-Herlin, 1997). In one of the earlier studies, it was demonstrated that grey squirrels (an invasive species in the UK) could not colonise new woodlands if they were more than a given distance from their current habitat. It was therefore recommended that if new woodlands were to be planted, it was better to plant a series of small woods rather than a single large one. The same type of argument has been used in the selection of Nature Reserves – 'SLOSS' – Single Large or Several Small. Here it actually depends on the objective of the reserve sites because some species e.g. raptors need large sites greater than a minimal size. Originally, sites were seen in isolation, but it is now being recognised in nature conservation planning that they must be seen in the context of the wider countryside – an important principle in the design of new or modified landscapes (Verboom and Pauwels, 2004). Whilst this principle has always been accepted for wetlands, many other habitats, e.g. urban areas, are now often included in the planning of networks.

The recognition of the importance of corridors and connectivity in the landscape has led to a series of policy initiatives from local to national and now to international scales. At a regional level in the UK, for example, local councils are now setting up their own networks of corridors and county councils such as Cheshire have used the Life EECONET initiative to examine connectivity in the county using a model which took basic information on distribution and then looked at the consequences for different species. At a national level, The Netherlands has set up the Ecological Main Network to encourage corridors between recognised key sites. Work is also progressing to develop a European network although this development may be restricted by recent financial problems. In some landscapes e.g. the *bocage* of western France (Baudry and Burel, 1998) there are already existing dense networks of linear features such as hedgerows providing connectivity within the matrix of agricultural land.

## Landscape description

Traditionally description of landscapes has been concerned with the ideas of appreciation of the appearance of the countryside which originated at the end of the eighteenth century and only recently have scientists and planners started to formalise the definition of different types.

Many of the methods and parameters used to describe landscapes in ecological terms originate from other disciplines which have not appreciated the need to understand the complex of elements which comprise a given area of land. For example, in the Czech Republic there are over 30,000 vegetation records, yet none are on linear features. Recording at the landscape level involves recognising mosaics and heterogeneity and is now covered by extensive literature (e.g. Forman, 1995; Bunce et al., 2008).

Landscape ecology not only seeks to incorporate this variability but actually recognises that mixtures and mosaics are needed to express the variable characteristics of landscapes (Forman and Godron, 1986; Hinsley et al., 1996). Again, as mentioned above, much mainstream ecology actually often specifically removes variation at the landscape level in order to simplify studies. In practice, in the partitioning of variation at the landscape level, there is often a greater proportion of variation explained between samples than within. This not only has important implications for description and analysis, but also in the strategies for monitoring and the assessment of changes discussed above. The present discussion is especially relevant for the relationships between landscape architecture and landscape ecology.

Characterisation of landscapes has in the past usually been carried out using descriptive expert judgement and most landscape classifications have been produced in this way (e.g. Milanova and Kushlin, 1993). Whilst this procedure has worked in practice, especially at a local level, problems arise across national boundaries because of the lack of repeatability and the actual parameters used in making the distinction between units (Figure 6.2).

Furthermore, the boundary between the aesthetic aspects of landscape and factual description has often been blurred. For example, Pedroli et al. (2007) provide a series of chapters representing a wide range of approaches to landscape description. Within landscape ecology the trend has been towards the use of scientific sampling and analytical procedures in order to define combinations of landscape elements in a statistically robust and reproducible way. Procedures have also been developed in the UK and elsewhere for landscape assessment to be carried out in a consistent way as a means of assisting countryside planning. The fact that intuitively derived classes can be related to objective strata and integrated with other data (e.g. on natural areas) has often been overlooked. However, this is the way in which national landscape classifications could be linked into a common framework; e.g. the different classes of England, Wales, Scotland and Northern Ireland could be compared using statistically robust environmental strata.

In parallel to the science-based landscape ecological approach there have been a series of developments since the 1970s to use more objective methods for landscape classification. Although these have not been widely recognised, there have recently been further developments using the Internet. Landscape ecological principles have also been applied in commercial forest design in the UK (e.g. Bell, 2003 and Bell and Apostol, 2008). Statistical methods are also under development but again have yet to be widely applied. However, the current situation has actually changed little since the late 1970s with virtually all policy driven landscape and consultancy characterisation at regional and national levels being carried out using intuitive judgement – this is in part because of cost but also because statistical classifications often draw boundaries that do not fit exactly with customers' preconceived ideas. On the other hand, the landscape ecological approach has been successfully applied at regional and national levels throughout Europe as described in Pedroli et al. (2007). This is an important conclusion for the present discussion as much

**Figure 6.2**
Landscape function:
a) Scythe-cut grassland near Poprad, in Slovakia. Traditional management of grassland maintains biodiversity, whereas intensive methods – using herbicides and pesticides – have major adverse impacts.
b) Abandonment is now taking place in European hills and mountains, on shallow soils and steep slopes. In the picture, taken in the hills near Castelnaudry in southern France, old machinery has been left on a former meadow, with tall grasses and invasive shrubs taking over. (Photos Freda Bunce)

consultancy work in landscape architecture uses the intuitive approach and it would probably not be very difficult to introduce landscape ecological principles to make the process more repeatable.

Most intuitive landscape classifications do not involve any data analysis and include cultural features such as the style of architecture and agricultural field patterns that change over time and cannot therefore be used for baseline studies of change. The alternative approach is to use the statistical analysis of relatively unchanging features, such as altitude and geomorphology, as base strata and then to relate the landscape features to these. The same applies to monitoring procedures and the two areas of work are closely interrelated. The first complete exercise of this approach for a region in Great Britain (GB) was for the county of Cumbria in the north-west of England. It was subsequently expanded into the GB Countryside Survey (Sheail and Bunce, 2003; Haines-Young et al., 2000). This approach has enabled the resources of landscape features, such as hedgerows and habitat complexes, to be estimated at regional and national levels. It has also enabled the important contribution of some features, especially those involving linear elements, to be determined in agricultural landscapes. A comparable approach has also been developed in Austria, using classes determined from satellite images. Originally, the construction of the strata was relatively time-consuming, but recent developments in database management and computing have enabled such procedures to be rapidly constructed, although the supporting field survey is still relatively expensive. The above procedure can also be integrated with the conclusions of the previous section, for example, in the role of linear features as corridors in the landscape. Urban ecology has not received sufficient attention within landscape ecology, and this topic could be an important area of collaboration with landscape architects.

Although not specifically described as a 'type' of landscape, the term cultural landscape has become progressively recognised and used over the last 20 years (e.g. Birks et al., 1988). Whilst not initially a term coined by landscape ecologists, it is now much used to describe regions that are rich in traditional agriculture and forest systems and a high degree of internal organisation e.g. the *bocage* landscapes in France and ancient terraced landscapes in Crete. Landscape ecologists have in recent years not only described such landscapes but especially, as discussed in a later section, examined the way in which they function.

Traditional landscape description has been central to rural policy but has not specifically involved landscape ecology. Recently however, there has been a trend to link such descriptions to nature conservation policy (Harms et al., 1998). The landscape ecological approach has been widely used in the UK, not only to define resources in the countryside, but also to identify landscape indicators linked to policy measures, such as hedgerow length, that can be determined at a national level in a statistically robust way (Haines-Young et al., 2000).

## Biodiversity at the landscape level

In recent years, the word biodiversity has been applied to a progressively wide range of parameters, from species to vegetation and habitats. However, in common with the first two topics discussed above, variation and distribution at the landscape level has often been ignored. Landscape ecology has been concerned not only with measuring biodiversity at the landscape level but also with providing precise definitions of what is involved in biodiversity measurement (Angelstam et al., 2001). In this process, landscape ecologists have collaborated with methods developed by other disciplines e.g. freshwater ecology (Armitage et al., 1983). As mentioned above, the importance of the landscape level has become increasingly recognised.

Biodiversity measurement needs to be considered at the following levels – each of which involves complexes of the tiers below them in the hierarchy:

1. Landscapes e.g. the *bocage* landscapes of Brittany and western France
2. Habitats e.g. hedgerows in England or Normandy
3. Vegetation types e.g. the plant communities of phyto-sociology
4. Species e.g. holm oak (*Quercus ilex*)
5. Ecotypes e.g. the Caledonian strain of Scots pine (*Pinus sylvestris*).

Underlying all these levels is genetic diversity, not only of native species but also of traditional races of animals. Recent approaches in landscape ecology have involved integration of at least the top four levels. Provided that a standardised framework is used for the initial data collection, other biodiversity groups can be subsequently added. Whilst the setting up of the Natura 2000 Network is a positive move, such sites can only ever cover a small proportion of the landscape. The wider countryside between these key sites has a major role maintaining biodiversity.

The studies on the measurement of biodiversity are for many floral or faunal groups now at a stage when they can be applied to obtain consistent population estimates e.g. the common bird census and butterfly walks; what is missing is coordination at the landscape level across Europe.

The statistical approach used in the UK for landscape features has also been used to identify biodiversity indicators at the national policy level e.g. the number of plant species per unit area. At the international level there are many initiatives involved in biodiversity e.g. the Rio Declaration and more recently, the Gothenberg Agreement to reverse biodiversity decline by 2010 and the statement following the Malahide Conference on biodiversity. Thus, it is now imperative to achieve more international collaboration in order to achieve such policy objectives but at the present time only limited progress has been made (Figure 6.3).

## Landscape function

Integrated studies of individual landscapes using information from all three topics above can be used in order to determine how landscapes function (Wiens *et al.*, 1993). Such studies take years of work before they fully understand the interactions between the different elements and have only been carried out for a few landscapes. The most detailed study has been of the *bocage* landscape in Brittany, western France, where landscape ecological studies have continued for over 20 years (e.g. Burel and Baudry, 1995). Such studies are continuing, for example of veteran trees and their role in the landscape. However, the function of many landscape elements, such as hedgerows and their role in maintaining a sustainable landscape, are now well understood (e.g. Hinsley *et al.*, 1996).

Among other well-studied landscapes are the open forested grasslands and croplands of Spain (*dehesas*) (Regato-Pajares *et al.*, in press). Detailed studies of energy flows and biodiversity have shown how these complex ecosystems function with many integrating elements and their importance as one of the reservoirs of highest biodiversity in managed landscapes anywhere in Europe. Elsewhere,

**Figure 6.3**
Landscape Description: A classic, mountain landscape in the High Tatras in Slovakia. The lake, waterfalls, scree and cliffs could be separated or treated as a single landscape. (Photo Freda Bunce)

few landscapes have been studied in such detail and the role of biodiversity in maintaining sustainability is not well understood.

An integral part of the functioning of landscapes is the understanding of the driving forces that not only maintain stability, but also lead to change (Petit et al., 2001). The understanding of key socio-economic features as well as global factors such as pollution and climate change has been the basis of many past and present landscape ecological studies, but again it is widely recognised that not enough is known about the role of driving factors, especially at a European level. The ever-increasing forces of globalisation and the influence of the reform of the Common Agricultural Policy (CAP) are all part of this process, especially in terms of how likely they are to influence different regions of Europe at the landscape level. It is also likely that, as mentioned above, these forces may combine to produce landscapes driven by factors other than those that have determined their function in the past, e.g. production driven agriculture and forestry (Figure 6.4). This is particularly true where there has been agricultural abandonment.

Understanding the functioning of landscapes is the key to developing appropriate policies for maintenance or enhancement. The role of policy initiatives such as Agri-Environmental Schemes remains in doubt because of limitations in the understanding of the processes involved and how individual managers are likely to respond to a given stimulus. The application of the DPSIR (Driving Forces, Pressures, State, Impact and Response) Framework in the Mirabel project (Petit et al., 2001) uses expert knowledge to assess the potential for change and shows the limitations in the data available at all levels across Europe as a whole. This framework has also been used to identify the key processes in the role of transhumance for monitoring ecosystems in mountain areas in Europe.

**Figure 6.4**
Biodiversity: A landscape near Clun in Shropshire, central England, with many elements; especially linear features, such as hedgerows; but also grassland, crops and small, deciduous woods. All the different patches need to be sampled in order to assess the biodiversity. (Photo Freda Bunce)

## Landscape ecological monitoring and the assessment of change

The assessment of landscape ecological change is fundamental to the understanding of the current state of the rural environment (Ihse, 1995; Lipsky, 1995). Whilst it is inherently involved in many historical studies of landscape, the need for quantitative studies of change is now widely recognised in order to support major policy initiatives on biodiversity as described above. Measurement of change is linked to the initial assessment of resources and consists of repeated surveillance (Bloch-Petersen et al., 2006). Such monitoring is essential if the effect of policy initiatives is to be fully understood. Hence, much of the earlier discussion is also relevant here but consistent recording is even more important, as repeated recording needs greater attention to detail and rigour, so that real changes are identified as opposed to those due only to background noise and not artefacts of the recording system (Bunce et al., 2008). Many descriptive studies, e.g. of vegetation, cannot be used to measure change, because the sites cannot often not be relocated and have inadequate quality control and assurance and therefore cannot be reliably repeated (Figure 6.5).

Despite the many meetings and conferences on monitoring, there is still no consistent approach across Europe, although the technical problems have now largely been resolved, except for full integration of *in situ* data with satellite imagery. The most complete integrated national programme has been carried out between 1978 and 2007 in the GB Countryside Survey (Haines-Young et al., 2000). The results of the 2007 survey are available on the internet (http://www.countrysidesurvey.org.uk/). This project demonstrates that integrated monitoring of landscape and its associated biodiversity including various *taxa* e.g. birds and aquatic fauna, is essential in order to appreciate the true landscape ecological changes over time. In this

**Figure 6.5**
Monitoring: Pre-desert in Almeria, in southern Spain. The procedure described by Bunce *et al.* (2008) is based on life forms such as low evergreen shrubs and palms. (Photo Freda Bunce)

respect, the rural/urban interface will need greater attention in future studies, especially at the European level, owing to the complex urbanisation taking place in many coastal areas e.g. Spain and Portugal, and the expansion of urban concentrations in the Netherlands and Germany.

Because of the inherent difficulty in obtaining sufficiently detailed data to measure change, good sampling procedures are essential in order to relate detailed records to the entire population or domain (Sheail and Bunce, 2003). Whilst the case study approach has often been used, these have rarely been selected using statistical procedures and cannot therefore be extrapolated. The most important principle is that the stratification of the samples must be independent of landscape ecological change, otherwise samples will change strata and confuse the identification of real change (Bunce *et al.*, 1996). Environmental strata have widely been used for this purpose e.g. in the integrated programmes in the UK and Spain. In both these projects, environmental classes are linked hierarchically to reporting procedures at a regional level. The use of such strata enables statistical measures of the reliability of the measurement of change to be made as well as changes in pattern. Strata have now been produced for Europe (Metzger *et al.*, 2005) and work is proceeding to develop a framework for biodiversity monitoring in mainland Europe.

Despite the recent advances in the resolution of satellite imagery, it has not yet been possible to produce reliable national figures for change at the landscape level because of the extent of background noise, although local studies have been successful e.g. of urban areas. Aerial photographs however, can be used to assess changes in major habitats (Ihse, 1995) but have limitations in the level of detail that can be obtained. In many situations they represent the only possible way of measuring historical change, because as stated above, most descriptive studies cannot be used. Some specific phyto-sociological studies can be used to measure vegetation change if there is enough information to relocate the sample sites sufficiently accurately. There is no doubt that an integrated programme involving a combination of satellite, aerial photograph and supporting ground survey is the optimal procedure. However, although technically feasible, this has not been undertaken, probably because such a project is multidisciplinary and crosses conventional boundaries of research initiatives. This is a central problem in landscape ecology – that traditional measurements of change, especially biodiversity, ignore the landscape level and take measurements of stock and change in isolation even although it is demonstrable that there is usually greater variation between than within points.

Repeated measurements at the same locations are essential in order to ensure that real changes are recorded, even although these may have high standard errors if they are rare in the landscape (Bunce *et al.*, 1996; Cooper and McCann, 2000). Quality control and assurance are required, both in the initial data recording and in the subsequent repeat measurements. In the UK such procedures have been used to follow changes in landscape elements, such as hedgerows, over a thirty-year period and have led to the passing of the Hedgerow Protection Act to halt hedgerow losses (Haines-Young *et al.*, 2000). However, many other features in common between landscape architecture and landscape ecology have not yet been covered e.g. vernacular architecture, landscape reclamation and gardens that could be included in an integrated programme. Repeat measurements of individual landscape elements are important because otherwise changes can be masked. For example, old hedgerows or forests can be removed and new ones planted in different places, involving a loss of ancient landscape elements of high quality with relatively uniform new features, although the survey may find the same amount of woodland in the landscape. Diagrams of flows between major landscape categories express these changes and are essential to identify changes in both quality and quantity of landscape elements.

A modern trend in government policy formulation is the identification of indicators to assess the effectiveness of policies e.g. the UK government now uses hedgerow length/$km^2$ as an indicator of landscape state. Whilst many of these indicators are not selected on the basis of research nor are adequately monitored, the procedures described above would enable these to be carried out and would then strengthen policy formulation. The reform of the CAP and its reported limitations make such an approach imperative at the present time. Similarly, the success or failure of agri-environmental schemes needs confirmation using correct statistical procedures in order to justify expenditure on such schemes.

## *Teaching methods in landscape ecology*

Many universities provide courses on landscape ecology, usually based on quite small groups and often stimulated by an outstanding individual scientist e.g. Pavel Kovar in the Charles University in Prague, Czech Republic and Francois Burel at Rennes University, France. These groups are often formed by PhD and MSc students who carry out research related to the specialisation of the leading scientist.

It would be useful to produce a directory of these courses (which could be carried out by IALE) and an initial list has already been informally compiled (Rob Jongman *pers. comm*). A summary of course contents and teaching procedures could also be produced to assist landscape architecture schools who wish to improve the knowledge in this area. It is probable that 'service teaching' is probably a good model for teaching in landscape ecology too. For example, at Nottingham, GIS teaching comes from the appropriate specialists and in Vienna, vegetation scientists contribute in a similar way.

PhD and MSc courses have been run in landscape ecology since the late 1980s – for example, by Jesper Brandt at Roskilde in Denmark. More recently, masters' courses have been run linked to the major IALE congresses in 2007 and 2009. These courses have been successful in sharing expertise and in the stimulation in international collaboration and the sharing of landscape ecological concepts between young scientists and experienced colleagues.

Visiting professorships have a long tradition in science and have a distinguished track record for stimulating collaboration and sharing experiences. They do however need to be for a sufficient length of time and to be focused on specific issues. For example, the issue of applying landscape ecological principles to planning and design could be an appropriate subject. An alternative would be to provide a framework for setting up collaborative teaching programmes between universities which have mutually complementary specialists. This would have the advantage that finding extra salaries would not be required. Another option is to lecture in a particular subject and then carry out project work to apply the principles learnt – this was done by the author in Madrid on the topic of 'quantitative methods for assessing stock and change at the landscape level'. Three of the projects carried out by the students were subsequently published.

## *Comparison of landscape ecological research areas and landscape architecture*

Landscape ecological research is not currently used to any great degree in this field of design. However, the principles of landscape structure and pattern and landscape function could be used in the design process. For example, corridors could be included in new projects and the spatial organisation of new elements could also be arranged according to landscape ecological principles. Such modifications in design could involve few extra costs but could add significantly to the quality of the final development.

Landscape ecological principles are already widely used in designing new developments as noted above. Ecological compensation areas, large-scale corridors and ecological networks are all examples of such applications. However the discipline is only minimally concerned with many of the topics relating to professional practice in landscape architecture at the present time.

However, the integrated landscape ecological studies discussed above are central to the formulation of countryside policy, which is also an increasingly important issue for landscape architecture. In this respect monitoring could be used to determine the trends following the development. Within Great Britain, the

Countryside Information System has delivered summary information at the regional level but elsewhere local information using the principles described above could provide integrated assessments of the landscape ecological resources at the required scale. For example, the state of a hedgerow network and the degree of disruption caused by a major infrastructure project would be an important contribution to a given Environmental Impact Assessment. The understanding and appreciation of cultural landscapes is also a shared interest.

Although restoration techniques are often discussed at landscape ecology workshops, this subject is mainly outside the remit of landscape ecology. There are however some exceptions – for example, the differential dispersal capacity of species and the degree of isolation of a given site are likely to have a strong bearing on the potential for recolonisation by local species. Another important topic is the selection of local stock and species for planting, which has important implications for subsequent biodiversity but this is more a basic ecological rather than landscape ecological principle *per se*. The communication of research results and the sharing of common experience would also be mutually beneficial.

Gardens, landscape management and urban open space planning are not generally regarded as landscape ecological subjects – although their contribution to the landscape ecology of urban areas has been little studied and has great potential for further research. Dispersal of introduced or invasive species, isolation of natural patches in an expanding urban fabric, road or railway corridors and the function of gardens as patches could all be important in determining the biodiversity resources of urban areas.

As discussed above, landscape ecologists have always been involved in planning and environmental policy issues, especially at the strategic scale. There is therefore potentially a high degree of synergism between the two disciplines. In practice, a number of landscape architects have been involved in various workshops organised by IALE and there has been much exchange of views. As discussed below, there is scope for further development of such collaboration.

## Future research opportunities

The section below presents some topics of mutual interest between the two disciplines where future collaboration would be particularly beneficial.

### *Characterisation and definition of landscapes including recognition of complexity and heterogeneity*

Currently, descriptions of cultural landscapes using quantitative measurements of the components are generally lacking, especially involving elements such as vernacular architecture, which add to the appreciation of landscape. This would not only be useful in education terms but also in terms of developing targets for landscapes – especially if the range was set to include both poor, average and good examples. These would need to be set in a European framework, as described above.

### Urban ecology
Whilst there have been an increasing number of relevant studies e.g. those provided in the World Congress publication of 2007, simple questions such as: 'What is the contribution of urban landscapes to biodiversity?' and 'Do they act as reservoirs of insects to restock the countryside?', still remain unanswered. Coordinated studies from the two disciplines could yield strong dividends.

### New European landscapes
There are currently many large-scale development projects and major rural structural developments, such as flood compensation areas, that would benefit from the two disciplines working together. Some good projects have incorporated both, but if the current trends continue there could be even greater opportunities.

### Landscape-scale monitoring
Integrated monitoring, including both rural and urban areas with elements such as vernacular architecture, farm buildings, trees and linear features would be valuable. A collaborative initiative could be set up involving coordinated PhDs supervised by both disciplines.

### Scenarios of change
Whilst useful, descriptive scenarios are limited in their applicability because they are difficult to validate and quantify. However, they can be used as a basis for formulating more quantitative approaches. This is especially true if they could be produced jointly by landscape ecologists and landscape architects as then they would have an added value for a wider audience. Quantitative scenarios involving rules e.g. of urban expansion, can be applied in an objective way to specific parcels of land, which can then be aggregated to regional estimates. This procedure could be used to link scenarios from landscape ecology and landscape architecture to be integrated and then be used to test policy options and to examine their impact in Europe.

## Concluding thoughts
Landscape ecology is primarily a research discipline and involves basic scientific principles. However, many of the results could be incorporated in teaching and practice in landscape architecture, in planning, design and management at various scales.

Collaborative projects could be set up that would be mutually beneficial but would need strong positive thinking and probably financial support. Demonstration projects would show the complementary roles of the two disciplines and could be used to promote joint actions.

Joint meetings between IALE and the IFLA could provide the stimulus to extend such cooperation and collaboration. This would also help in increasing the understanding between both disciplines, which are each interdisciplinary in their own way, with many overlapping concerns and interests.

Joint teaching between universities and institutes, including both PhD and MSc supervision, could also provide a basis for future collaboration and promotion of common ideals.

# Bibliography

Angelstam, P., Breuss, M. and Mikusinski. G. (2001) 'Toward the assessment of forest biodiversity at the scale of forest management units – a European perspective'. In A. Frank. O. Laroussinie and T. Karjalainen (eds) *Criteria and indicators for sustainable forest managment at the forest management unit level.* European Forest Institute, Joensuu 59-74.

Armitage, P.D., Moss, D., Wright. J.F. and Furse, M.T. (1983) 'The performance of a new biological water quality score system based on macroinvertebrates over a wide range of un-polluted running-water sites'. *Water Research* 17: 333–347.

Arnaud, J.-F. (2003) 'Metapopulation genetic structure and migration pathways in the land snail Helix aspersa: influence of landscape heterogeneity'. *Landscape Ecology* 18: 333–346.

Bailey, S.-S., Haines-Young, R.H. and Watkins, C. (2002) 'Species presence in fragmented landscapes: modeling of species requirements at the national level'. *Biological Conservation* 108: 307–316.

Baudry, J. and Burel, F. (1998) 'Dispersal, movement, connectivity and land use processes'. In Dover, J.W. and R.G.H. Bunce (eds) *Key Concepts in Landscape Ecology.* IALE (UK), Preston. Pp. 323–339.

Bell, S. (ed.) (2003) The potential for applied landscape ecology to forest design planning. Forestry Commission, Edinburgh.

Bell, S. and Apostol, D (2008) *Designing Sustainable Forest Landscapes.* Taylor and Francis, London.

Bellamy, P.E., Hinsley, S.A. and Newton, I. (1996) 'Factors influencing bird species numbers in small woods in south-east England'. *Journal of Applied Ecology* 33: 249–262.

Birks, H.H., Birks, H.J.B., Kaland. P.E. and Moe, D. (eds) (1988) *The Cultural Landscape – Past, Present and Future.* Cambridge University Press, Cambridge.

Boatman, N. (ed.) (1994) *Field Margins: integrating agriculture and conservation.* BCPC Monograph No.58, Thornton Heath.

Bloch-Petersen, M., Brandt J. and Olsen, M. (2006) 'Integration of habitat monitoring based on plant life form composition as an indicator of environmental change and change in biodiversity'. *Danish Journal of Geography* 106: 41–54.

Bunce, R.G.H. and Howard, D.C. (1990) *Species dispersal in agricultural habitats.* Belhaven Press, London.

Bunce, R.G.H., Barr, C.J., Clarke, R.T., Howard, D.C. and Lane, A.M.J. (1996) 'Land classification for strategic ecological survey'. *Journal of Environmental Management* 47: 37–60.

Bunce, R.G.H., Metzger, M.J., Jongman, R.H.G., Brandt, J., de Blust, G., Elena-Rossello, R., Groom, G.B., Halada, L., Hofer, G., Howard, D.C., Kováŕiř, Mücher, C.A., Padoa Schioppa, E., Paelinx, D., Palo, A., Perez-Soba, M., Ramos, I.L., Roche, P., Skånes, H. and Wrbka, T. (2008) 'A standardised procedure for surveillance and monitoring of European habitats and provision of spatial data'. *Landscape Ecology*, 23: 11–25 Countryside Survey 2007. Field Manual, Vegetation plots 60pp.

Burel, F. and Baudy, J. (1995) 'Species biodiversity in changing agricultural landscapes: a case study in the Pays d'Auge, France'. *Agriculture, Ecosystems and Environment* 55: 193–200.

Cooper, A. and McCann, T. (2000). *Northern Ireland Countryside Survey 2000: Field Handbook.* University of Ulster, Coleraine.

Dramstad, W.E. and Fry, G.L.A. (1995) 'Foraging activity of bumblebees in relation to flower resources on arable land'. *Agriculture, Ecosystems and Environment* 53: 123–135.

Dramstad, W.E., Olsson, J. and Forman, R.T.T. (1995) Landscape ecological principles for landscape architects and land use planners. Postgraduate School of Design, Harvard University, Harvard, USA.

Forman, R.T.T. (1995) *Land mosaics.* Cambridge University Press, Cambridge.

Forman, R.T.T. and Godron, M. (1981) 'Patches and structural components for a landscape ecology'. *BioScience* 31: 733–40.

Forman, R.T.T. and Godron, M. (1986) *Landscape Ecology.* New York: Wiley.

Fry, G. (1995) 'Landscape ecology and insect movement in arable ecosystems'. In D.M. Glen, M.P. Greaves and H.M. Anderson (eds) *Ecology and integrated farming systems.* John Wiley and Sons, Chichester.

Fry, G. and Särlov-Herlin, I. (1997). 'The ecological and amenity functions of woodland edges in the agricultural landscape; a basis for design and management'. *Landscape and Urban Planning* 37: 45–55.

Harms, W.B., Smeets, P.J.A.M. and Werner, A. (1998). 'Nature and Landscape Planning and Policy in NW Europe: Dutch and German examples'. In: Dover, J.W. and R.G.H. Bunce (eds) *Key Concepts in Landscape Ecology*. IALE (UK), Preston. Pp. 265–279.

Haines-Young, R.H., Barr, C.J., Black, H.I.J., Briggs, D.J., Bunce, R.G.H., Clarke, R.T., Cooper, A., Dawson, F.H., Firbank, L.G., Fuller, R.M., Furse, M.T., Gillespie, M.K., Hill, R., Hornung, M., Howard, D.C., McCann, T., Moorcroft, M.D., Petit, S., Sier, A.R.J., Smart, S.M., Smith, G.M., Stott, A.P., Stuart, R.C. and Watkins, J.W. (2000). *Accounting for Nature: Assessing Habitats in the UK Countryside*. DETR, London.

Hills, G.A. (1974) 'A philosophical approach to landscape planning'. *Landscape Planning* 1: 339–371.

Hinsley, S.A., Bellamy, P.E. and Newton, I. (1995) 'Bird species turnover and stochastic extinction in woodland fragments'. *Ecography* 18: 41–50.

Hinsley, S.A., Pakeman, R., Bellamy, P.E. and Newton, I. (1996) 'Influences of habitat fragmentation on bird species distributions and regional population sizes'. *Proceedings of the Royal Society of London*, Series B 263: 307–313.

Ihse, M. (1995) 'Swedish agricultural landscapes – patterns and changes during the last 50 years, studied by aerial photos'. *Landscape and Urban Planning* 31: 21–37.

Lipsky, Z. (1995) 'The changing face of the Czech rural landscape'. *Landscape and Urban Planning* 31: 39–45.

Metzger, M.J., Bunce, R.G.H., Jongman, R.H.G., Mücher, C.A. and Watkins, J.W. (2005) 'A climatic stratification of the environment of Europe'. *Global Ecology Biogeography*, 14: 549–563

Milanova, E.V. and Kushlin, A.V. (1993) *World Map of Present-day Landscapes. An Explanatory Note*. Dept of World Physical Geography and Geoecology. Moscow State University, UNEP.

Olff, H. and Ritchie, M.E. (2002) 'Fragmented nature: consequences for biodiversity'. *Landscape and Urban Planning* 58: 83–92.

Opdam, P. (1990) 'Dispersal in fragmented populations: the key to survival'. In: Bunce, R.G.H. and Howard, D.C. *Species dispersal in agricultural habitats*. Belhaven Press, London. Pp. 3–17.

Pedroli, B., van Doorn, A., de Blust, G., Paracchini, M.L., Wascher, D. and Bunce, F.M. (2007) *Europe's Living Landscapes: Essays Exploring the Identity of the Countryside*. LANDSCAPE EUROPE. Wageningen/KNNV Publishing

Peterken, G.F. and Game. M. (1984) 'Historical factors affecting the number and distribution of vascular plant species in the woodlands of central Lincolnshire'. *Journal of Ecology* 72: 155–182.

Petit, S., Firbank, B., Wyatt, B., and Howard, D.C. (2001) 'Mirabel: models for review and assessment of biodiversity in European landscapes'. *Ambio* 30: 81–88.

Regato-Pajares, P., Castejon, M., Gimenez, S. and Elena-Rossello. R. (In Press) 'Recent landscape evolution in DEHESA woodlands in Western Spain'. In: S. Mazzoleni (ed.), *Dynamics in Mediterranean Vegetation Landscape*. Gordon and Reach Publishers, Reading.

Sheail, J. and Bunce, R.G.H. (2003) 'The development and scientific principles of an environmental classification survey in Great Britain'. *Environmental Conservation* 30: 147–159.

Turner, M.G. (1987) *Landscape Heterogeneity and Disturbance*. Springer Verlag: New York.

Verboom, J. and Pouwels, R. (2004). 'Ecological functioning of networks: a species perspective'. In: Jongman, R.H.G. and Pungetti, G. (eds), *Ecological Networks and Greenways Concept, Design and Implementation*. Cambridge University Press, Cambridge Pp.56 –72.

Vermeulen, H.J.W. and Opdam, P.F.M. (1995). 'Effectiveness of roadside verges as dispersal corridors for small ground dwelling animals'. *Landscape and Urban Planning* 31: 233–248.

Webb, N.R. (1990) 'Changes in the heathlands of Dorset, England between 1978 and 1987'. *Biological Conservation* 51: 273–286.

Wiens, J.A., Stenseth, N.C., Van Horne, B. and Ims, R.A. (1993) 'Ecological mechanisms and landscape ecology'. *Oikos* 66: 369–380.

Chapter 7

# The past speaks to the present

## Historical geography and landscape architecture

*Klaus-Dieter Kleefeld and Winfried Schenk*

### Introduction

Time and space are primal constituents of life. They both determine human existence as well as social and governmental structures. As a result, geographical research always entails a temporal dimension. Within that part of geography that deals with the spatial aspects of human existence (human geography or 'anthropogeographie'), three fields of research have emerged, which programmatically link the temporal to the spatial dimension:

- Historical geography studies human activities and resulting spatial structures in a historical perspective in order to deduce laws of temporal spatial differentiation. This requires describing, differentiating and explaining the scale and quality of economic, social, political, demographic and natural processes. It also includes the reconstruction of past landscapes.
- Genetic cultural landscape research seeks to explain present spatial structures and processes in terms of the past. It centres on humans as agents of landscape development and projects back into history as far as connections between the past and the present exist and can be uncovered. The human impact on landscape is most discernible in settlements and their surroundings.
- Applied historical geography aims to implement the results of the branches described above in regional planning and environmental education. When sustainable development takes centre stage, it is called cultural landscape conservation.

If all three aspects are considered, most people speak of the 'historico-genetical' approach in geography. However, this term is rather idealistic, as in fact most geographers define themselves by their methodological approach (e.g. as geomorphologists) or their socialisation within one of the manifold sub-disciplines in the field (e.g. as social geographers), even if their questions are 'historico-genetical'. Even those who define themselves as historical geographers cannot and do not want to pursue historical geography exclusively. In practice, they are integrated into the curriculum of geography departments, which also brings about a certain breadth in research.

What is meant by historical geography is defined by the cross-sectional character of the historico-genetical approach, which intends to analyse spatial structures *and* spatially orientated patterns of behaviour. Consequently, historical geography always contributes to environmental issues as well. Its principal objectives lie in:

- Fundamental research, since 'historico-genetical' research reveals the background and causes of economic, social, demographic, mental and ecological processes and conditions. This leads to the geographical fundamentals of history. Another aim is to enlarge the period under consideration in order to get new insights that contribute to the understanding of natural and human developments.
- Contributions towards environmental education that show the history of our environment and demonstrate that it has changed and is still changing. Historico-genetical research, as with every other kind of historical research, seeks to reveal historical mistakes and aberrations in order to avoid them in the future.
- Applied studies and active cooperation in planning processes to protect the cultural heritage in our landscapes and to emphasise its potential for future development. This includes aspects such as biological and cultural diversity, regional identity or aesthetic quality. In this context, administrative provisions serve as starting points for the formulation of concepts of development and protection.

Questions posed by historical geography and environmental history are quite similar, though geography highlights aspects of landscape change more strongly. This also distinguishes historical geography from applied geography, although the latter also carries out investigations at the medium scale. Furthermore, the history of the discipline demonstrates that historical geography is a sub-discipline of geography in spite of its proximity to history.

## *History and development of the discipline*

The objectives and priorities of historical geography have changed several times over the years. Up to the beginning of the twentieth century, historical geographers focused on the reconstruction of ancient battle fields. The campaigns of Alexander the Great were a classical subject, and in some Israeli departments geographical approaches to the Bible are still very important. As there is no relation to the present, proponents of this branch of historical geography consider themselves within the science branch of history.

Since the time of Otto Schlüter (1872–1959) and Robert Gradmann (1865–1950), the precursors of historical geography in Germany, many academics have been concerned with historical settlement research or processes of cultural landscape development, especially in regional studies. Here, two strands of development have been emphasised: processes of colonisation and processes of settlement desertion.

Research on the large-scale abandonment of settlements at the end of the Middle Ages led to the notion that settlement processes do not proceed linearly. This means that the possibilities to infer knowledge about the past from more recent circumstances are very limited. However, this was exactly what research did around 1900. At that time, it was a common objective to trace back spatial structures as far as possible in order to substantiate ethnic continuities in the settlement system. This type of research was connected to a traditional concept of landscape; it was bound to the object and largely morphographic. Its close relationship to the concept of cultural landscape development led to various typologies, used to describe and explain the developmental processes of field or settlement layouts, often in terms of distribution maps.

Until the end of the 1960s, when early social geographers as well as applied geographers began to criticise it for being unscientific, the 'historico-genetical' approach remained at the core of geography. At that time, many geographers turned to new questions and fields of research. Several chairs of 'historico-genetical' geography were changed into those of spatial planning or economic and social geography. In the former East Germany (DDR), where geography was primarily concerned with planning issues, historical geography was completely marginalised.

An international forum for the remaining historical geographers is the 'Permanent European Conference for the Study of the Rural Landscape' (PECSRL), which has held its meetings every two years since its foundation in 1957. Since 1974 the 'Arbeitskreis für Genetische Siedlungsforschung in Mitteleuropa' (Working Committee for Genetic Settlement Research in Central Europe, since 2005 the Working Committee for the Study of Historical Landscapes in Central Europe) has organised annual and thematic conferences mainly for archaeologists, settlement historians and geographers from the German speaking part of Europe (Zeitschrift Siedlungsforschung). Owing to its interdisciplinary orientation towards settlement geography and archaeology historical geography partially separated itself from the new approaches within human geography. In Germany, this led to the dissociation of the approach, field of research and organisation of historical geography. As a result, important developments within the historical geographical research of the last 25 years have often been overlooked from the outside. Most frequently it has been ignored because of a perception that modern 'historico-genetic' research is conceived as process analysis, which centres on social groups as actors and on questions of environmental history.

In Germany, the development of cultural landscape management started from the morphogenetic perspective, which is in contrast to the Anglo-American situation, where historical geography is characterised by its social and economic approaches. The development of application-oriented research and training was stimulated by influences from Dutch and Swiss historical geographers. In this context, the department for historical geography at Bonn served as a focus for many activities. These activities were bundled in the 'Arbeitsgruppe für Angewandte Historische Geographie' (Working Group for Applied Historical Geography), which is affiliated to the committee for genetic settlement research. In summary, two developmental trends of historical-geographical research in Central Europe can be discerned:

- Since the 1960s, historical geography has developed in parallel to geography in general. This development was characterised by a continuous emphasis on the morphogenetic approach. In terms of its thematic and institutional organisation historical geography was not only geared to geography, but also to settlement archaeology and settlement history. Consequently, historical geography was neither fully integrated into the overall development of geography or environmental history in Germany, nor into the development of Anglo-American historical geography.
- Applied historical geography is a characteristic of Central European geography. It emerged from the application-orientated implementation of its morphogenetic approach.

## Basic concepts of historical geography

### Documentation of landscape history

Evidence of past landscapes can be drawn from the interpretation of different types of written and cartographic documents. Important sources for environmental history are manorial registers of property ownership or rights with different regional names (e.g. Urbare, Lehn- and Salbücher), tax registers and bills as well as ministerial files (e.g. privy councils, forest administration, building commission). Documents that derived from conflicts are particularly valuable, because they often regulated access and utilisation of spatial resources such as soils, waters or forests. However, arguably none of these sources has been written down in order to pass on environmental information. As in every other historical discipline, it is the task of the historical geographer critically to interpret and acquire this information.

Most sources provide information which is limited to a local or regional scale. Furthermore, the data can only be understood against the backdrop of the very specific context in which they were written. Hence, increased efforts are now being made to build up 'long rows' in order to detect large-scale trends in a comparative perspective. Homogeneous and quantifiable rows of data are compiled from archival files, e.g. bills.

For that purpose it is necessary to extract quantitative information such as frequencies of occurrence, hierarchical grades and codifications in order to build up groups, periodisations and rankings by statistical means. The usage of computers offers the possibility to build up historical environmental databases. Meanwhile a multitude of procedures including approved catalogues of criteria and questions has been developed. Such databases provide good starting points for objective comparisons, e.g. to reconstruct past climates or the development of landscapes as

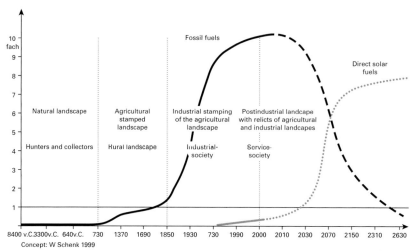

**Figure 7.1**
A diagram of historical development based on the concept of the constraints of the solar energy system and the possibilities released by the use of fossil energy sources. (Source Winfried Schenk)

a result of energy availability (Figure 7.1). Their adoption makes it possible to detect breaks and continuities in certain developments. Nevertheless, in many cases area-wide information is missing. Moreover, it is often impossible to go back much further than the High Middle Ages. Thus it is required to analyse a preferably great number of different sources of diverse quality and temporal coverage in a combined manner.

Historical geography uses maps as both sources of knowledge and means of communication, as does geography in general. Therefore two different types of maps can be distinguished:

- old maps (hand drawn plans and sketches, early official surveys such as first cadastral maps, so-called *Urkataster*) as sources and
- historical maps, which are used to illustrate facts and results.

In research and teaching both types of maps are closely related, since old maps are not primarily addressed from a historico-cartographical perspective, but as an important source which allows the reconstruction and illustration of past landscapes. The map of cultural landscape changes aims to display the succession of multiple maps in different cross-sections so that several phases of the cultural landscape development are merged in a single map sheet. It is based on the method of backtracking, described later in more detail: the combined interpretation of cadastral maps of the nineteenth century and manorial registers of early modern times allows a step-by-step approach to identify earlier stages of field and settlement layouts. In this instance the names of sites and places are important sources of information (meadows, forests and waters). Furthermore, settlement places are linguistic entities, which are important sources for the history of the landscape, namely because of:

- the spatial clustering of certain types of settlement names, which can have different causes, e.g. geographical, historical, linguistic, social and ethnic, among others

- the typology of the names, which frequently relates to their meaning (e.g. 'Rotacker'/red field as a name for the colour of the soil or the type of clearing)
- the temporal stratification of the names, particularly in regard to their endings. In certain periods some suffixes appear to have been in vogue, e.g. -ingen or -heim in the early medieval settlement phases of south-west Germany.

The history of art tells us that landscape paintings have emanated from the poetical mind of the painter. Yet landscape paintings include references to reality. Thus, if art historical source criticism is kept in mind, they are well suited to illustrate past landscapes. Under very favourable conditions it is even possible to draw information of landscape conditions (e.g. the state of glaciers in the Alps) from them. Photographs may be used to infer information on more recent landscapes.

## Landscape as an archive

For the historical geographer, the cultural landscape is a book written over thousands of years by hundreds of generations of people. The origins of the words 'landscape' or 'Landschaft' in themselves convey meanings which both confuse and enlighten (Figure 7.2). It is like a palimpsest with all the relevant information on past environments. Whoever consults landscapes as archives will find several elements and structures of the past in them. If they still exist, often without any function and therefore devoid of context, they are called relicts. The addition of the word 'historical' means that these elements can no longer be authentically constructed. Some of them, like 'rigg and furrow' systems, were purposely created. Many others are the by-products of human actions. The same holds true for their current ecological effects and significance (e.g. sunken tracks).

The notion that human actions affect the relief led to the term 'human variance', which is parallel to the term of 'climate variance'. Anthropogenetic geomorphology and soil science is concerned with the formation and ecological consequences of the 'inventory of quasi-natural features'. One prominent field of research is forest clearings and their effects on the landscape, particularly in regard to the deposition of alluvial clays which extend along the banks of Central European rivers. These late Holocene alluvial clay deposits closely correlate with the main periods of forest clearings and reorientations of agricultural production systems, especially with those of the Roman period, the Early and High Middle Ages and the nineteenth century.

Historical relicts bear information on past systems of production, on patterns of thought and ways of living. They are the 'historical constants' of the cultural landscape. The landscape is individualised by the dominance and peculiarity of single elements. Think of the landscapes of Northern Germany (Knick-Landschaft) or of Normandy/Brittany (Bocage), both characterized by hedgerows. These historical relicts are intrinsic elements of the cultural landscape. They are places of remembrance or ecological importance, which affect individual and societal decisions. When the material or immaterial heritage affects current actions, e.g. by re-use or protective measures, historical geographers speak of persistence.

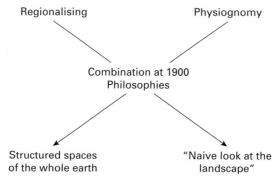

**Figure 7.2** A scheme showing the complex etymology of the word "Landschaft". (Source Klaus-Dieter Kleefeld)

Scientific methods give important insights into processes of landscape development. Palaeoecological investigations provide information on past vegetation and land-use conditions as well as absolute (radiocarbon dating, dendrochronology) or relative (soil stratigraphy, archaeological correlations) chronological data. In addition, methods like phosphate analysis enable abandoned settlement sites to be located. The adoption of soil profiles, pollen diagrams, territorial features and archival sources makes feasible a critical comparison of pollen analytical results with residues from charcoal kilns. Very often, historical geography does not carry out such investigations by itself, but integrates scientific results into its landscape research. Therefore, the ability to critically evaluate scientific sources is a core competence of the historical geographer. The crucial question is: what does the specific method contribute to landscape research?

At present, historical geography is in a process of re-orientation due to a partly accomplished generation change. Should it move towards a science-based landscape history, participate in the discourse of historical and cultural disciplines or, with the needs of students in mind, concentrate on questions of application? This 'trilemma' is reflected by the scope of historical-geographical fields of research.

Open land and forests are in permanent interaction. No matter whether they are of natural or anthropogenic origin, they influence the ecosystem as well as the possibilities and strategies for the use of landscape resources. Today, physical geography demands from historical geography the provision of quantitative data on the proportion of open land and forests at a mid-scale level. This data is needed to analyse processes like soil erosion, occurrences of landslides and the effects of past land-use on chemical cycling (e.g. in rivers). To define broad-scale models the effects of land-use changes on reference areas are reconstructed with the help of Geographical Information Systems (GIS).

The integration of the historical dimension contributes to central tasks of climate research: on the one hand, calibrated climatic records are too short for long-term reconstructions; even in favourable cases they only reach back to the nineteenth century at the most. Historical climate research is able to push the limits much further. Thus, with regard to the high variability of climatic developments, quantitative reconstructions become more sound. This helps to understand extreme changes much better. On the other hand, historical climate research uncovers analogies which give clues to the effects of climatic fluctuations. Without consideration of the causal chain: 'climatic fluctuations – agrarian crisis – famine – social crisis', the changes and crises of pre-industrial societies cannot be understood. Hence, historical climatic research expands the area of historical and climatic investigation. The contribution of geography lies at the regional level. It collects historical climate indicators – ice levels, flooding or the yields of certain products such as wine, hay or crops – in a database, since this kind of data can only fruitfully be analysed by the use of computers and in a European-wide network of research. Methods of interpretation range from simple correlations of the quality of wine with the mean summer temperatures to multiple regressions or complex factor analyses. If these questions can be answered, 'long rows' can be used to deduce future flooding probabilities or to explain single flood events and their effects more precisely. However, it remains one of the most important tasks to improve the spatial and temporal resolution of historical climate reconstructions.

At the moment, research into feudal landscapes is of certain importance within historical geography. The feudal influence on settlement development becomes apparent in the regional differentiation of inheritance customs. While inheritance of undivided property tends to have a conserving influence on the structure of farm acreages, *gavelkind* appears to be the cause and driving force of a more dynamic cultural landscape development, which can be characterised by the causal chain 'diversion of property – multiplication of small holdings – population increase – settlement growth'. Other immediate consequences are changes to buildings, the fragmentation of land-holding, the intensification of land use (particularly by means of viticulture) as well as – after undershooting the subsistence limit – the transfer of rural population to craft and finally to industrial production. At least in Altwürttemberg, the rise of industry was based on the supply of labour in the areas of *gavelkind* and Protestant ethic.

The early modern formation of quarters of the rural underclass belongs in this context as well. Studies on the Cistercians have proved particularly instructive to illuminate the correlation between feudalism and landscape development. During their heydays at about 1400, more than 400 male and 900 female monasteries belonged to the Cistercian order. As the so-called 'Grey Monks' were mystically committed to alter the cultural landscape, their economy had strong and long-lasting effects on many regions. Very often these effects are still visible in the landscape, which should prompt us to treat the specific relics carefully.

Due to their geographical extent and the length of production processes involved forests are important objects of study in historical geography. However, historical geography not only looks at forests as a form of vegetation, but also at processes in landscape development occurring in areas determined by forest production. This might require analysis of areas and processes located far apart from each other. The early modern development of the 'dark woods' – a term that refers to the coniferous woods of the Black Forest and the Franconian Forest – can only be understood with regard to the markets of the target areas of Northern Germany whence large timber rafts were sent via the rivers Rhine and Main. It is a matter of studying the complex network of cause and effect between woodland usage, forest conditions and regional developments in pre-industrial times. In accordance with this specific interest historical geography participates in environmental historical discussions, e.g. on the reality of timber deficits in the eighteenth century. By the analysis of different sources (e.g. forest bills as a cross reference to other documentary sources such as forest orders) concrete forest conditions can be reconstructed against the background of regional supply and demand structures. Thus historical geography literally returns some abstract theories of historical woodland usage back to the earth. In a similar way Hans-Jürgen Nitz tested the feasibility of Wallerstein's model of the 'Early Modern World-System' by historical-geographical analysis at a regional level.

Historical geographers increasingly contribute to another domain of environmental history, the history of water and air. They mainly focus on specific spatial structures. The study by Andreas Dix on the cloth mill of Ludwig Müller in Kuchenheim (Dix, 1997) has to be seen as a contribution to the effects of industrialisation on the cultural landscape, particularly on smaller watercourses. This approach provides the opportunity to study the steps of past industrial substance flows – extraction and production, conversion and processing as well as usage and consumption – in a systematic and spatially differentiated way. Thus it tries to grasp the phenomenon of industrialisation in a system- and process-oriented manner without writing the decadent story of environmental stress and destruction right from the beginning.

## Applied historical geography and cultural landscape conservation

Applied historical geography is the response of historical geography to the paradigmatic shift of geography towards an applied science. Owing to the conditions of mission-orientated research (deadline constraints, limited budgets, addressee-focused) basic research is often neglected, which very often brings about a low temporal depth to papers and new methods of report presentation (e.g. expertises). Public duties and contracts set the framework; the re-use of results of basic research and appealing presentation techniques dominate. The criteria for evaluations and scales are constrained by the project. The following fields of activities are considered to be established:

- In Bavaria, a standardised method for the recording of historical structures has been developed in connection with village renewal programmes and land clearance projects ('Denkmalpflegerischer Erhebungsbogen'). The recording technique is based on the method of backtracking. Recent village structures are compared to cadastral maps of the nineteenth century and formative structures are documented in texts and maps.
- The inventory of historical traffic routes in Switzerland is an applied project, too. It benefits from the protection of the natural and cultural values of the homeland and includes a survey of historical traffic routes that are considered necessary to be protected. Furthermore, it provides insights into Swiss history by means of publications and the reconstruction of old paths and tracks, which are made accessible to tourists.

When the planning efforts refer to historical elements and structures of an area, one speaks of cultural landscape conservation. This concept picks up the ideas of the regional development agendas of the German Federation and the European Union, which regard cultural landscapes as a basic resource for future developments. The Upper Middle Rhine Valley, which was added to the UNESCO's World Heritage list in 2002, relates to the global dimension of this discussion (von Droste zu Hülshof *et al.*, 1995). Cultural landscape conservation combines methods and results of historical-geographical research with the intention of applied historical geography, which is to measure and direct regional development regarding to the model of sustainability. Ideally, the process of spatial planning in cultural landscape conservation has three steps:

1  Cultural landscape inventories or cadastral inventories register, describe and explain cultural landscape structures and elements. Such inventories or registers are accompanied by large-scale maps associated with standardised recording methods for the description and evaluation of the cultural landscape. In this context, geographical information systems have become increasingly important.
2  Due to the complexity of cultural landscapes the evaluation of historical elements and structures is based on a mixture of criteria. These criteria derive from nature ('the beauty, specific character, and cultural diversity' according to the 2002 edition of the German federal law of nature protection, namely § 2, Abs. 1 Nr. 14 including the term 'historical landscape') and monument conservation (historical importance, age, landscapes as stages of historical events) as well as regional planning concerns (in particular the regional specificity according to the principle of the Federal Regional Planning Act which says that 'grown

landscapes' should be protected in their characteristic features including the monuments of nature and culture close to them).
3 Conservation measures allow for the fact that very often even old historical landscape structures can only be conserved to a very limited extent because of high costs or insufficient public acceptance. Thus, unless potentials for future developments in terms of sustainable development are not destroyed, cultural landscape conservation has to come to terms with continuous changes as a characteristic of cultural landscapes. The discussion on values, overall concepts and measures should ideally take place in open dialogues with the parties involved. Guides to the cultural heritage in our landscapes, heritage trails and landscape museums should draw people's attention to the assets of the cultural landscape.

## *Cultural landscape and cultural landscape management from a historical geographical point of view*

This also provokes the question, what does cultural landscape really mean, then the meaning of cultural landscape is not clear at all and it is important for the relationship between historical geography and landscape architecture. It doesn't mean a collection of museums or cultural activities. Cultural landscape is in the first place a geographical term, which should also be used in the understanding of the UNESCO, Council of Europe and the European Union. From the point of view of applied historical geography the historical dimension of the present landscape is the most important research object. Its genesis and development have to be described, analysed and evaluated, so that the cultural landscape can be regarded as a legitimate object of public interest in spatial planning and preservation decisions at different levels of importance (Figure 7.3).

European civilisation changed the natural landscape into a cultural landscape even in the furthermost parts of the continent. It is therefore necessary to research all parts and zones of the landscape, not only the monuments and historical highlights but also the structure of settlement, transportation, trade and industry and, in particular, land-use forms. Within the past cultural landscape development the older cultural landscape elements and structures were not replaced by new ones as they tend to be today, but they mainly changed or expanded – evolution, not transformation.

Historical geography has defined cultural landscape as:

> The cultural landscape in a geographical view means the human-shaped, transformed and adapted natural area from the point of view of existential, social, economic, religious, cultural and aesthetic needs, which developed in the course of time with increasing dynamics and had/have respectively been continually changed and transformed caused by changing conditions. The cultural landscape today represents a spatial, functional and process orientated system, of which the visually observed and structured artifacts comprise point elements, connecting linear elements and summarising area elements.
>
> (Burggraaff 1996, pp. 10-12).

**Figure 7.3**
This diagram structures the different levels of significance and value of historical elements in the landscape. (Source Klaus-Dieter Kleefeld)

**Cultural landscapes as heritage of the world**
*Tentative list based on the UNESCO-criteria

**Cultural landscapes of European importance (European Landscape Convention)**
*Assigning of cultural landscapes

**Cultural landscapes of national importance**
*Dividing into cultural landscape units

**Cultural landscape state program**
*Marking and short characterisation of historical landscape regions

**Regional cultural landscape management concept**
*Inventory of regional individuality (uniqueness)
*Structural analyses and assigning cultural landscape zones

**Cultural landscape management in regional landscape planning**
*Substantial and structural analyses
*Arrangements and measures

**Cultural landscape management in local (municipal) planning, natural and monument conservation**
*Inventory
*Substantial analyse (elements)
*Concrete measures

Design Drs. P. Burggraaff and Dr. K.-D. Kleefeld

**Research steps:**
*Study of written sources, literature and historical maps
*Inventorying, mapping and describing
*Marking assign interpretation and valuation

The definition embraces the whole present landscape with its natural processes and includes industrial zones, agglomeration areas and also the contemporary agricultural and reafforestation areas. In terms of their function, the use and the appearance could vary with regards to the natural potential and natural environment conditions between different types of cultural landscapes. Concerning the morphology, for example, it is possible to differentiate between high or low mountain or lowland areas and coastal areas. Functions such as agriculture, forestry, industry mining, trade, commerce and traffic have also shaped cultural landscapes which take on a regional identity.

On the other hand non-concrete shaping phenomena such as religion, politics, defence, social, economic, cultural and aesthetic value systems, processes, utilisation and management systems, traditions and customs also have to be considered. They have been the processes which have produced a particular cultural landscape. Furthermore, the visual and non-visual archaeological heritage can also be regarded as part of the cultural landscape.

The term cultural landscape principally has a neutral meaning to applied historical geography. Any valuation with regard to conservation, management and a careful future development has to take place after the inventory, mapping, characterising, evaluating and associating of cultural landscape areas, structures and elements. Thus, all parts of the cultural landscape, its local and regional characteristics and features, have to be considered. The most important purpose is to support responsible further development by respecting and preserving the cultural heritage (Figure 7.4). According to the characterisation and valuation of the cultural landscape and cultural landscape areas, there are grades of future development ranging from preservation to new landscape creation:

1. Preservation: areas designated as a nature conservation area or even as a cultural landscape conservation area. These cannot be used and managed in a modern fashion and they will probably be conserved like a museum.
2. Conservation management: To work out relationships between the present demands by appropriate forms of use and management, in which the conservation of the historical substance and structures will have a priority.
3. Sensitive future development: Development has to be harmonised with the core values of the area. Concepts have to be developed with continuing regard

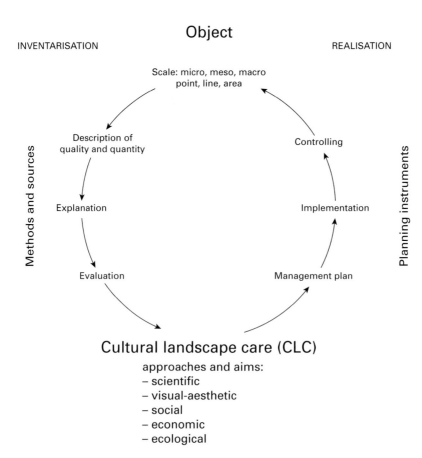

**Figure 7.4**
A model of the process of understanding, valuing and caring for cultural landscapes. (Source Winfried Schenk)

for the natural potential, the specific regional outlook (presentation) values and characteristics of the landscape and its historical development.

4   Replacement and even destruction of historical cultural landscapes (areas). In some cases this is necessary due to the needs of national and international political or economic objectives or interests, for example open cast brown coal mining in Germany, hydro-electric dams or large-scale transport infrastructure. In these cases the documentation of the cultural landscape which will be lost is necessary.

For future development it is very important to conserve the cultural landscape values at the different levels: global, European and national – both in rural areas with low development pressure and urban agglomerations and industrial zones with high development pressure – in order to conserve the values they contain without hindering the necessary dynamic of development. This is only possible by formulating guiding principles for spatial planning and landscape development policy at the national landscape level, which contains cultural, historical, ecological and aesthetic values. These have to be worked out in spatial development plans at the regional level and in specific measures at the local level.

In Europe since the 1950s we have had a new quality in changing, transforming and shaping landscapes: we have tended to destroy historical structures and elements and therefore the historical dimension or the cultural heritage of the different historical epochs to replace them by a single present monotony. However, in the development process of cultural landscapes, historical structures and elements have built diverse (objective), characteristic and beautiful (subjective) cultural landscapes in Europe with many regional (individual) differences and values. This is important for tourism and recreation, but also for the regional identity of the inhabitants, who are living in and identifying with their regions. This illustrates the important role of the cultural heritage of Europe and European countries.

The present development process cannot be stopped, because landscapes have to be developed in the future. Every generation has the right to carry out its own development and add its own traces to the landscape. However, we need to know what is important for the cultural identity of the different European regions in order to ensure that important values are protected. The vision is a cultural landscape management which understands our European cultural heritage as valuable capital and which can be considered as the roots on which future development has to be based.

The rural and urban cultural landscape can also be considered as a form of economic capital for the vitally important recreation and tourism sectors. In this way cultural landscape and cultural heritage management can be seen as having a major economic significance. It is necessary to present a new vision: cultural heritage (structures and elements) of the present landscape in European regions have to be analysed and have to be integrated in the European spatial policy and landscape development.

Klaus-Dieter Kleefeld, Winfried Schenk

## Cultural landscape research and analysis

Applied cultural landscape research and analysis in Germany has a research tradition going back some 25 years. One of the first important historical geographical research projects was the cultural landscape inventory of lower Franconia for the purpose of farmland consolidation, which was carried out by Thomas Gunzelmann of the University of Bamberg (Gunzelmann, 1987). Gunzelmann inventoried mainly rural cultural landscape elements in unsettled areas such as canals, allotment forms (cultivation), the development of land use, hedges (bocage), old frontiers etc., which had to be protected within the land consolidation. Peter Burggraaff carried out research in 1982 in the former duchy of Cleve (Lower Rhine area), examining the development of the settlement pattern since 1730 on the basis of a land register map of 1730, to support spatial planning. Approaches can be defined as progressive, retrospective or retrogressive, depending on which layers in the landscape are the starting point – the earliest looking forward or the present looking backwards, for example (Figure 7.5).

The 'bureau for the study of historical towns and landscapes' (Burggraaff and Kleefeld, 1998) started in 1990 with several research projects. In co-operation with the Archaeological Monument Office of the 'Landschaftsverband Rheinland', the District of Cleves and the Department of Historical Geography of the University of Bonn the bureau carried out an interdisciplinary research project about the cultural landscape development and archaeological survey in the Lower Rhine area between 1990–1992. Since 1990 this bureau has carried out about 60 cultural landscape research projects on different levels.

Cultural landscapes contain elements of many different kinds. They can differ in material (like built objects) and biological elements, which have been influenced by man through agriculture, forestry and special land use forms respectively. Much older cultural landscape elements have lost their original functions. Some of them assumed new functions and others became redundant. These can be regarded as fossil elements or relics. The second group of elements is threatened to disappear, because they have no functions and so no reason for them to be maintained.

Within the field of research into the cultural landscape it is possible to consider two aspects – cultural landscape elements and cultural landscape structures. Cultural landscape elements are mainly built structures, such as settlement, agricultural and industrial structures. The structures, which evolved in the past, now contain a variety of cultural landscape elements, which have developed in different historical periods. Cultural landscape elements may have been changed and lost their original appearance or they may even have disappeared over the course of the past development. Cultural landscape analysis is a very important instrument in researching the (historical) substance and structures of the present landscape for informing spatial planning and planning decisions, which may threaten historical landscape elements and structures. With exception of the protected monuments many cultural landscape elements such as fields, draining fields, old roads, walls, canals etc. as well as historical land use forms and structures cannot easily be legally protected.

An important characteristic of the history of the European cultural landscape since Neolithic times is the dynamic of its development. Every historical period

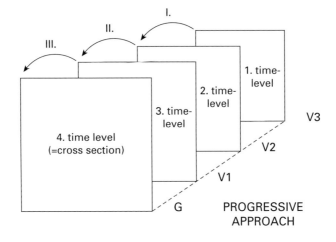

**Figure 7.5**
A model showing different approaches in historical geographical research. (Source Klaus-Dieter Kleefeld)

can be shown to be dependent on the different cultural and technical capabilities leading to specific use and settlement forms. In addition to these direct shaping measures indirect and non-directed processes have also taken place, which influence and change the flora and fauna and other different natural factors such as hydrology or local climate.

Cultural landscape analyses follow different methodological steps. The first step is the study of history of cultural landscape, then the present cultural landscape has to be diagnosed by working out the state of preservation of persistent structures. In such a case a cultural landscape geographical information system (GIS) is desirable, but the installation and management of a cultural landscape register requires much work and care. Therefore the structural analysis is a first method for assembling regional landscape maps and associated spatial relationships. The method of cultural landscape analysis contains the following research steps:

1   Research into the cultural landscape development with

- the description of the cultural landscape development on the basis of literature, interpretation of old maps and selected written and printed sources, information from administrative departments and offices such as the monuments preservation and nature conservation boards, statistic boards etc.
- mapping the land use on precise historical maps (land register and land surveys)
- mapping cultural landscape changes over the last 150 or more years on the basis of precise historical maps (land survey maps and topographical maps).

2   Analyse and diagnose the present cultural landscape:

- mapping and inventory of persistent cultural landscape elements and structures
- evaluation of the cultural landscape elements and structures and
- mapping the cultural landscape structures and the selection and recording of different cultural landscape units and regions.

3   Evaluation of the cultural landscape units and regions and the formulation of guiding principles and development purposes for future planning
4   Management plan for the cultural landscape and public presentation for residents, landscape-user and politicians.

The development of the cultural landscape up to the beginning of the nineteenth century can mainly be described using literature and the study of some selected written and printed sources. From the early nineteenth centuries in many areas of Western and Central Europe the French introduced the land register with cadastral maps. After the defeat of Napoleon, in most European states since 1813 the land-register has been completed and continually updated since then. This also includes the making of small scale land surveys (1:10 000–1:100 000) and early topographical maps (1:20 000–1:200 000). An important example is the French 'Tranchotmap' of the Rhinelands (1801–1813) and the eastern parts of Belgium and the Netherlands or the first Prussian land survey (1837–1850). The second Prussian land survey of 1890–1898 is the basis of the present topographical maps, for example in Nordrein-Westphalia, which have been continually updated. Exact land use maps only can be made on the basis of precise historical maps. Exceptionally we can use older maps, like the land-register maps of the former Duchy of Cleves from 1730. In the UK the Ordnance Survey started systematic and accurate surveys in the nineteenth century with updating cycles of some 40 years.

This method has been specially developed for cultural landscape research. The scale can vary, depending on the research area, from 1:1 000 to 1:500 000. It is important for the analysis of cultural landscape to present the extent of older or earlier land use forms. The maps contain the extent of the following land use forms: built up settlement, industrial areas (including roads and railways), open cast mining areas, fields, pastures (green land), forests (deciduous, coniferous and mixed forests), heaths and waters (canals, lakes, rivers etc.). It is possible to follow the development of the land use since the early nineteenth century and through the following times. The common standard is to make land use maps of the early nineteenth century until 1850, 1895/1900, 1950/55 and of the present situation, which cannot be presented in the map of cultural landscape change but which only shows the present state of the landscape as a reference. The series of land use maps show how in the course of time the extent of the different land use forms has changed. For example heaths greatly reduced or even disappeared in many regions and were replaced by new forms of use such as the cultivation of new arable land or afforestation. With these maps one can compare the state or situation of the landscape at different fixed time periods. Abandoned settlements and other vanished cultural landscape elements can also be located and presented. This only matters for maps with scales up to 1:50 000. In the basis of the contents of these maps it is also possible to work out proposals for an adapted revival of older use forms within the context of cultural landscape management.

A second very important mapping instrument for the presentation of the change of the cultural landscape since the early nineteenth century is the so-called cultural landscape change map. This method was developed for the research of landscape change in the lower Rhine area for the Historical Atlas of the Rhinelands in 1992. The method has already been used successfully in about 35 projects. The most important condition for making this map is the presence of exact land surveys and land register maps and early topographical maps in Western and Central Europe (about 1800–1850). The map should be presented on the basis of the current edition of the topographical map. The presentation scale can vary from 1:5 000 to 1:100 000 (depending on the research area).

The cartographical sources for the presentation of cultural landscape change between the oldest available exact map and the present map are historical land surveys or old editions of the topographical maps. The selected periods have to correspond with the historical topographical maps, which are newly made or completely revised. The maps used also have to be critically considered and evaluated for their accuracy or precision.

The following procedure is to compare the oldest map with the newest or completely revised and published maps from after 1850. The map summarises the historical dimension of the landscape development, by which the different cultural landscape elements and land use forms after their development period are mapped for the different periods with different colours over the present edition of the topographical map.

The map of cultural landscape change in fact summarises and shows the development of the cultural landscape by comparing the topographical map of 1850 with the maps from, for example, 1895, 1950/55 and 2000 in one map. In this way historical cultural landscape elements and land use forms can be chronologically registered, presented and described. It is possible to mark the development, change and persistence for about the last 150–200 years. The elements which vanished between 1850 and 2000 are not presented. They can be presented in special maps after comparing the land use maps.

These maps are very important in reporting the cultural landscape development as part of the applied (historical) geographical research for cultural landscape inventory and analysis to be used in spatial planning and by conservation boards for landscape policy and programmes, landscape plans, inventories of cultural heritage, nature and landscape conservation areas, cultural management, tourism, recreation and landscape guides. They also can be made for almost all European countries.

The cultural landscape change map does not show the conservation state of cultural landscape elements, but only the location of a farm or another building or the course of roads. This means for example, that today the building can be a modern or even new one (on the old site) or the road appears as a modern traffic route. The location or the course are decisive. These maps differ between strongly changed (dynamic) and less changed (persistent) areas, between landscape areas with a large number of small historical parts. The purpose of this map is to work out historical cultural landscape aspects to be used as guiding principles for national spatial and landscape planning, landscape programmes and landscape development plans.

An important step in cultural landscape analysis is inventory and mapping the historical cultural landscape elements and areas in the present landscape. These elements and areas have to be presented on current topographical maps with a scale which can vary from 1:1 000 to 1:50 000 (mainly depending of the research level and area and the purpose of the research). In this case the meaning of 'historical' has to be defined or fixed. We consider more or less modern cultural landscape elements as historical ones; when they are not built or constructed anymore, then they become the expression of a finished period that has already passed.

The important sources for this map are: land use maps (1850–2000), cultural landscape change maps (1850–2000), historical maps from before 1850, information of the monument protection and nature conservation boards, literature from some selected and important written and printed sources, and photographs. The most important source is the landscape itself, which has to be checked by field work. The following aspects will be examined:

1   The form and shape of the cultural landscape elements. They can be divided into:

    - Point elements. They can differ between farms, houses, churches, chapels, castles, residences, mills, factories, public buildings, road crucifixes, memorial stones, archaeological objects, solitary trees, graveyards and so on. These elements and relics are catalogued as in terms of their state of conservation: nearly intact, somewhat changed, completely changed (new building on the old location) or former location.
    - Linear elements. These consist of roads, avenues, railways, rivers, canals, drainage canals, dykes, hedges etc. An important linear element is the historical road, even if it has a modern structure, but where the course of the road itself is historical and original. Such a road has to be mapped as a historical cultural landscape element. This also matters for waters and canals. Many rivers and brooks today have been straightened and changed. However, the historical changing measures such as straightened and changed parts will also be mapped.
    - Area elements are, for instance, historical villages and town centres, historical industrial places, woods and forests, vineyards, fields, pastures or common land. The area of historical settlement can be mapped precisely after the comparison of historical maps and the inspection of the allotment and road structure.

2   The function of the elements. It is very useful to relate cultural landscape elements to functions which are important for the state of the element and for cultural landscape and land use development. The following functions can be defined:

    - Religion (cult): churches, chapels, monasteries, crucifixes, graveyards
    - Defence/military: forts, castles, town walls, defence walls, gun emplacements
    - Administration/jurisdiction/law: residences, prisons, frontiers
    - Agriculture/forestry: farms, forester's houses, fields, forests, parcelling forms
    - Mining/industry/trade/service: mills, potteries, mines, factories, markets
    - Traffic/transport/infrastructure: stations, bridges, tunnels, canals, dykes, market places
    - Living/settlement: houses, villages, towns, quarters
    - Education/health: schools, universities, hospitals
    - Culture/recreation/tourism: hotels, museums, yards, recreation parks.

    Many historical elements have kept their original function, but in the course of the technological and industrial revolution up to now many elements have changed their function or lost their function (a fossil relic). In the last case many of these elements have already been destroyed or disappeared and replaced

by new ones. When vanished elements were not replaced by new ones, then the location should be mapped for archaeological reasons.
3   The present state of the cultural landscape elements is also an important indication. We differentiate between well preserved, preserved and changed and vanished elements. This is also important for the evaluation of the elements.

Summarising these maps presents the historical dimension of present cultural landscapes. They are very important for spatial planning and landscape policy (landscape plans) at regional and local level for districts and municipalities, for monument, nature and landscape conservation, cultural management, tourism and recreation. In this way the historical cultural landscape elements and structures (the historical dimension of the present landscape) can be regarded the same as other public interests in spatial planning. With the information from these maps it is possible to decide consciously about their existence or even their destruction.

After mapping the historical land use forms, cultural landscape change and the mapping and collection of historical cultural landscape elements in the present landscape a database should be assembled, which has to be classified and evaluated within the analysis of the present landscape. The first step in classifying cultural landscape elements is to work out a summary of the elements within a structural framework which relieves the decision about the marking of valuable historical cultural landscape areas. Cultural landscape structure maps are somewhat controversial within the discussions of geographical basic research, because the various valuation schemes related to the complex formation which is 'cultural landscape' are not suitable. Within the present dynamic development of the landscape, with expanding requirements of areas, decisions have to be made. Therefore a map with cultural landscape structural zones is necessary.

Structural information has to be presented on current editions of topographical maps. The scale can vary from 1:25 000 to 1:500 000, depending on the extent of the research area and the purposes of the study. The structural map is the result of the cultural landscape analyses and marks and divides the present cultural landscape into different structural orientated zones such as: historical agriculture dominated zones, younger cultivation zones, older and younger forest zones, historical settlement zones, historical industrial zones and strongly changed zones, in which no special historical structures can be identified. Within the framework of working out the effects of large scale planned crucial changes, such as open cast mining, on the cultural landscape a substantial cultural landscape analysis mostly with solitary elements is too expensive. In this case the method of generalised structural analysis has been developed.

Analogous to the main units of the classification of natural landscapes, guiding principles can also be worked out on the basis of the cultural landscape analysis, so that cultural landscapes units at a national level or cultural landscape zones or section at a regional and local level respectively can be distinguished and recorded. This method is also suitable for the European level.

## Concluding thoughts: historical geography as a bridge

It is often claimed for geography in general and historical geography in particular that it acts as a bridge between different natural and cultural sciences in both research and teaching – with landscape architecture being an example of a subject which uses a lot of historical aspects as well as developing ideas about the future of a given landscape. With regard to its methods and questions historical geography occupies a very distinct position in the field of science:

1. It is a sub-discipline of geography as a whole, in particular of historical-genetical geography with its close ties to the questions posed by physical geography regarding landscape history and those questions posed by somewhat 'soft' spatial sciences such as climate history or landscape ecology. Historical geography is also part of applied geography as well.
2. Historical geography is an environmental-historical bridge between historical and cultural sciences (in particular of economic, social and regional history as well as archaeology and folklore studies) because of similar methods and complementary questions. In this interdisciplinary network historical geography is characterised by its great historical depth reaching far back into prehistoric times. Irrespective of this last remark the recent tendencies in historical-genetical research can be circumscribed by three comparatives, being greater proximity to environmental history, more presence-oriented and more planning-oriented.

In the context of landscape architecture there are many points of contact and in fact, with the advent of the European Landscape Convention and the need to understand cultural landscapes in a broad and deep way, historical geography can help in uncovering the essential time-depth of any landscape. The methods developed for this may be specialised but the application has much to offer.

## Bibliography

Burggraaff, P. and Kleefeld, K-D. (1998) *Historische Kulturlandschaft und Kulturlandschaftselemente* Teil I. Bundesübersicht. Teil II: Leitfaden. Bonn-Bad Godesberg = *Angewandte Landschaftsökologie* 20.

Burggraaff, P. (1996) 'Der Begriff "Kulturlandschaft" und die Aufgaben der "Kulturlandschaftspflege" aus der Sicht der Angewandten Historischen Geographie'. 10-12 = *Natur- und Landschaftskunde* 32.

Burggraaff, P. (2000) *Fachgutachten zur Kulturlandschaftspflege in Nordrhein-Westfalen: Im Auftrag des Ministeriums für Umwelt, Raumordnung und Landwirtschaft des Landes Nordrhein-Westfalen.* Mit einem Beitrag zum GIS-Kulturlandschaftskataster von Rolf Plöger. *Siedlung und Landschaft in Westfalen* 27. Münster.

Von Droster zu Hülshof, B., Plachter, H. and Rössler, M. (eds) (1995) *Cultural landscapes of universal value. Components of a global strategy.* Jena, Stuttgart, New York.

Dix, A. (1997) *Industrialisierung und Wassernutzung. Eine historisch-geographische Umweltgeschichte der Tuchfabrik Ludwig Müller in Kuchenheim.* Rheinland Verlag, Köln.

Fehn, K. (1996) 'Grundlagenforschungen der Angewandten Historischen Geographie zum Kulturlandschaftspflegeprogramm von Nordrhein-Westfalen', *Berichte zur Deutschen Landeskunde* 70. 121–130.

Gunzelmann, T. (1987) 'Die Erhaltung der historischen Kulturlandschaft. Angewandte Historische Geographie des ländlichen Raumes mit Beispielen aus Franken'. *Bamberger Wirtschaftsgeographische Studien* 4. Bamberg.

Jäger, H. (1987) *Entwicklungsprobleme europäischer Kulturlandschaften. Eine Einführung.* Darmstadt.

Jeschke, H. (2001) 'Vorschläge für ein europäisches Konzept, Kulturlandschaft'. In: *Kulturlandschaften in Europa* (a.a.O.), 181–224.

Job, H. (1999) *Der Wandel der historischen Kulturlandschaft und sein Stellenwert in der Raumordnung. Forschungen zur deutschen Landeskunde* 248. Trier.

Rössler, M. (1995) 'Neue Perspektiven für den Schutz von Kulturlandschaften: Kultur und Natur im Rahmen der Welterbekonvention', 343–347 = *Geographische Rundschau* 6.

Wagner, J.M. (1999): *Schutz der Kulturlandschaft – Erfassung, Bewertung und Sicherung schutzwürdiger Gebiete und Objekte im Rahmen des Aufgabenbereiches von Naturschutz und Landschaftspflege.* Saarbrücken = *Saarbrücker Geographische Arbeiten* 47.

Chapter 8

# Trees shaping landscapes

Links between forestry and landscape architecture

*João Bento and Domingos Lopes*

**Introduction**

It is well-known that trees play an important role in producing oxygen and absorbing carbon dioxide and that could be the starting topic of this chapter. However, as we intend to reflect about the links between forestry and landscape architecture, we start by looking at trees as one of the most important components of the natural and, for that matter, cultural landscape. Trees are the living element that best identifies the link between forestry and landscape architecture. The presence of a tree takes a hegemonic form in forests, and it is also a striking feature in landscape architecture.

The main forestry activities focus their attention on trees as major landscape elements in various combinations and scales, so that trees become the essential part of the process of forest growth. The establishment phase of new forest stands involves introducing young seedlings or saplings that will be the new trees in the future, so that time plays an important role as forests grow and develop over spans of many decades. The other elements of these forest ecosystems such as shrubs, even in forests managed for multiple purposes, are frequently treated as passive or secondary components, in which a set of interventions will be carried out in order to restrict their competition with the arboreal layer.

Silviculture is the art and science of growing trees and managing forests and in that respect it is a little like landscape architecture which also combines art and science. The forester manipulates the composition and structure of the forest canopy layers in order to produce different effects and can in some senses be seen as a landscape designer, constantly thinking about the future of the forest and how it develops and changes. Much like a landscape architect, the forester rarely lives to see the final fruits of his or her work which takes more than a human lifetime to come to maturity.

João Bento, Domingos Lopes

Taking into account the concerns that differentiate these two activities (forestry and landscape architecture), it is clear that there is usually a more individualized and located intervention in the way trees are used in landscape architecture, when compared to the more extensive processes in forestry practices. Landscape interventions can occur at an individual level or in individualized small patches. As the size of the areas for forestry management is obviously larger, the scale of the project is disproportionately higher in forestry, although it may be manifested in lots of smaller scale or stand-level interventions which aggregate together to form a larger, 'landscape' scale.

We do not want to argue in favour of 'small is beautiful', but the previous references may reflect such an interpretation. Nevertheless, the work scale in forestry results plainly from the presence of forests being highly visible in the wider landscape all over the world, to different degrees in different places.

While the authors' professional forestry experience is mainly concerned with forests growing in Mediterranean regions, in these environments, forest management represents a stimulating challenge since adequate conditions have to be created to make it possible for trees to survive in adverse hydrological stress and intensive wildfire hazard. The whole of the following chapter certainly reflects these characteristics. In other parts of Europe, such as Finland, the forest is so extensive that it forms the essential landscape matrix within which the cultural landscape is set, imparting a very different character to the nation and the culture from the Mediterranean. In north-western Europe forests were almost removed from the landscape so that they occupy a relatively minor place in the scene. As a result of these observations, it is worth starting the chapter by examining exactly what and how much forest there is as a resource as well as a landscape component.

## Forest distribution

Forests are one of the major representative vegetation types around the globe. Occupying about 4 billion hectares (Table 8.1), forests cover nearly a third of the surface of the continents, and three times the area of cultivated lands. The temperate

**Table 8.1** Global distribution of the main types of forest

|  |  | Area billions ha | Percentage of the globe | Percentage of land area |
|---|---|---|---|---|
| Oceans |  | 36.1 | 70.8 |  |
| Continents | Deserts | 4.8 | 9.4 | 32.2 |
|  | Forests | 4.0 | 7.8 | 26.8 |
|  | Grasslands, steppes, savannahs, farmlands | 6.1 | 12.0 | 41.0 |
|  | Total | 14.9 | 29.2 | 100.0 |
| TOTAL |  | 51.0 | 100.0 | 100.0 |

Source: Arnould et al., 1997

and boreal forests (2.2 billion ha), occupying more than a half of the global forest area, are mostly located in the northern hemisphere; whereas tropical dry and rainforests, about 1.8 billion hectares, are mainly present in the southern hemisphere (Arnould *et al.*, 1997).

With nearly 50 countries and representing some 17 per cent of the land area of the entire globe, Europe contains approximately a quarter of the world's forests, with more than 1 billion hectares or 81 per cent of this total found in the European part of the Russian Federation west of the Urals (FAO, 2009).

When assessing the importance of European forests, excluding the areas belonging to the Russian Federation, it can be noted that forests extend over more than 205 million hectares (FAO 2009, FAO 1995). Forests in Europe occupy on average over 30 per cent of the land area, only surpassed by the Latin America region (EUROSTAT, 2008) – and this in a continent we usually think of as highly agricultural and urbanised. In fact the forest area of Europe is actually increasing as a result of many factors such as land abandonment as well as afforestation programmes.

Forest is the climax vegetation of Europe – remove the cities and abandon the farmland and forest would return. Both the development of agriculture over the millennia and the land use changes that occurred during and after the industrial revolution led to a reduction of European forested areas, more so in some countries than others. Since the middle of the nineteenth century, the importance of forests for providing many values has strongly increased, mainly in Central Europe, and the need for their preservation, expansion and management has also become unquestioned. This new point of view has led to the need to implement several policies, based on the concept of sustainable forest management – the original use of the concept of sustainability! During the twentieth century, the forested areas continuously expanded in most European countries, especially in the period after the Second World War. From 1950 onwards forests have gradually increased by more than 55 million hectares in area (FAO 2009, FAO 1995).

Analysing the areas covered by forests in European countries, it is important to stress that the largest proportions can be found in Finland (77%), Sweden (75%) and Slovenia (65%), in contrast with Malta (1%), Ireland (10%), the Netherlands (11%) and the United Kingdom (12%). The largest producers of paper, timber and other wood products are currently Sweden (65 million $m^3$), Germany and France (62 million $m^3$), Finland (51 million $m^3$) and Poland (32 million $m^3$), which together represent almost two-thirds of the European production (EUROSTAT, 2008).

## Basic concepts of forestry
### Forest planning and management models

Forests have always been of key importance in the development of human societies. They have long been used as a source of energy through fuel-wood consumption, as a feeding place for wild animals and for the original production of round-wood used both for building and mining, sawn wood for ship building and food for grazing or foraging domestic animals. Forest lands have also functioned as a reserve whenever farmland had to be enlarged. Small areas of woodland relicts in agricultural lands usually survived because they retained an importance in the local economy. The empirical management of forests based on the accumulated experience of generations has for centuries allowed them to maintain their support function and to sustain their production.

Though varying according to different geographical conditions, the continuous increase in population has caused the over-exploitation of forest in several regions. In Central Europe this pressure was felt in the early Middle Ages with the consequent and progressive separation of agricultural and forest landowners and the identification of them as individual, not integrated land use types. Scientific forest management was the answer to provide a regular and sustainable supply of wood, an essential raw material during the growth of the industrial economy. This was started in the eighteenth century in order to regulate and increase yields of timber (Bell and Apostol, 2008). By the eighteenth and mainly in the nineteenth century, with different policies all over Europe, the amount of forest cutting, especially clear cutting, progressively increased. The reason for the decreasing forest areas was in part a consequence of the more intensive management for wood production, but its effects could also be seen in unbalanced flows of water and increased soil erosion. In the nineteenth and twentieth centuries, concerns over the effects of these activities led to strengthened public institutions with responsibilities in forest management, such as state forest services. Remnant forests were sometimes protected as National Parks, following the earlier North American experience. Moreover, this was also a way to develop and promote new forest stands, thus leading to management tools capable of considering the various influences of forest ecosystems. Early management planning models were concerned with ensuring the successful regeneration of the forest and these evolved into sustained yield models where the continuous flow of timber over time was the main objective. This can be seen as a precursor to sustainable management where a wider range of benefits or values are needed by society.

The year 1960 was remarkable in the general acceptance of multiple use as a guide for forest management – the first movement beyond management for a single objective, timber production. This year saw both the approval of Multiple Use Sustained Yield Act by the US Congress and the 5th World Forestry Congress held under the theme of Multiple Use of Forests. With these came the recognition of forest management in such a way that, while conserving land as a basic resource, it would yield a high level of its major timber and non-timber products – water, forage, recreation and wildlife – for the benefit of the greatest number of people in the long run. The idea of forest management for competing or multiple purposes was no more than a belated recognition of all the traditional practices long exercised over the generations. However, its application was in contradiction with the formal development of the scientific approaches supported by the arithmetic rules of timber production prevailing in Europe applied through the concept of the 'normal forest', where stands of all ages coexist in a forest unit so that as one matures and is cut another takes its place. Multiple use management has continuously evolved and incorporated subsequent developments in forest planning.

Recent decades may be characterised by the involvement and contribution of several scientific disciplines, which have resulted in more accurate approaches to particular aspects of forest management across progressively larger regions. An example of this is the application of operations research, using linear and multiple objective programming, in order to optimise harvest scheduling and land use management. Landscape ecology has been a powerful support for forest planning, since it gives the opportunity to move not only from the forest stand to the forest landscape but also to incorporate better understanding of forest ecology, wildlife management and stand and landscape dynamics. Finally, the general availability of geographical data bases organized as information systems (through the use of GIS) allows an easy and rapid appraisal of the consequences resulting from different alternatives and solutions.

## Forest functions

Because of wars between countries and many types of pressure on them throughout the centuries, most European forests have significantly changed in area, composition and structure, mainly as the result of human intervention. Many forests are managed to provide a wide range of values that include not only timber but also many non-timber products and services associated with recreation, protection and nature conservation. Management for these activities results in forests with structure and composition which leads to higher biodiversity and scenic qualities than those managed just for timber. It is increasingly recognized that forests play an important role in carbon sequestration, as a result of the conservative practice of cutting and regeneration (FAO 1995).

The various functions performed by forests can be divided into five main groups, comprising several sub-functions (DGF, 2001; DGF, 2004; Pardal *et al.*, 2000). A single forest management unit may be managed wholly for one function or, more likely, include zones managed with different balances of functions:

- *Production* – Forests contribute to the material well-being of rural and urban societies through production of timber, resin, cork, biomass for energy, fruit and seeds, berries, mushrooms, fodder and other organic materials.
- *Protection* – Forests contribute to maintaining ecosystems and anthropogenic infrastructures – through the protection of river systems (the hydrological balance) and against erosion or damage caused by floods, avalanche, mudslides and wind; microclimatic and environmental protection and safety, notably by contributing to carbon sequestration.
- *Habitat conservation* – Forests contribute to the maintenance of biological and genetic diversity and geomonuments at site and landscape scales – conservation of protected habitats; conservation of flora and fauna; geomonuments conservation; conservation of genetic resources.
- *Pasturage, hunting and fishing in inland waters* – Forests contribute to the supply of game and wildfowl for hunting as well as fishing in inland waters and pasturage by domestic animals and also beekeeping.
- *Recreation, aesthetics and spiritual values* – Forests provide opportunities for the physical and mental health and well-being, aesthetic enjoyment of the landscape and spiritual and social development of citizens especially where

**Figure 8.1**
A recreational area inside a forest, one of the main areas where landscape architecture intersects with forestry. (Photo Diogo Bento)

forests are located close to where people live or in areas with high values for outdoor recreation and nature tourism (Figure 8.1).

## Main aspects of forest management

These general rules include a set of principles that should be applied in a wide range of situations. However, it is important to stress that they are not universal but depend on the circumstances. In many places forest management starts with existing forests of natural origin which are opened up for some kind of exploitation, such as timber harvest. This to some extent still occurs in northern Canada but in most places, especially in Europe, management involves the continued interventions into forests which have already been affected by human activities for centuries or millennia and, increasingly, restoration or reafforestation of landscapes which lost their forest cover in the recent or distant past. The principles are based on sustainable forest management and are organized by themes, from planting to harvesting (and thence to replanting). It can be used as a guide to provide, in each situation, the justification for adopting a particular procedure, and to reflect on the advantages and disadvantages of each choice. While some of these principles are likely to be recognisable to landscape architects, others will be more specialised. All involve manipulation and change to the landscape at a small or large scale.

### Selection of sites for plantations and/or reforestation

One of the tasks of forestry in countries with low percentages of forest cover is reforestation which contributes to the development of new landscapes and is often nowadays carried out in conjunction with landscape architects. The process of afforestation starts by investigating the legal constraints of converting land into forest, although in some cases policies to expand forest cover are important – in North

Western Europe for example. In afforestation projects, the maintenance and recovery of ecosystems with a high conservation value should always be considered. The selection of the best locations for afforestation is based on soil and site analysis and on the evaluation of the growth rates of the existing vegetation and potential for forest species. Valuable existing trees, shrubs and splendid specimens of native species ought to be maintained and preserved as a kind of nucleus.

Increasingly, the visual appearance of forests in the landscape needs to be considered and here the forester interacts directly with the landscape architect. This trend has been especially important in countries where new forests, especially using non-native species, have made significant changes to the scene in hilly or mountainous countries. A whole branch of 'forest landscape design' has arisen as a unique collaboration between foresters and landscape architects (Bell and Apostol, 2008).

## Selection of forest species

The opportunity to use local natural regeneration should always be considered if there is a seed source available. Emphasis should be focused on either native or non-native species, possibly classified as being naturalized or appropriate for afforestation, which are regulated by law. The adaptability of species to each place must always be considered in the afforestation process. The profitability of selected species, for example, should be assessed according to the objectives initially proposed and landscape impacts have to be predicted from the use of these selected solutions.

In afforestation processes, the ecological value of each species has to be taken into account, mainly for the native species, which produce important food for wildlife. The benefit of associating species should always be considered, avoiding pure stands, which are, in general, poorer in terms of biodiversity. The option for installing mixed stands (broadleaves and conifers) should depend on ecological conditions and management goals. The use of non-native species has been controversial in many places – such as the extensive use of Eucalyptus species (*Eucalyptus globulus*) in Portugal or Sitka spruce (*Picea sitchensis*) in the UK – because of their effect on the landscape as well as biodiversity.

## Forest composition

Natural plant communities are usually mixtures of different plant species, and forests are no exception. Even forest stands that are initially planted as monocultures are often subsequently invaded by other tree species (Gobakken and Næsset, 2002). Under a paradigm where forest management should emulate natural patterns and disturbance regimes, increasing criticism has arisen against forest practices that result in widespread single-species plantations that radically differ from natural forest ecosystems (Chen and Klinka, 2003).

Compared to what happened a few years ago, people nowadays regard forests in a very different way. The production perspective of forest lands, which dominated a decade or two ago, is now changing into an ecological and multiuse perspective. Even keeping in mind the simplistic production perspective, mixed-species forest stands are generally considered as being more productive than single-species stands (Chen and Klinka, 2003). Mixed-species stands are not only more diverse in respect of tree species but also provide more diverse habitat. Besides, they may also have greater resilience to natural disturbances such as fire or insect attack. A mixture of species also provides a more varied range of products, such as fruit and wood, as well as services (Kelty, 1992), and could be an essential tool for producers to have a diverse source of production.

Mixed forest stands are more resistant to pests and diseases and also more resilient to forest fires due to their increased biodiversity. Forest fires are especially problematic in the Mediterranean area, above all in monoculture forests of pine and Eucalyptus. They are also a means of rehabilitating degraded soils which have lost nutrients, as well as improving the soil quality. They allow wider exploitation of non-timber products and also provide some intermediate income between planting and final felling.

The management of mixed stands tends to be more complex than of pure stands, owing to a need for a greater knowledge of growth patterns and rates of different trees and their competitive characteristics. Mixed stands, being more similar to natural landscapes, can be considered as being closer to the philosophy of landscape architecture than pure stands, which can be seen as simplified landscapes. The complexity and beauty of a mixed forest increase the potential of these areas for recreational purposes.

## Site preparation

It is desirable to maintain existing shrub and the herbaceous layers whenever they do not increase the risk of fire or compete too strongly with the planted trees in order to reduce soil erosion. Cultivation of small spots for tree planting is usually advisable. In extreme situations of poor nutrients it is desirable to adopt schemes for soil improvement, such as the use of nitrogen-fixing plants (such as lupines), which also protect the soil against the rain impact, reduce erosion, prevent surface overheating, reduce water losses by evaporation, and additionally help control weeds and reduce the incidence of pests and diseases.

## Plantation, sowing and natural regeneration

The best forest establishment process should be chosen considering the initially defined objectives for the forest. Natural regeneration should always be the first option, if there is a good seed source from trees of good form and genetic origin. The most suitable technique for forest establishment must be carefully chosen, and the main physiographic characteristics of the place (aspect, slope, etc.) should be taken into account. The minimization of the losses of organic matter and soil compaction should also be taken into consideration during cultivation operations.

If natural regeneration is the option, the harvest planning for the old stand should be done carefully, so as to assure a good distribution of suitable seed trees over the area. In the planning process, the estimated mortality rate after planting should be considered, according to the species and the local characteristics, in order to prepare for filling in the gaps over the next couple of planting seasons.

## Weeding

The need to reduce water and light competition from vegetation should be evaluated, identifying the cost/benefit ratio of the methods to be used for controlling it, such as cutting or controlled use of herbicides. The weeding can be carried out only around each tree or along rows of trees or across the entire area. If fire risk is low, the best approach is the localized removal of this competing vegetation or by using partial strips. These approaches ensure better soil protection, higher rates of water retention, better protection against harmful biotic agents and reduction of the chance of developing other new plant communities, which are sometimes more difficult to control.

## Stand tending

Once the forest stand has been established and the crowns of the trees have coalesced to form a continuous canopy the next stage of management is stand tending. Interventions in forests, such as tree pruning to improve form, and thinning to reduce competition among the trees and to focus the timber growth on selected trees, should be planned according to the objectives. The species of trees and the type of forest products to be produced should also be taken into account, using cost/benefit ratios for decision support. Forest management has to consider not only the aspects related to forest production but obviously also environmental and ecological issues.

Pruning must be done with moderation, in order to shape the crown and its growth and stimulate the production of fruit, or as a response to health control. It aims to improve the quality of timber, and should only be applied to the best trees. When this intervention targets the prevention of forest fires, it should only be applied to the lowest branches, covering a larger number of individuals (known as brashing).

Thinning ought to be done periodically as a way to improve the main stand, but always trying to avoid drastic changes in the stand density by opening the canopy too much. Thinning represents the possibility to advance some financial revenue, while maintaining the process of achieving larger trees with greater eventual value and quality. The type and intensity of thinning depend not only on the tree's physiological characteristics but also on the use and objectives of each forest. As a rule, working with light demanding trees (such as *Pinus* or *Fraxinus* species) forces the forester to choose the dominating ones to be thinned, whereas alternative solutions are available for shade tolerant species (such as *Fagus* or *Abies* species).

Fertilization may be used to achieve higher yields and/or increase or maintain the soil fertility, while avoiding the contamination of waterbodies. The quantities of fertilizer to be applied depend on the soil characteristics, on the needs of the stand and on management goals. This knowledge can be obtained by visual observation of trees and by chemical analysis of soil or leaves. This may be used at the establishment phases as well as during stand tending.

## Timber harvesting

Timber harvesting operations form one of the ultimate tasks of forest management when timber production is one of the main objectives. It can have major impacts on the landscape and ecosystems so it should be carefully planned at a landscape and site level. It is necessary to be particularly cautious with the management of the residual biomass, branches, tops left behind after the timber has been removed. The management of forest residues, including their removal and recovery, or chipping and crushing them, may be done in order to avoid increasing fire risk and reducing the susceptibility to diseases. The timber extraction process should be planned so as to prevent soil degradation and erosion, especially in places with higher ecological sensitivity. The harvest planning should take account of the size and shape of the patches to be logged and their possible impact on the landscape (Bell and Apostol, 2008). Other impacts on ecosystems have to be considered in order to avoid serious degradation of biological and geochemical processes. This is another area where foresters may work closely with landscape architects and design solutions which must be technically and economically practical as well as aesthetically and ecologically sensitive.

Timber harvesting in plantations is often still carried out by complete clearance of the trees in the stand as opposed to selection systems which only take a proportion of the trees – sometimes known as 'continuous canopy silviculture'. These usually have much less visual or ecological impact but may suit mixed stands of native species better than more intensive ones.

## Forest roads and tracks

Forest areas should be well organized and facilitate access to all areas for management, fire protection and timber harvest operations. Thus a network of roads and tracks of varying kinds is needed. In some places such as Mediterranean countries, linear fire breaks, networks of water points and fire watching towers are needed, all linked into the road and track network. Their impact on the landscape may be high, although they also permit recreational access and thus perform important secondary functions (Figure 8.2).

**Figure 8.2**
A forest road also acting as a fire break, part of the infrastructure which can have a significant landscape impact. (Photo Diogo Bento)

### Road network
Forest roads are essential in forest management for a large number of reasons. The road network gives access to the forest, and the number of entries into a forest is essential for conducting intensive and efficient forest management. As the road system is a connection network, it can be divided into different types of roads according to their importance: i) main roads, used by all types of vehicles throughout the year; ii) secondary roads or forest dirt roads, with limited movement (by all-terrain vehicles), used to give access to operations inside the forest and also for forest fragmentation; iii) trails, which are routes with an ephemeral existence, used for the sporadic passage of tractors and forest machines.

### *Preventive silviculture*
The term 'preventive silviculture' means a set of standards and activities that aim to reduce the progression and intensity of forest fires, thus avoiding or limiting tree damage. Preventive silviculture has to include norms that may improve the forest capacity to protect itself against fire, mainly in the Mediterranean region where forest fires are highly problematic.

Silviculture activities should address their attention to the forest structure, in order to create and maintain vertical and horizontal discontinuities between different fuels inside the stand. A variety of techniques is required based on mechanical and manual clearance, manual pruning, restricted burning, prescribed fire and controlled pasturage, among others. These preventive measures should be guided by two principles: the first one is related to the fact that different species present different resistance to fire; the second one is that fire resistance should be seen in the forest vegetation as a whole.

## Silvicultural models by function
Some basic approaches for forest management, which can be related to the main forestry activities summarized above, can be used to manage forests for different combinations of objectives. These are not exclusive to a single management unit but can be applied at different scales, for example in various zones across a forest area.

### *General production model*
The species usually planted in productive forests are well known for their fast growing capacities, the quality of their wood (for biomass, energy production or pulp) or for other non-timber products, such as resin and cork, or even fruit and seeds and other organic products. Production cycles are in general shorter and the stand densities higher than in the other models. In this silvicultural model, technical interventions are increasingly important in order to manage stands and get products of higher quality. As a result, such forest stands are usually of less interest for recreation or biodiversity and may look unattractive in the landscape.

### *General protection and conservation model*
Forests are extremely important to preserve high rates of biodiversity. Among other functions, they purify the air, help sustain the quality and availability of freshwater

supplies, regulating the environmental conditions locally and globally and constitute the habitat of a huge number of animals and plants. The species used in this silvicultural model provide protection conditions for more degraded forests or ones vulnerable to biotic and non-biotic factors such as forest fires. They also protect ecosystems against weather, wind, snow and water damage or erosion.

Several features such as the density and composition of stands are crucial to the achievement of the protective function. Because of many physical conditions of the deteriorated environment, the densities of these stands are usually lower than in productive forests. In terms of forest exploitation and composition, this silvicultural model is managed over longer periods and, whenever possible, in mixed stands, allowing a better protective effect.

In order to achieve conservation purposes, species must be carefully chosen according to their performance. As a rule, native species present more ecological advantages than non-native ones. In this type of silvicultural model it is very important to preserve and improve soil and water and also to guarantee biodiversity. The selection of harvest trees should be made at an individual level, never at an intensive grade, in order to minimize the disruption of the ecosystems that are intended to be maintained. In fact timber harvest may never occur in this type. It usually offers a continuity of ecosystem functioning and may be a long-term element in the landscape.

## *General pasturage, hunting and fishing model*

In this silvicultural model, the selection and treatment of species should consider that this forest will support both domestic and wild animal life. Topics such as the period of time for the rotation of forest stands, the adoption of mixed composition and the choice of low densities are essential to maintain and improve pasturage, hunting and fishing. Longer rotations are desirable to avoid disrupting the systems we want to protect (in the case of hunting and fishing) or to keep pastures with overhead cover. It is also desirable to increase floristic diversity in order to attract more diverse fauna. Lower tree density helps the movement of grazing animals. The selection of tree or shrub species capable of producing fruit, seeds and fodder should be encouraged to support hunting and pasturage (Figure 8.3).

## *General recreation and landscape model*

For this silvicultural model, the species selection needs to bear in mind that the main function of forests is recreation and landscape enjoyment. In this particular model, topics such as the use of mixed species and of single elements are decisive and the switching/mixing of species by ages is vital to achieve the objectives of this silvicultural model. Special attention should also be given to density since forest areas with spaced trees with large canopies are suitable for parks, picnic sites or walks. On the other hand, the adoption of screening by dense areas of trees may be an element of appreciation of the landscape in places where less desirable aesthetic elements are present.

**Figure 8.3**
Pasturage activities in Mediterranean ecosystems – traditional management practices which create important cultural landscapes. (Photo Diogo Bento)

## Forestry education

Forestry and landscape architecture have many points in common, especially in terms of the impact of their activities in outdoor spaces. In both cases, the use of vegetation is extremely important. Thus, it is not surprising that in both activities, part of their educational background partially coincides. One of the main differences is the scale of activity which is often larger in forestry, while both need to consider the effects of time and to be able to visualise how the landscape will be over many decades.

There is a set of subjects that can claim to belong to the forestry educational universe. In this way, we can refer (Figure 8.4) first, to the central disciplines of forest establishment, its management and harvesting; and second, to those related to the evaluation of individual trees, stands and forests.

The forestry educational curriculum is based on ecology and plant biology. As forestry deals with large plant populations, whose elements interact with one another in complex ways over time, it is natural to recognise the increasing importance of mathematics, modelling and statistics. The bases of these subjects are essential for achieving the main objective in forestry: the successful establishment and management of forests over time.

Thus, the subjects in the silviculture curriculum (with several different approaches and references), such as forestry engineering, forest protection and harvesting guarantee a scientific and a practical background that is necessary to implement adequate interventions that will promote the persistence and reproduction of forest stands in healthy conditions.

Moreover, the monitoring and control of the forest growing stock are based on procedures that are developed and explained in forest mensuration and forest inventory, using direct measurements or indirect approaches, like the use of remote sensing (aerial or satellite photographs). The distribution and the spatial and

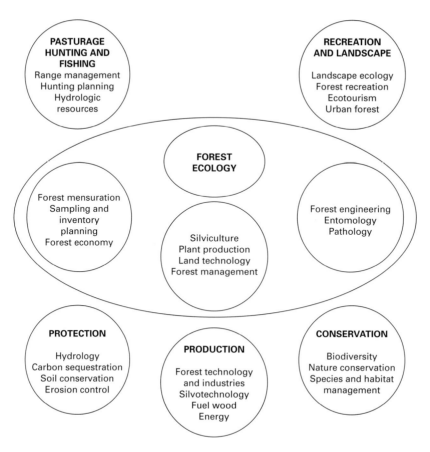

**Figure 8.4** A diagram showing the different aspects which make up the forestry educational curriculum. (Source Bento and Lopes)

temporal evolution of forests are studied in planning and forest management, obviously using Geographic Information Systems (GIS) nowadays. In every case, all the interventions and management of these areas are supported by numerical information, obtained from the forest inventories. What to do and how to act can only be defined after knowing what exists.

Traditional inventories are essentially related to the characteristics of individual trees (diameter at breast height (dbh), total height, volume, dominant height, among others), as components of stands with certain parameters (number of trees per hectare, basal area per hectare, volume per hectare, among others). Until recently, periodic forest inventories had tended to simplify the reality and had often only cared about the production information. Even nowadays, only a few forest inventories are more specific and detailed, being also concerned with the pathological and health problems found in stands, and with biodiversity and its complexity. The new context of increasing ecological awareness has allowed and forced foresters to understand forests as a complex and important ecosystem that does not allow simplistic analysis. Forest multifunction perspectives are increasing and imposing this more complete concern on society. As noted in this chapter, forest uses for recreation are important enough to justify maintaining special forest areas, mainly close to the urban areas. The new philosophy for the delineation of forest inventories reflects this new kind of context. For example, the last National Portuguese Forest Inventory evaluated, for the first time, biodiversity in each sampling plot.

The connection between landscape architecture and forestry has also helped to change this new paradigm of forest inventories, because the previous simplistic approach, mainly focused on the production rate of these areas, is not supported by landscape ecology, where ecosystems are considered in their global amplitude, nor the effect on the scenery of such mechanistic approaches.

## Educational interconnections

Besides the subjects briefly discussed in the previous section and according to the historical, social and economic context of the different forestry educational programmes, greater emphasis can be given to contributions related to timber production, to conservation, hunting, fishing and pasturage, or to recreational purposes and landscape, among others. The philosophy of each course always needs to be adapted to each specific context. The balance between a more productive or a more ecological perspective is always related to each context and this in turn suggests where interconnections between forestry and landscape architecture can be strengthened.

Briefly, some of the courses and modules, representing a more specific training for each function, are:

- Production: technology of forest products; silvo-technology; forest industries (wood, fibre, resin, cork, energy)
- Protection: hydrobiology; river rehabilitation; soil conservation
- Conservation: biodiversity, species and habitat conservation; nature conservation
- Pasturage, hunting and fishing: pasturage scheme; hunting planning; hydrological resources
- Recreation and landscape: ecotourism; landscape ecology, urban forestry, forest landscape planning and design.

Considering their links with landscape architecture, the diversity of contents that generally constitute the educational background in forestry can be included in three main groups:

- Disciplines only slightly associated with landscape architecture
- Core contents for forestry, where at least part of their curriculum is important to landscape architecture
- Disciplines simultaneously relevant to both backgrounds.

We may also add a fourth group connected with the topics that will be part of a core educational programme on landscape architecture and that should receive greater attention in the curricular structure of forestry.

The first of these groups will include issues related to harvesting, forest products technology, forest management planning and forest economics. It may occasionally be useful to clarify some issues relating to the maintenance of wood in outdoor spaces, and the operation and use of certain forest equipment employed widely.

As previously noted, the second group can include subjects whose contents are also useful for landscape architecture. Some of the possible examples are: forest establishment techniques, seed collection and preservation; planting practice; tree measurements; general sampling procedures; production appraisal; spatial structure of forest areas; water management and soil protection. Even within this group, we can identify some issues, such as health protection (entomology and pathology) or plant nurseries and production techniques, which require more intensive knowledge by landscape architecture students.

The third group corresponds to important aspects for both landscape architecture and forestry, like recreation, nature tourism, forest landscape planning and design, management of animal resources and nature conservation.

Lastly, a fourth group of subjects can be added in order to list those areas where forest training has traditionally been lacking, and where the exchange of points of view in common with landscape architecture would be desirable. The incorporation of aesthetic awareness and the increasing demand for visually more interesting forest composition structures should be a gradually increasing concern for forest management. Use of social science methods of understanding how people use places, how they perceive the landscape and its changes over time are also aspects which would be quite new to most foresters.

Finally, we recognize and strongly encourage the development of innovative activities for students from both backgrounds, joint activities which could, for example, combine design and project development. This would enable the sharing of training, knowledge and affinities from both forestry and landscape architecture. It could also provide an experienced learning group and identify the advantages and difficulties of inter-disciplinary experiences. Furthermore, at this level, field trips with students from both backgrounds may bring confrontation of different views and perspectives about landscape analysis and interpretation. Joint projects looking at landscape character, landscape change and conservation, linking cultural and natural landscape processes, would also be useful.

## Innovative challenges in research

The connection observed in the educational background between forestry and landscape architecture can also be reflected in their main investigation topics and in the complexity of the teams working together. Let's take, only as an example, the Forest Landscape Ecology Lab, from the University of Wisconsin (USA). As their designation indicates, they primarily focus on studying ecological aspects of forested ecosystems. Most of their work tries to understand the basis of sustainable forest landscapes. Their projects include forest ecosystem field studies, forest landscape modeling, GIS and remote sensing applications, analysis of the effects of forest management practices and natural disturbances on ecosystems, the influence of landscape-scale factors on wildlife populations, and the effects of land use and historical factors on current landscape patterns in ecosystem processes (FLEL, 2010).

These are the topics of greatest concern all over the world, regarding every kind of forest, with specific problems and challenges, in this general context. Any other similar groups that could be reported would not change the topics of their investigation, as they are actually universal to the scientific community, and their composition would be similar, with the recognition that forest and landscape researchers must have essential presence within these teams.

Taking into consideration only the scientific areas noted in Figure 8.4, and being aware that this report cannot be exhaustive, it is recognized, for instance, that the studies developed by Zavala and Oria (1995) are examples where several issues are connected, mainly the relations between the protection and production perspectives of forest ecosystems and their conservation goal. These investigators stress the point that biological diversity is a complex concept that includes many different aspects of ecosystems. In addition, its relationship with stability and resilience has been a subject of long debate in ecology and it still remains confusing. Investigators have reviewed the existing approaches for the conservation of biological diversity and have realized that most of them derive from population biology and emphasize the species richness in biological diversity. These methods do not account for ecosystem processes and economic constraints, which makes them inappropriate to be implemented in managed lands. They propose four directions for further research that would help to couple forestry and ecology showing the Pacific Northwest example, in which considerable research efforts have been made making forestry and conservation compatible and also the Mediterranean basin, where a certain level of management is required for the preservation of biological diversity.

Another relevant project is known as 'Fire Paradox', which has been an important attempt to combine all the research on forest fires. Humans need fire to regulate the dynamics of natural ecosystems for the benefit of stability and productivity, but the uncontrolled use of fire often leads to ecological and humanitarian catastrophes that threaten biodiversity and result in economic loss. The participants in Fire Paradox want to contribute to develop concrete, operable contributions in order to reduce the social, economic and ecological impacts of large-scale or highly severe forest fires, such as those which occurred in Portugal, Spain and France in 2003 and 2005. The aim of this project is to provide the scientific and technical means to 'learn to live with fire' and will examine four fire components: Prescribed Burning, Wildfire Initiation, Wildfire Spread and Backfire (FIREPARADOX, 2010). This project has enabled participants to share experiences from the various geographical regions of Europe, mainly from the European Mediterranean countries most affected by fires: Spain, France, Greece, Italy, Portugal, Slovenia and also from two Maghreb countries – Tunisia and Morocco.

Finally two scientific and technical co-operation projects in the European Union COST programme: Cost Action E33: Forest Recreation and Nature Tourism (COST E33, 2010) and Cost Action E39: Forests, Trees and Human Health and Wellbeing (COST E39, 2010) represent an effective way to exchange professional experience and institutional discussion about two very important emerging aspects of forests and their management. They reflect the recognition of how important forests and other open spaces can be to individual citizens and collective organizations in

**Figure 8.5**
Trees defining paths in the landscape, experiences and play of light and shade. (Photo Diogo Bento)

European countries. The development of forest spaces in order to make them attractive to receive an increasing number of people, looking both for tranquility and for exciting physical activities, is only possible with the collaboration of several professional experts, including foresters and landscape architects. Both represent good examples of multi-disciplinary approaches as well as the collaboration of researchers, managers, policy makers and representatives of the public and private sectors (Figure 8.5). The materials to emerge from both Cost actions have significantly increased the volume of knowledge available about the subjects (Bell *et al.*, 2009; Proebstl *et al.*, 2010; Nilsson *et al.*, 2010).

## Opportunities to develop collaboration

Forest investigation has allowed the use of new technologies in forestry, such as biotechnologies, information management and communication, ensuring automatic control procedures, together with an effort towards better understanding of the functioning of forest ecosystems. FAO (2009) considers the following areas to be key subjects for forest research: forest management, planted forests and wood production, agro-forestry, harvesting and processing of wood products, non-wood forest products, wood for energy, provisioning of environmental services. We will only concentrate on the ones we foresee as having more potential for close collaboration between foresters and landscape architects.

*Forest Management* – From an early age, forest management practice has considered sustainability as one of its objectives. More recently, these concerns have extended to the effects on economic, social and environmental issues, with forest certification being a response to consumers' demands and to these new viewpoints. The ability to introduce new indicators that may characterize the forms and conditions of developing forest landscape is an obvious area of collaboration between landscape architecture and forestry.

*Plantations* – The new forest plantations, in particular with fast-growing species, allow slower levels of exploitation of natural forests. However, the concerns about the environmental consequences of this more intensive practice have led to cooperation with different scientific backgrounds, in order to understand their effects better and more deeply. The need to convert large areas of monoculture tree plantations into more diverse structures or the restoration of degraded systems has created challenges for finding better solutions for their use. The creative potential of these solutions has clearly required the close involvement of foresters and landscape architects.

*Agro-forestry* – In developed countries the topic of grazing in forest areas is now becoming more and more important, not only as a response to the demand for food quality (using traditional sustainable techniques with good animal welfare) but also as a way to control the underground biomass, because of the danger of forest fire. If we add the use and dissemination of wind-breaks for amenity purposes, or the progressive demand for agricultural areas to be used for round-wood production, we have a number of issues likely to attract researchers from different academic backgrounds.

*Energy* – Using wood as fuel is one of the earliest benefits derived from forests. With the growing demand for solutions to obtain energy from renewable resources, wood is likely to be part of the answer to those requests, although research in appropriate technologies is still required. The expansion itself, the planting of woody species for energy, the recognition of the best solutions and methods to be implemented are still areas that require technical cooperation and research. Effects on the landscape of large-scale energy plantations need to be evaluated.

*Environmental Services* – The contribution of forests to biodiversity and nature conservation or to ameliorating the urban environment (such as heat island effects, storm water interception and flood control) are some of the multiple opportunities for further study bringing together foresters and landscape architects.

## Concluding remarks

Generally speaking and all across Europe, forestry will continue to be a topic of discussion and concern in the near future. In spite of the effects of the global economic crisis, European countries will keep environmental concerns in their political agenda, with special focus on forest sustainability and on conserving and protecting their natural resources.

The current trend for increasing forest cover in Europe will be maintained in the next years as a result of afforestation. In addition to environmental concerns, the decreasing pressure of other types of activities, including intensive farming with high levels of industrialization, creates conditions for increasing natural forests. Furthermore, the general reduction of population outside the most important urban areas also releases extensive areas making them available for afforestation.

This new century will probably be known among other things for its emergent environmental consciousness, felt more intensively than at the end of the previous one. Under the pressure of continuing population growth, natural plant communities remain under pressure. They are also reducing in area because of the expansion of agriculture, urbanization, dam construction, forest fragmentation and road building, among other indirect human impacts, such as the invasion of exotic species.

The recognition of the importance of forests in the process of carbon sequestration and the increase in their biomass use as a source of energy constitute enough reasons for recovering forests and recognizing their value triggering an effort to expand new plantations.

As societies are now confronted with intensive levels of industrialization, they desperately require spaces close to natural patterns, located in urban or peri-urban forests. These urban forests offer a wide range of well-known environmental, social and economic benefits like: filter of air pollutants and particulates; energy conservation as a consequence of transpiration and cooling reduction, shade and wind protection; water flows regularization; noise buffering; support for wildlife habitat; aesthetics and psychological comfort (Konijnendijk et al. 2005).

In a variety of land uses, where forests have acquired a significant importance, urban residents and urbanized populations need spaces where recreation, contemplation and other activities close to nature are possible. In this way, nature- or ecotourism has been consolidated. Developing countries have begun to recognize the economic benefits of ecotourism. This is a means of simultaneously getting economic return with relatively small exploitation or extraction of resources, and also of increasing the ecological conscience of our societies, which are currently facing important global challenges and problems.

The threats that are nowadays recognized against European forests and mostly against the forests of the whole world, are: i) health problems, associated with pests and diseases, mainly at the interface of industrial activities; ii) cyclic occurrence of forest wildfires, more problematic in the Mediterranean countries and those with similar climates such as Australia. These aspects constitute sufficient motivations for deeper understanding of forest ecosystems, not only in their own details but also at the interface of other uses and occupations which are recognizable in the global landscape.

Modernizing forms of knowledge permanently, monitoring and intervening in forest ecosystems, and also ensuring the practical implementation of sustainable forest management are some of the challenges for the convergence of a broad range of skills and academic background, which should concern both landscape architects and foresters. The concept of sustainability is simple to understand because it represents the capacity to endure; but it is very difficult to implement, requiring some solid ecological foundation, if we can be able to guarantee that biological systems will remain diverse and productive over time. All these challenges need to be understood as a multidisciplinary issue (see Figure 8.6).

This effort to share concerns and responsibilities is already reflected in the structure of some educational institutions with responsibilities in training both foresters and landscape architects. Sometimes this is reflected in the structure of some university departments, like the one at Universidade de Trás-os-Montes e Alto Douro, with its Department of Forest Sciences and Landscape Architecture. This organization allows the simultaneous sharing of application domains and learning

**Figure 8.6**
Trees shaping landscapes or trees presented as singular elements? (Photo Diogo Bento)

facilities and laboratories, or the use of the same areas for experimentation and research. This cooperation and exchanges between forestry and landscape architecture require a commitment between the two backgrounds, which can only be enriching and fruitful.

## Acknowledgments
The authors are grateful to:

Leonor Figueiredo and Maria do Rosário Cristovão for their helpful comments and revision of the English manuscript;
Diogo Bento (diogobento@gmail.com) for use of photographs;
Project FCOMP-01-0124-FEDER-007010 (previously designated by PTDC/AGR-CFL/68186/2006) for scientific background support.

Special thanks are due to Simon Bell for his critical and detailed revision of the original manuscript, comments and suggestions for chapter improvement.

## Bibliography
Arnould, P., M. Hotyat and L. Simon. 1997. *Les Forêts d'Europe*. Fac. Géographie. Nathan. Paris.
Bell, S. and D. Apostol. 2008. *Designing Sustainable Forest Landscapes*. Taylor and Francis. London.
Bell, S., M. Simpson, L. Tyrvainen, T. Sievanen and U. Proebstl (Eds). 2009. *European Forest Recreation and Tourism: A Handbook*. Taylor and Francis. London.
Chen, H. and K. Klinka. 2003. 'Aboveground productivity of western hemlock and western redcedar mixed-species stands in southern coastal British Columbia'. *Forest Ecology and Management*, 184: 55–64.
COST E33. 2010. 'Forest Recreation and Nature Tourism'. *European Cost Action*. (http://www.openspace.eca.ac.uk). Accessed in December 2010.
COST E39. 2010. 'Forests, Trees and Human Health and Well-being'. *European Cost Action* (http://www.e39.ec). Accessed in December 2010.

DGF. 2001. *Guião para a coordenação e harmonização dos PROF*. Direcção-Geral das Florestas. Lisboa.

DGF. 2004. *Macrozonagem funcional dos espaços florestais em Portugal Continental*. Direcção-Geral das Florestas. Lisboa.

EUROSTAT. 2008. Eurostat newsrelease. 146/2008 (http://www.ec.europa.eu/eurostat). Accessed in December 2010.

FAO. 1995. *State of the World's Forests*. Food and Agriculture Organization of the United Nations. (http://www.fao.org/DOCREP/003/X6953E/X6953Eo5.htm). Accessed in December 2010.

FAO. 2009. *State of the World's Forests*. Food and Agriculture Organization of the United Nations. Rome.

FIREPARADOX. 2010. Fire Paradox. (http://www.fireparadox.org/). Accessed in September 2010.

FLEL. 2010. Forest Landscape Ecology Lab, Department of Forest and Wildlife EcologyUniversity of Wisconsin-Madison. (http://www.forestlandscape.wisc.edu/). Accessed in September 2010.

Gobakken, T. and E. Næsset. 2002. 'Spruce diameter growth in young mixed stands of Norway spruce (*Picea abies* (L.) Karst.) and birch (*Betula pendula* Roth B. *pubescens* Ehrh.)'. *Forest Ecology and Management*, 171: 297–308.

Kelty, M.J. 1992. 'Comparative productivity of monocultures and mixed-species stands'. In: Kelty, M. J., B. C. Larson and C.D Oliver (Eds). *The Ecology and Silviculture of Mixed-Species Forests*. Kluwer Academic Publishers, Dordrecht.

Konijnendijk, C. C., K. Nilsson, T. B. Randrup and J. Schipperijn (Eds). 2005. *Urban Forests and Trees*. Springer. Heidelberg.

Nilsson, K., M. Sangster, C. Gallis, T. Hartig, S. de Vried, K. Seeland and J. Schipperijn (Eds). 2010. *Forests, Trees and Human Health*. Springer. Berlin.

Pardal, S., J. Pinho and P. Bingre. 2000. 'Espaços silvestres'. *Normas Urbanísticas, vol. IV. Planeamento integrado do território. Elementos de teoria crítica*. Direcção-Geral de Ordenamento do Território e Desenvolvimento Urbano / Universidade Técnica de Lisboa. Lisboa.

Proebstl, U., V. Wirth, B. Erlands and S. Bell. 2010. *Management of European Forests for Recreation and Nature Tourism*. Springer. Berlin.

Zavala, M. A. and J. A. Oria. 1995. 'Preserving biological diversity in managed forests: a meeting point for ecology and forestry'. *Landscape and Urban Planning*, 31 (1–3): 363–378.

## General silviculture

Daniel, P., U. Helms and F. Baker. 1979. *Principles of Silviculture*. McGraw-Hill. New York.

Shepherd, K. 1986. *Silviculture*. M. Nijhoff Publishers. Netherlands.

## Forest management

Davis and Johnson. 1987. *Forest management*. 3rd edition. McGraw-Hill, Series in Forest Resources. New York.

## Forest and land use planning

Nagendra, H. and J. Southworth. 2010. *Reforesting Landscapes, Linking Pattern and Process*. Springer, Landscape Series. Dordrecht.

# Part 3

# Economics, cultural anthropology, regional planning and cultural geography

Chapter 9

# . . . and how much for the view?

## Economics and landscape architecture

*Colin Price*

**Introduction**

The subject matter of economics is the use of scarce resources to satisfy the competing wants of humans (and arguably of other sentient beings). Wants are taken to embrace both physiological needs and psychological desires: economists do not on the whole seek to distinguish the two categories. Among these wants are aesthetic ones, and among the scarce resources are the land and sea and sky, which are referred to as *natural* resources, although all are now pervasively modified by human activity. If the subject is seen in this way, discussion of landscape inevitably has an economic dimension.

In the landscape ecologists' sense of landscape, as a territory occupied by living things, within which interactive processes occur, the economics of landscape bears upon everything that occurs – on the scurrying human condition of getting and spending (and, by analogy, the transactions and decisions of other creatures), with landscape as the arena in which their actions and interactions are played out. This descriptive function is what economists call *positive economics*.

In the more customary, perceptual understanding of landscape embraced by the author (Price, 1978), it is explicit that landscape is imbued with economic meanings: 'Landscape, it can be said, is the perceived environment which results from the interaction of the earth's resources and humankind's needs.' The definition of landscape by the European Landscape Convention, adopted under the auspices of the Council of Europe in 2000, commands most consensus. According to that definition 'Landscape means an area, as perceived by people, whose character is the result of the action and interaction of natural and/or human factors.' This equally clearly puts landscape within the direct interests of economists.

In so far as economists merely concern themselves with recording the preferences that are rooted in perceptions, and the trade-offs that people choose to make when confronting them, economics remains a descriptive, value-neutral subject. But when it claims that the data favour one course of action or another, it becomes a normative subject, impinging on the territory not only of landscape architects, but that of political scientists too.

When landscape simply *is*, no resources are apparently devoted to it, no actions are indicated: but landscape architecture entails *design*, and that inevitably introduces resource use as it is applied to landscape change or maintenance. And, whether or not this change has a primary aesthetic purpose, aesthetic wants will be more or less satisfied, left unsatisfied, or made dissatisfied in consequence of the change. Even leaving landscape undisturbed, as a conscious abstention from action, has implications for the landscape's ability to meet other wants, and so has economic consequences that ought to be a mediated part of the conscious decision.

This chapter discusses economics, as the subject is generally presented in introductory economics texts, and in introductory school and university courses. There is, appropriately, a bias *away* from the financial institutions that so often grab the economic headlines and occupy long chapters of texts, and *towards* the natural resources which are the basis of all real production. It considers what economists have to say about resources and wants, with emphasis on aspects of the subject that have particular importance for landscape architects.

It recognises also, however, that relevance does not equate to consonance. Economists and landscape architects do on the whole have different ways of looking at things. I once asked the late Dame Sylvia Crowe whether she regarded economists, or philistines, as being her greater enemy. Her reply 'is there any difference?' might have been a crowd-pleaser; but in being so it revealed a perception of what *would* please crowds: that is, an identification of economics as a hostile discipline, especially one that is anti-aesthetic, and anti-spiritual. Within the conventional prejudices about economics, the discipline is characterised not only as philistine, but also ultimately self-destructive, as the processes it puts in place undermine the resource-base on which production is founded. The belief may be that, if the indications of mainstream economics are followed, the pursuit of growth at the expense of resource stocks and environmental quality will continue until *the economy as well as the ecosystem* is starved or smothered.

Not only do landscape architects have suspicions about economists' attempts to place cash values on aesthetic conditions, but for their part economists distrust the subjectivity that appears to pervade landscape architects' judgements. And nor do professionals in other fields share the viewpoint of either of these two protagonists. In the introduction to the monograph *Landscape Economics*, published back in 1978, I wrote:

> But examining the viewpoint of the academic economist has convinced me that economic purism, pressed too far, simply prevents the achievement of any useful result. The customary plea for an interdisciplinary approach does not ask enough; for, if advances are to be made, economists, landscape architects and political scientists must come prepared not only to collaborate, but also to abandon some of their cherished

> preconditions for analysis. What I have written has assumed that such flexibility is acceptable, and it will no doubt on that account be considered trivial by purists.

These were hard words then, a premonition of how subsequent conference presentations having a transdisciplinary nature would be received. They are no less relevant in the third millennium.

> Professionals guard territory aggressively. Unwritten codes forbid making concessions to the perspectives of other professions. Thus economists are uncomfortable with aesthetic judgement; landscape designers revile quantification and monetisation; and planners trust only in their intuition, or in public meetings. To compile a method of valuing trees which draws on the expertise of all these professionals might be an obvious course in theory: but those who take it need not expect a welcome in anyone else's house.
>
> (Price, 2007a: 12)

Mindful of this, in what follows economics is presented as a potential contender as well as a potential collaborator. No effort is made to gloss over the differences that exist between professional viewpoints, because understanding those differences is a first step towards resolving them, and towards developing a constructive synthesis. Hence the chapter also deliberatively contemplates what often seems to be an interdisciplinary chasm, in the hope of discovering ways in which interdisciplinary bridges may realistically be constructed. It does this in the following section, by considering the subject as a discipline with numerous points of relevance to landscape architecture; and, in a later section, the possibilities of research collaboration are explored, drawing on the techniques which economists deploy.

These possibilities of conflict and of conflict resolution are also the theme of the concluding section, which offers some insights from personal experience of interdisciplinary collaboration, and makes proposals for interdisciplinary research programmes.

## The intellectual territory of economics: a guide for outsiders

Formal economics in the sense already discussed has early origins: there is much in ancient sacred texts – *The Bible*, *The Laws of Manu*, *The Quran*, for example – that encourages, prohibits or regulates particular modes of production, distribution, taxation and trade. The secular texts of Greek and other classical philosophers (such as Plato (*c*4th century BC) and Aristotle (*c*4th century BC)) address economic issues too.

The primogenitor of modern economics is often considered to be Adam Smith, whose *Wealth of Nations* (1776) formalised an overview on common issues of the day: industrialisation of production, and liberalisation of trade, for example. But Adam Smith's academic title was 'Professor of Moral Philosophy'. And later economists – Hume, Bentham, Malthus, Ricardo, Marx, Pareto, Dupuit, Allais – are as readily classified as philosophers, as economists. Economics, perhaps, *was* then regarded as part of a broad way of thinking about resources and their capacity to satisfy human wants. It is with Jevons (1871) and Marshall (1890) – both having the title 'Professor of Political Economy' – that the synthesis representing neoclassical economics came into the technical form in which, essentially, the subject is presented today. Macroeconomics – the economics of industries and nations seen as something other than a simple summation of the acts of individual firms and consumers – took shape in the twentieth century with the works of John Maynard Keynes (1936) and, on the contrary side, Milton Friedman (1962).

Even introductory economics texts (classically, Lipsey (1989), and Samuelson (1980)) run to many hundreds of pages, and it is not the purpose of this chapter merely to provide a highly condensed version of them. Instead, it identifies some of the key concepts with which such texts deal, emphasising those that have particular relevance to landscape architects, and that should, perhaps, feature in their education.

## *Supply*

Two concepts, *supply* and *demand*, and their interaction in *the market* are often seen as the basis – and sometimes, erroneously, as the entire subject – of economics.

Economics is concerned with the production of goods and services, and with the conditions that affect that production. Economists generally divide productive resources into three, four or five *factors of production*. These are conventionally termed land, labour, capital, raw materials, and enterprise. Sometimes raw materials are regarded simply as intermediate products, and sometimes enterprise is considered to be a specialised form of labour. The introductory texts invariably consider these factors in relation to industrial production, and sometimes by reference to agriculture. But they are equally relevant to the creation of new landscape.

For landscape architects, land will provoke the greatest empathy. It is, after all, the canvas on which they practise their art, and, according to the economic perspective, is the primal factor of production in the creation of landscape. In fact land is but one cluster of entities among natural resources, the others, water bodies and sky, also providing important, though more transient ingredients of an aesthetic package.

Especially in pre-industrial times, labour – the productive effort of human beings – was also required in massive amounts to construct features – pyramids, stone circles, chalk mounds and downland figures – which still stand tall and distinctive in landscapes surviving from the prehistoric era (Figure 9.1). In pursuit of other objectives, labour also remodelled the landscape of the industrial revolution in heroic ways that are admired today. The visually inspiring bridges of the canal and railway ages might conventionally be attributed to famous engineers such as Telford, Brunel and Stephenson: they were equally the product of the tens of thousands of 'navvies' who worked on, and the hundreds who died in, constructing them. And until the advent of major earth-moving equipment, labour remained the key factor in transforming landscapes dramatically into parks for pleasure, more fastidiously in

**Figure 9.1**
The Ring of Brodgar, Orkney – a Neolithic ritual landscape wrought with much labour. (Photo Colin Price)

intimate garden maintenance and, more subtly, through evolving the cultural landscapes of grazing, *bocage* and wood pasture. Urban parks, too, remain to this day as heavily labour-dependent resources.

The mental processes of landscape design might also be considered as labour. Some economists would see these skills as part of *enterprise* – the organising factor of production. But there is a further, risk-bearing function in enterprise, and it is the landowners who financed the creation of landscaped parks on a grand scale who risked mightily in order that their vision might be realised.

Raw materials are things that pile in the yards of contractors and local authorities' works departments: sand and gravel, fencing posts, bags of seed and fertiliser. Characteristically, they are of little use in their present condition: they need to be transformed by further human effort to achieve that useful status.

This introduces capital. Contrary to the common perception, by 'capital' economists do not mean *monetary wealth*, but rather a collection of raw materials to which human labour has given continually productive form. Capital is machinery, for manufacturing goods or moving earth; it is buildings as working and residential locations, and as the follies and temples of aesthetic intrigue; it is vehicles for the transport of resources and of products and to bring visitors to outstanding landscapes; it is the individual know-how (the intellectual capital of knowledge and understood design techniques); it is the social capital of the institutions needed to gestate major projects for constructing landscape; and some would say it is the constructed landscape itself.

The growth of capital, in the form of earth-moving machinery and devices for controlling or maintaining vegetation, is what has enabled large-scale landscape creation and maintenance to survive the loss of cheap labour.

Capital, its nature and its role, has been an issue among economists throughout the modern era. Marx's remembered preoccupation (1867, 1885, 1894) was with unmerited reward to productive capital, as much as with the factor of production itself. Debate has long raged among economists about the *measurement* of capital (Robinson, 1956). In truth, capital's physical quantum is impossible to assess meaningfully, because of its relentlessly miscellaneous nature. The *value* of capital, however, is simple to assess in principle: it is the value of that continual flow of services to which it gives rise over a period of time. How the value of the flow is to be summed over time is something that we shall briefly revisit under the heading 'Investment'. In this light, it can be said that anything giving rise to a flow of aesthetic services is landscape capital. For example, many consider wind turbines, especially those located offshore, to be a positive and dynamic part of landscape – as most present-day viewers certainly regard ancient onshore windmills. Because of the durability of landscape, the capital element in the production process is of key concern. Again, we shall consider this further under 'Investment'.

Cost is conventionally interpreted in terms of financial outlay. Economists however also recognise the importance of 'opportunity cost' – the loss of potential revenue or benefit incurred when a factor of production is withdrawn or withheld from some alternative form of economic activity. Of particular importance is the loss of material production that may follow the dedication of land to primary aesthetic purposes.

That all these factors of production may be combined to meet aesthetic needs is not controversial. But economists are further concerned with the conditions under which such combination occurs, and with the quantitative detail of relationships between inputs and outputs, technically known as *production functions*. In theoretical texts this is often represented as an equation which allows continuous variation in quantities of factors of production. In the real world there may be only a few, discrete possibilities of production.

A feature of standard economics texts is a model known as 'the theory of the firm'. This describes how an individual economic entity – a small factory, a farm or whatever – increases its production, with progressively increasing cost of producing an extra unit of the product (marginal cost). The process continues to the point at which the marginal cost equals the extra or marginal revenue achieved by selling one more unit of the product, at which point the profit of the entity has been maximised. The assumption is that producers will seek to maximise profits within certain legal and ethical constraints.

This process, possibly combined with cost increase as decreasingly efficient manufacturers or farmers are drawn into production as the offered price rises, leads to a positive relationship known as the supply curve between price of product and the total quantity supplied.

Diminishing returns to such factors of production as fertiliser do not only furnish standard textbook examples, but also explain why (for example) one cannot create 'instant' soft landscape by very large doses of fertiliser to trees and shrubs. And, reaching back into history and prehistory, the rising opportunity cost of labour as it was withdrawn from increasingly important tasks to do with subsistence would have curtailed the production of ever-more-grandiose schemes for landscape modification.

Scale economies exist when larger production units (factories, farms) have cost advantages over smaller ones, through specialisation of labour force, better utilisation of capital, superior geometrical configuration and ability to exploit bulk purchasing power or cut margins for error. Such economies may also be achieved at the level of the whole industry (hence the concentration of particular forms of manufacture in one location), or even the whole economy (hence the growth of large cities). Diseconomies exist too, curtailing the competitive advantage of the largest productive entities.

Scale economies may underlie the nature of the land resource with which landscape architects deal, such as increase in farm and field size, and concentration of manufacturing on industrial estates. Labour specialisation is also the reason for why there are economists and landscape architects.

Whether these phenomena are important to landscape architects' understanding of their own activities in a given economic context is another matter. At the level of the firm, clearly large practices have advantages in labour specialisation, but disadvantages in communications problems and (often) rising overheads; marginal costs may rise for a practice of given size, as long hours of working reduce the productivity of individual 'workers'. But at the project level, landscape architects are not really deciding a minutely optimal *level* of output, but choosing among discrete schemes, which may differ rather in design quality or concept. Nonetheless, the point *is* always reached when further input to improving the detail of a particular design offers diminishing returns in enhancement of aesthetic quality: this point is determined by reduced productivity as exhaustion increases, or by the opportunity cost of time, which could otherwise be productively devoted to defining the broad characteristics of incoming schemes. In such senses economics produces an account of the sensible input of time, and has the capacity to deliver sensible decision criteria, though the difficulty of quantifying increments of design quality compromises its capacity to do so in this situation. Thus traditionally, decisions on the optimal allocation of labour resources among active projects are instead based on intuitive judgement informed by experience.

## *Demand*

It is said by economists that 'the end of all economic activity is consumption'. By consumption is denoted not just digesting food or using up material goods, but also the enjoyment of sold and unsold aesthetic products, whose nature may mean that enjoyment does not entail depletion of the resource. As to being 'the end' of economic activity, consumption is seen as both the ultimate objective of production, and chronologically the last act in the economic drama. The latter interpretation, now, might be disputed by environmental economists, who would see the disposal of waste as the key later stage in the economic process. Creative activity *as an end in itself* also challenges the stylised structure of economists' views, while manipulation of consumers' desires by producers undermines the vision of consumers as the ultimate driving force of the economic process. With these reservations, the adage provides an insight into economists' perspectives on the world, and is needed to justify their belief that markets do, within limits, provide a beneficent mechanism for increasing the good of the world.

The upward slope of the conventional supply curve is underlain by diminishing returns to increasing inputs and the summation of outputs from firms of different efficiency: in the same way, the downward slope of the demand curve reflects the diminishing marginal utility of consumption and the summation of demands from consumers of different tastes and income. Diminishing marginal utility is among the key concepts of economics. It means that the value (utility) derived from the consumption of an extra (marginal) unit of a product diminishes as total quantity of consumption of the product increases. At an individual level, this arises from hierarchies of wants (a product is first used to meet the most important wants that it can satisfy); and from satiation (within a given time period, consumers weary of repeated acts of consuming the same product). At the social level of aggregation, increased consumption brings less marginal value, as individuals with lesser taste for the product join the circle of engaged consumers. From these phenomena it is inevitable that as price decreases, and *only* as it decreases, so a larger volume of sales will be achieved.

Diminishing marginal utility gives a formal account of why people will visit a landscaped property once during a year, but not at every available opportunity. For manufacturers and for the managers of visited landscapes alike, it is important to know the strength of the relationship between price charged and quantity of sales, characterised by the *price elasticity of demand*. Where sales are highly responsive, total revenue is increased by price cuts: where they are unresponsive, total revenue is increased by raising prices.

Quantity of sales is also affected by the position of the demand curve, which in turn reflects population, taste, income levels and the availability of competing products. All else being equal, sales rise in proportion to population.

Income level influences demand statically, among sections of a contemporary population: not everyone in ancient Babylon could afford a hanging garden! It also acts dynamically as income levels change through time. Crucially, as incomes rise, so a smaller proportion of income is devoted to the basic needs of survival, and more to the pleasures of aesthetic experience. This phenomenon, coupled with the ability (examined later) of landscapes to provide more experiences with little increase in the total cost of provision, led to a spectacular rise in the value attributed to landscapes during the twentieth century (Krutilla and Fisher, 1972). This has balanced the demise of landscape owned by extremely wealthy resident landowners, and the loss of cheap labour for landscape creation and maintenance. Income also varies cyclically with boom and recession in economies, and this makes enterprises which rely on selling aesthetic services vulnerable to the sensitive dependence of demand on income.

The influence of taste upon demand for a certain style of product is something with which the history of landscape architecture can identify closely: while landscape in a general sense may be increasingly prized, particular configurations of landscape have had waxing and waning appeal. This is notoriously evident in the succession of preference from formal 'landscape gardening' to naturalistic and romantic idealisations, to the current 'ecological aesthetic', under which taste is meant to be conformed to ecological primacy and its sustainability (Sheppard and

**Figure 9.2**
Versailles, product of the formal taste in landscape that was superseded by the naturalistic preferences of the Romantic Movement. (Photo Colin Price)

Harshaw, 2001) (Figure 9.2). The counter-argument, that taste is linked to evolutionary functionality, is also increasingly heard, as a justification for taking people's tastes for what they are: a valid expression of landscapes' capacities to satisfy their felt needs (Parsons and Daniel, 2002).

## Markets and the lack of them

The market is the conceptual space (not necessarily a physical location) where supply and demand meet. The upward sloping supply curve and the downward sloping demand curve intersect at the market price, the price at which the quantity willingly supplied by producers equals the quantity willingly demanded by consumers: there are no shortfalls and no surpluses.

Economists accord to the market a special importance, and special powers and virtues. It is through market competition among producers that the product is made available at the lowest possible cost (in money terms and in terms of the scarce resources expended). The market brings production and consumption into balance without the action of despots or bureaucrats, neither of whom would be sensitive to the cost structures of producers nor the varied wants of consumers. The market adjusts to changes of technology and taste without the intervention of committees of politicians, consumer watchdogs or indeed of economists. Under certain really quite plausible assumptions, the market creates the best of all possible worlds, where 'best' relates to human satisfaction, and 'possible' refers to what the availability of scarce resources allows. Interference with the market has been seen, since the time of Adam Smith, as a force for inefficiency of production and a blow against the interests of the consumers.

The market, however, does not accomplish all good, and an outline knowledge of the ways in which it fails gives a better basis for proposing alternatives, than an ideological opposition or an intuitive distaste.

The market mechanism has created inefficient monopolies, tolerated inhumane treatment of children and animals, and, as discussed later, generated distributions of income that are certainly unequal and arguably inequitable, all in the pursuit of self-interest and profit.

But for landscape architects the most significant defect of the market is its failure to exist in relation to the products that most concern them. Economics has traditionally dealt with *private goods* – products which exhibit the following features: rivalry in consumption (if one consumer consumes the item, another may not do so); a positive marginal cost in production; and excludability in transaction (those who do not pay, do not get the product). Such are the goods that fill supermarket shelves. But there are other categories of product to which economists have turned their attention in more recent times (Lifran and Oueslati, 2007). Toll goods such as bridges are not rival in consumption: the bridge is built once for all, and all who pay may cross it, without incurring significant marginal cost, and without interfering with the possibility for others to do so (until such time as congestion sets in). Yet proprietorship allows exclusion of those who do not pay the toll. To obtain the greatest sum of societal net benefit from the bridge, all who wish to cross should be permitted to do so, and hence many bridges have been freed from toll, rather than allowing their owners to 'charge what the market will bear'. Even when the proprietor, as of a commodious landscaped garden, does not seek maximum profit, but only to cover the costs of upkeep, the requisite entry charge excludes some potential visitors, despite the fact that their presence would incur trivial marginal cost in additional upkeep, or in loss, through crowding, of potential benefit to other users.

Common property resources and open access resources, by contrast, *do not* allow (legally or physically) exclusion of potential users, yet use *is* rival, and significant marginal costs *are* incurred by additional use. Thus deep sea fisheries are over-utilised by self-interested users, who cannot be excluded yet who impose costs on others through the lowered quality and productivity of the resource, even to the point of extinction.

Public goods, of which lighthouses are the classic example, neither allow exclusion of potential users, nor entail significant marginal cost, nor exhibit rivalry in consumption. If the service is provided for one, it is provided for all without the possibility of excluding non-payers: if it is denied to one, it is denied to all. And, once the cost of its provision has been laid out, maximisation of benefit would anyway entail free access to the service. Any landscaped property visible from places of free public access falls into this category of resource.

For these three categories of product, a case can be argued for public provision or regulation, to ensure both that full benefits are obtained, and that account is taken of costs imposed by consumers on the provider or on other users. This indeed has been done for many products of this nature.

There are cases similar to the public good (or bad), in which beneficial or detrimental effect on environment or society arises that is incidental to the purpose driving an economic activity. Telecommunication masts, for example, constitute a public good (better television reception) which also produces a public bad (intrusion on wild or organic landscapes). Such phenomena, known as *externalities*, are also attendant on the provision of private goods (e.g. biodiversity depletion as a concomitant of tropical timber products); and on use of toll goods (vehicular pollution from bridge traffic), and of common and open access resources (most spectacularly, use of the atmosphere as a free dump for sulphur dioxide and other pollutants).

An externality exists when a product or the act of its production generates benefit or cost without a matching financial transaction. In such cases, the market neither promotes creation of sufficient benefit, nor conversely sufficiently penalises the generation of cost. Once again, public intervention seems justified, in order to encompass the full consequences of economic activity. But the further questions now arise: what form should intervention take? and what is the appropriate level of regulation (of negative externalities), or promotion (of positive externalities)? With regard to the latter, a question of key interest is, what equivalent to the market price can be used for externalities, which exist in no market, and therefore have no market price?

A common question levelled at economists is 'How can you possibly put a money value on the song of the nightingale, the sight of a majestic sunset, or deer on a morning-dewed meadow beyond the ha-ha, or the peace and seclusion of a deftly contrived *plaisance*?' Since the assumed answer 'you can't' is an excuse to disengage from economics, as a discipline that has nothing to say about the finer things of life, it is important that landscape architects are at least acquainted with the actual answer that an environmental economist would give. Thus they avoid exposing themselves as ill informed by asking such rhetorical questions. Because environmental economists may have an answer such as 'well, there are basically eight methods: which one would you like me to explain first?' A less political and more practical reason for having at least an outline of this knowledge is that increasingly such valuations, of ecological as well as aesthetic services, pervasively arise in discussions of land use change. It may not be the role of landscape architects to *undertake* them, but it may be expedient that they *understand* them.

There are differences among environmental economists as to where the boundaries between evaluative methods lie, and as to the most appropriate terminology. Some methods depend on actual market transactions arising elsewhere in the economy: others depend on answers to hypothetical questions posed experimentally. The methods in current use are subject of much present research in environmental economics, both to derive case-specific values, and to refine techniques in response to criticisms – from within and from outside the economics profession. For this reason they are outlined and discussed in the research section. Alternative accounts are given in Pearce and Turner (1990), and in innumerable papers on environmental economics that have appeared since. Not all are equally valid.

## *Investment*

Investment is at the heart of capitalism. In its financial form, it came into bad public repute at the end of the third millennium's first decade, being characterised as the product of short-term greed, and seen as the source of the catastrophic collapse of the world economy. As provider of the productive machinery of a modern economy, however, it is the main pillar of the unparalleled growth in production and income, experienced worldwide since the industrial revolution. And therein has lain a deeper, more lasting potential problem. Investment has powered the pillage of the world's natural resources, the impoverishment of its life-forms and the taming of its wild landscapes. In so doing, it has come, like economics itself, to be reviled as an evil force, hostile to the finer goals of humanity.

In fact, to natural resources economists, the concepts surrounding investment provide also the means of analysing the problem, and of identifying and achieving an appropriate balance between the desirable aspects of growth (alleviation of poverty and enrichment of life experience) on the one hand; and, on the other, sustaining the earth's productive capacity. In its real sense, investment means incurring costs early in order to increase revenues or benefits later. This is generally seen to be accomplished by combining other factors of production into making new capital or productive machinery – not just buildings and machines and vehicles, but the human capital of know-how, skills and productive social arrangements. Financial investment is merely the means of providing for this beneficent forgoing of present indulgence, in order to improve the condition of later periods and generations. Investment includes moving earth and laying paving stones and creating lakes and building bridges and planting trees, to delight the senses and uplift the spirit. The monumental mounds and megalithic circles of prehistory may not have been without these elements, and they were intrinsic to the cathedrals and pleasure gardens of the second millennium.

Investment may also be seen in terms of opportunity cost: forgoing present consumption in order that future consumption may be enhanced. For natural resource economists, the obvious examples are moratoria on depleted fisheries, and refraining from logging over-cut forests, so that their full productivity may be restored. The same is seen in restrictions on recreational access, when paths cannot sustain the existing degree of trampling and when the deterioration of the paths has adverse visual effects.

As the beneficiaries of all these processes, it is not for us to dismiss investment wholesale, as the agent of the devil.

Landscape architects need not concern themselves deeply with the processes by which banks raise and lend investment funds – except to note that when banks get it wrong it can have disastrous consequences for real investment. What is more to the point is ability to appraise a particular investment in a given context, accepting that heavy financial outlays incur interest payments (or forgoing of interest on self-financed projects), and that humans are impatient for quick results.

When there were no recognised procedures for quantifying aesthetic values, and when landscape was constructed primarily for the benefit of an individual's family, an intuitive assessment of the worth of investing in landscape capital sufficed. Now that methods exist to give a monetary equivalent to aesthetic pleasure, and now that multiple stakeholders are involved in providing funds and obtaining benefits from projects, intuitive appraisal of investment by a single stakeholder is not the only option, nor is it reliable in balancing interests. Moreover, the aesthetic externalities – positive and negative – of projects with other purposes, can now, potentially, be included in cost–benefit analyses.

Cost–benefit analysis – a comprehensive assessment of the advantages and disadvantages of projects, normally in a common, monetary unit of account – has its procedures, derived from those used in financial appraisal, for balancing future benefit against present cost. They involve discounting – an inversion of compound interest, such that future benefit (or cost) is treated at less than parity with present cost (or benefit).

$$[\text{Present value}] = \frac{[\text{Cash equivalent of future benefit or cost}]}{(1 + [\text{interest (or discount) rate}])^{[\text{time lapse}]}}$$

where [time lapse] is the interval of time between the present and the time when the benefit or cost is expected to be realised.

The reasons for this apparently bizarre process include the greater productive value of money or resources if they are available immediately, and human preference to have good things early. The debate which questions the relevance of these reasons, and their compatibility with the declared ethos of *sustainability*, is alluded to in the research section, because it is an active and unresolved debate. But, irrespective of the theoretical validity or ethical acceptability of the process, it is something that in practice *is done*, and landscape architects need to be able to follow such appraisals, if not undertake them.

It is important to understand that discounting has nothing to do with monetary inflation: it would be equally apposite (or, some would argue, equally inappropriate) in periods of stable money prices. Investment appraisal is only patchily and unsatisfactorily covered in introductory economics texts. Fuller accounts of the topic may be found in many kinds of text on financial or social appraisal, e.g. Hawkins and Pearce (1971), Mishan (1975), Price (1989, 1993).

## *Distribution*

There is no reason to suppose that the market assures a just distribution of income: factors of production are rewarded according to their scarcity, as well as to their industry, and such scarcity is largely an accident of how society chooses to arrange production. The landscape architect of rare talent and flair may be more useful to society than the merely competent professional, but is not necessarily more morally worthy nor more personally efficient in converting income into happiness. As for members of the unskilled forces who, as navvies, accomplished tremendous works in the landscape, they were not necessarily insensitive to pleasure, just on account of their abundance as a factor of production.

While the neoclassical account of payment to factors of production is generally orientated to their marginal contribution to output, there are some more explicit focuses. Malthus (1798) foresaw that gains would accrue to wealthy landowners, when population outstripped the capacity of land to feed it. Ricardo (1817) declared that unskilled labourers were doomed, by the fecundity of their class, to an eternal subsistence level of wages. Marx (1867, 1885, 1894) believed that all value should be attributed to labour, and that payment of interest and other rewards to capitalists and landowners represented expropriation. And it is certain that those with monopoly or other powers are able to extract a greater reward from the market system than their contribution to it seems fairly to warrant.

For landscape architects *as landscape architects*, the important consequences of distribution lie in how it has affected the funding of projects. The famous landscape architects of the past had access, or aspiration, to fabulous wealth – Capability Brown, Fürst Pückler, Le Nôtre. A case might be argued that the landscapes democratically enjoyed today owe their existence to the past's great inequalities of wealth. That may be true in the sense of formally created landscapes. Yet it is also true that the patient acts of winning a sustainable livelihood, by those at a more modest level of living, formed the cultural landscape, within whose matrix the more dramatic aesthetic *tours de force* could be set. In a sense, Kent and Repton and Brown merely reproduced in a stylised and idealised form the landscape that could be seen before them. More lately, the well-to-do have affected the aesthetics of enclosed and often private patches, rather than spread their visions across public landscapes.

Fabulous wealth still exists today, but its exercise in publicly visible works is much restrained by social intervention. Hence the role of past aesthetic entrepreneurs has been taken over by public oversight of landscape change, but more often through mitigation of the malign effects of public works pursued for other purposes, than through creation of landscapes of delight.

## *Intervention*

The sundry failures of the market mean that governments intervene to adjust the allocation of resources. They may do this by legislation or by taxation or subsidy or by provision of public services. In the natural resource arena, it may be by nationalisation or public purchase of land; or by exhortation of private persons to serve the common interest; or by restricting individuals' rights to make adverse changes (especially within the urban landscape); or by compelling individuals to institute beneficial changes – perhaps through the trade-offs embodied in planning gain. Within a strongly market orientated system, the preferred intervention often appears to be financial incentives to comply with public interest. Inside the European Union, agri-environment schemes have commonly attempted to promote aesthetic improvement by targeted grants, for example by creation of water features or reinstatement of traditional field boundaries (Whitby, 1994).

Much has also been said about intervention failure, when government measures, in pursuit of an objective like food security, have caused adverse environmental consequences such as loss of trees and hedgerows and fields rich with colourful wild flowers (Bowers and Cheshire, 1983).

## *Spatial economics*

The development of modern economics, with an orientation to industrial production, left spatial aspects relatively neglected. However, geographer–economists, such as Von Thünen (1875), Christaller (1966), Isard (1956), Alonso (1964) and Evans (1973), emphasised the importance of space in terms of locational separation, as well as space for occupation. Their ideas became influential in urban and regional economics, as taught to town planners, and are to an extent important to landscape architects.

Because of their extravagant demand for comfort, and because of the opportunity cost of their time, human beings are the most expensive of all commodities to transport. This has important implications for location of residence and amenities (Price, 1982; Wu and Irwin, 2008). Both resource depletion and pollution from fossil fuels have led to increasing concern about spatial separation.

## Macroeconomics

This, the branch of economics that deals with the mode of functioning of nations' economies, and the world economy, less evidently demands understanding by landscape architects. However, it really does bear upon them, because when economic downturns come, private works of pleasure are naturally sacrificed before those of survival. And, inversely, when public works are sought to alleviate distress, to use a work-force for investment in and creation of amenities is to create a flexible form of employment, less reliant on interaction with other parts of the productive system. Keynes advocated government intervention in the face of cyclical economies, through taxation and public expenditure (fiscal policy). This can be productive of aesthetic works, replicating at public level the humanitarian projects that private landowners sometimes initiated in the hard times of the past. The dominant alternative model of managing the economy (monetary policy) advocates disengagement from 'excessive' government, together with minimal control of the economy through the interest rate and limitation on public sector borrowing. This leaves less scope for public aesthetic works.

Whether landscape architects need, *as landscape architects*, to know much of the intricacies of macroeconomics is less clear than the case for engagement with microeconomics.

## Accounting

Accounting is often seen as a separate discipline and usually constitutes a distinct university degree course. Their own concerns, with day-to-day and year-to-year financial management, may be seen by landscape architects as more helpful than knowing the theoretical models of how firms maximise profits, or of how competition leads markets to maximum profit – and fails to lead to maximum contribution to society. But landscape architects, like other professionals and business-people, employ accountants' services. What they need is not a detailed knowledge of accountancy practice, but the terminology in which accounts can be discussed: liquidity, cash flow, net worth, capital (in a different, accounting sense), cost centres, enterprises. Conventions for allocating overheads over projects, or revaluing a practice's physical assets in inflationary times, need to be understood, if financial viability is not to be imperilled.

## How is economics taught?

The two branches of economics, micro and macro, are what are normally taught today in undergraduate economics courses. Positive economics covers the behavioural science, which seeks to record, systematise, understand and predict the manner in which humans use resources to meet their wants. Normative economics embodies a more political perspective, making assertions about what *ought* to be done to meet wants, and how, if at all, governments and societies should interfere in the activities of free individuals pursuing their own interests. This perspective is commonly identified with a socialistic and interventionist outlook: however, it embraces just as much the standpoint of proponents of the free market, who contend – with an argued case – that the market is best left to its own devices.

The emphasis, however, varies across the subject matter and among political leanings. Some courses – no doubt much the majority these days – lean towards a market-orientated account of resource allocation; to a greater or lesser extent they note reservations about markets' efficacy, and offer remedies for what are seen as the most damaging deficiencies. Others, according to the inclinations of lecturing staff, may give more weight to alternative models: Marxist, Malthusian, environmental, radical. While some courses focus on the relationship of economics to societies and political economies, others are more technical, teaching the techniques of statistical analysis of market and other data, within the sub-discipline known as econometrics.

For landscape architects it is important that serious allusion is made to the limitations of the market with respect to the provision of public goods and the regulation of externalities; and that time is not frittered away on revealing the intricacies of stock markets and their financial instruments. Students should be conversant with the means by which the deficiencies of the market might be corrected, and be encouraged to reflect on how landscape architects, in their roles of public servants or private consultants, may bring about a more appropriate allocation of resources.

Approaches to teaching that have proved effective (to foresters, agriculturists, environmental scientists, town planners, and not a few straying landscape architects over the years) have had the following characteristics.

- Engagement with economics by starting from the questions that these neighbouring disciplines need to answer. It comes as a revelation to students of applied land use, that economics is a way of thinking about things, which embraces a range of philosophies, some of them markedly sympathetic to environmental and social concerns, and all of them bearing on the real-world decisions of real-world professionals.
- Practical exercises drawn from the arena of expected professional activity. Forget about the theory of the firm and the profit-maximising number of cars or sweatshirts to produce. Students of land-related disciplines are interested in how an operation can be done at least cost, or whether an investment in a new facility will eventually pay for itself, or what the outcome was of a particular cost–benefit analysis of a visual intrusion in the countryside.
- Simplifying without patronising. Not even an economist can be an expert in all the specialist areas within the subject. Students of landscape architecture have little time to devote to the whole discipline. Therefore its bare bones need to be presented, with an acknowledgement that time constrains the search for detail, and *without* any implication that deeper aspects of the subject would be beyond students' comprehension. It seems really to be appreciated, when it is

stated that a particular presentation has taken the class to somewhere near the frontier of current knowledge.
- Enthusiasm – this guy really *believes* in this stuff! Rather revealingly, from an economist's perspective, students often think that lecturers are paid by the hour of teaching delivered. That lecturers choose to devote more time to teaching than their contracts require – that they knowingly incur the opportunity cost of forgone participation in high-prestige and high-reward research activity – and that they can do this with a smile – seems to count for something. Yes, economics is concerned with the busy human condition of getting and spending in the pursuit of self-interest, but that is not *all* that it is concerned with. It seeks also to increase the sum of good within the limits of resources: and economics lecturers, as scarce resources, can be a living allegory of what they preach about.

Such an empathising approach is not always successful, however. Some students bring to courses the attitude 'I don't know anything about economics, but I know I don't need to know anything about it'. Such recipients do not *want* their comfortable prejudices and facile distinctions to be undermined, and are resentful of efforts to present economics as being relevant. Attempts to elicit what alternative means of allocating resources are acceptable to such students do not succeed either.

However that might be, economics needs to be engaged with by landscape architects, because it does have an influence on how resources are allocated in practice, whether for good or ill. At minimum, curiosity, or a desire to see into 'the mind of the enemy' should motivate a wish to understand economists' way of thinking. And economics seen from an empathising perspective should be known to trained landscape architects, as a potential friend and ally in the search for rational and pleasure-giving use of resources. They should also have some insight into the less benign by-products of economic mechanisms, so that they can identify with greater precision, not just evil outcomes, but the processes by which those outcomes are mediated. From a basis of this understanding, it becomes easier to argue convincingly for whatever is needed to override adverse consequences of markets.

## Research methods and approaches

The previous section reviewed the basic ideas of modern economics, as they have relevance to landscape architects. This section explores the application of those ideas to design practice, with respect both to information needed for individual projects, and to research uncovering more general prescriptions for land uses which have significant aesthetic elements.

Although much common ground exists between the subject matter of economics and that of landscape architecture, there has been less sharing of research techniques between the subjects than this would indicate. Indeed, as suggested in the introduction, it is probably true to say that each profession has had an instinct to pull up the drawbridge to its own intellectual castle, and lob ordnance at the other, similarly encastled professionals, as the principal means of interaction with them. In particular, there is a fundamental difference of approach to appraising the aesthetic merit of alternative designs, or of specific land use changes. This reflects a more general stand-off in the environmental sciences, in which economists often stand beleaguered by a motley collection of biological and molecular scientists, geographers, planners, and academics in the fine arts and humanities and even in other social sciences. No doubt it is sometimes their own fault, that they are regarded as unpromising subjects for collaborative research.

If the topics presented at the first international conference on landscape economics in Vienna in 2009 (http://www.ceep-europe.org/workshop.php?id_workshop=48&view=programme ) are anything to go by, research on the interface between economics and landscape architecture falls under the following main headings:

- costs, especially opportunity costs, of landscape preservation and enhancement;
- the role of landscape in regional development;
- issues concerning relative location of residence and landscape amenities, and spatial configurations of land use elements via spatial modelling;
- revelation of explanatory variables for aesthetic preference;
- monetary evaluation of aesthetic pleasure (although several evaluative papers took landscape as a setting for evaluating services which were not aesthetic ones);
- environmental and land use policy's effect on landscape, and landscape's impact on environmental policy;
- governance of protected landscapes, and of countryside generally;
- sustainability issues.

Some of these belong more properly to politics or environmental psychology than to economics, so are little dealt with below. To engage with the discourse of regional development, landscape architect researchers should know something of the macroeconomic techniques of input–output analysis, multipliers and econometric approaches to disentangling trends from other causal factors.

## Engaging with economic arguments

We begin with matters pertaining to the supply side, where other professionals will readily agree the claim that economics has relevance.

Landscape architects become involved in development projects, as designers purely of the landscaping elements, and as witnesses in public debate. Or that is how their involvement could be officially perceived. Their work may be seen to inform *only* that aesthetic element in which aptitude, training and experience make them experts. They are there to mitigate adverse impact on landscape, or if possible to turn aesthetic intrusions into aesthetic assets: they probably claim to do no more than that (Figure 9.3). Yet their involvement has an element of advocacy for the whole scheme: explicitly by the public defence of its merits, or implicitly by the facts of having engaged with the scheme, and having put their names on it. How often have proposals for private and public works – ranging from commercial afforestation to nuclear power schemes – mentioned in passing that the services of this or that well-known landscape architect have been retained 'to ensure that the scheme will fit

**Figure 9.3**
Mounds and tree planting designed by Sylvia Crowe partly screen and partly soften, but cannot turn Wylfa Nuclear Power Station on Anglesey in Wales into an aesthetic asset. (Photo Colin Price)

into [or even enhance] the landscape'! It would be naive to suppose that such a professional seal of approval can only, and therefore *will* only, legitimately be interpreted as being confined to aesthetic aspects. The implicit argument, whether it is legitimate or not, will often be that 'such and such people or practices would not involve themselves in – would not allow their names to be attached to – the project, were they not persuaded of its overall merit'. Without their input, perhaps, the project's claim for support would be doubtful: with their expert design, its overall merit may be assumed . . .

It is not as though the narrow economic case will be uncontroversial and uncontested. When major public works such as motorways or nuclear power stations are proposed, conflicting economic paradigms are referred to, to assert the merits of both development and of conservation as the means of best satisfying human wants with limited resources. It is therefore helpful to be aware of the physical nature of the works intended. And at the least to be able to read a project's economic appraisal, not only understanding the language and following the arguments, but exploring what appear to be disagreements *between* economists, and weaknesses in what appear to be consensuses *among* them. This may require a certain familiarity with technical issues and economic ones, when for example an expert team is being assembled to oppose a large development. Obversely, defending or retaining high quality landscape *against* a particular development proposal is most persuasively done, if the countervailing merits of the proposal can be systematically appraised.

Awareness of economic arguments and engagement with valuation of other environmental aspects may also help in positive advocacy of aesthetic improvement or conservation: in favour of national park designation, creation of peri-urban recreation areas, or green aspects of urban renewal. There is an implicit opportunity cost of restricting land uses that might otherwise have proceeded on the land in contention, and it does the reputation of a professional body no good if it is seen to be blind to the potential benefits of competing uses.

## *Pushing out economic understandings*

The 'intellectual territory of economics' section discussed the basic background of knowledge with which landscape architects ought in general to be equipped, in order at least to have a sense of the economic mode of arguing, in the roles just outlined. But it may go further, into finding new information, whether by research in the sense of adding to the knowledge base, or by that which is relevant to a particular policy or project case. For larger issues, a landscape architect would not work alone, but might expect at least to hold a dialogue with economists, informing the argument with aesthetic considerations, while weighing the merits of financial ones. Often enough, however, the matter will be a small and local one and no such team will be formed. Here, even if motivated by no more than personal curiosity about the merits of a proposal, a landscape architect might move out of the comfortable zone of aesthetic appraisal, into that of new economic arguments.

I myself initially engaged in forestry because of concern for its aesthetic effects on the UK's countryside (Price, 1966). And yet I was inexorably drawn into firstly the conventional wisdom of commercial foresters (Hiley, 1956), secondly into the conventional wisdom of investment appraisers (e.g. Hampson, 1972). And, finding a chasm between the two positions, I felt eventually obliged to understand them (Price, 1971a), then to develop them (Price, 1971b), then to engage in wider research that eventually took in the whole foundation of human exploitation of natural resources (Price, 1977, 1984, 1993). So may grow a passing, amateurish interest in landscape design! And – the point really is – having entered these fields in search of what was known on the subjects, I found much that should have been known but was not, and hence stood in need of researching.

## *Accounting and costing*

The procedures of accountancy are well laid down, and landscape architects may expect to participate in them only as recipients of established methods.

Cost accounting offers more scope for collaboration. In a sense, landscape economics as a subject to be researched has a long history, which began here. From the first time when investments were made in them, builders of highly visible features have been moved to count the cost. As the Biblical account has it:

> For which of you, intending to build a tower, sitteth not down first, and counteth the cost, whether he have sufficient to finish it? Lest haply, after he hath laid the foundation, and is not able to finish it, and all that behold it begin to mock, saying, This man began to build, and was not able to finish.
>
> (Luke, 14, 28–30)

Whether the tower was strategic or aesthetic in purpose is not recorded, but later manifestations of aesthetic equivalents are not rare. But the Bible also refers to the pearl of great price – for its aesthetic qualities no doubt – for which a merchant sold all he had in order to possess it.

With the same single-mindedness, those who reconstructed Seville cathedral said, according to various translations: 'let us build on such a scale, that men will think us mad to have attempted it'; 'let the church be so beautiful and so great that those who see it built will think we were mad'; 'we shall have a church of such a kind that those who see it built will think we were mad'; '. . . such a building that future generations will consider us to be lunatics'. And the same might have been said of the giant enterprises of the Neolithic ritual landscapes, of Versailles or Stourhead or Bad Muskau and Branitz (Figure 9.4). In some cases the owners or initiators of such enterprises *did* ruin themselves or came close to it (Pückler had to sell Bad Muskau). We may have an instinctive admiration, sneaking or open, for such extravagance of ambition. Yet we may also recognise a voice of reason which says: had *they* better not first have counted the cost, in financial terms, in terms of the human resources diverted, in terms of the productive possibilities forgone on the land which they occupied?

Now the computation of cost may be, and perhaps most frequently is, addressed by a recipe book approach. In this, reference to standard prices for elements of jobs (Langdon, 2009) saves time enormously, compared with the alternative of constructing costs from first principles. But the underlying research agenda is not just the patient digesting of figures. There is also innovation in the modes of costing. At its broadest, this extends to the opportunity cost of land resources devoted or diverted to aesthetic purposes, which in its turn entails deeper probing of the long-term global benefits and costs of any alternative land use. In this sense the areas of economic research of interest and importance to landscape architects embrace the whole of economics. No enterprise is an island entire of itself;

**Figure 9.4**
Fürst Pückler's Pharaonic aspirations, and their Romantic parkland setting at Branitz, nearly ruined him. (Photo Colin Price)

there is no consideration of any, without consideration of all. The ripple effects of projects spread out, endlessly it might seem; though there may be, should be and legitimately *are* limits to how far they are to be explored (Price, 1988, 1990). The first research question is: to what extent should landscape architects trust to economists to define these limits appropriately, and to evaluate dispassionately, knowledgeably and wisely within them? And to what extent is it incumbent upon landscape architects to know enough, such that they appreciate when the conventional judgements of economics should be held accountable in a wider review?

There may also be a subtle research programme which seeks to identify, quantify and monetise the individual determinants of the cost of works: the gradient of the site, the nature of soil and bedrock, the climatic conditions which the works must withstand. Such investigations are the bread-and-butter work of econometricians, who engage in the statistical analysis of data of an economic kind. Their favoured techniques are the standard ones of statistics: multiple regression analysis, time series analysis, with autocorrelation and multiple lags of response forming their peculiar interest. Uniqueness of site and of place and of the nature of intervention, however, means that a perfect predictive model of cost-of-works will never be available.

## *The demand side*

Preferences are not stagnant, and much work on the border between economics, environmental psychology and landscape architecture has striven to explain their nature and how they change through time. As far as econometricians are concerned, taste is often regarded as residual variation (Stigler and Becker, 1977), unaccounted for by other factors, and its uncertain future shifts are sometimes left implicitly for the discount rate to represent. But even simple, replicated studies (Price, 2011) show that preferences are far from random.

In an exercise that has been repeated many times over 20 years, the individual landscapes viewed (by untrained students) were themselves overwhelmingly the most significant differentiator of scores, with one particular view being scored the highest by the majority of respondents on most occasions. Yet objectively measurable characteristics of each view were only weakly correlated with subjective scores, and the most valued view was distinguished, not by its individual components, but by the way in which they combined into an appealing composition. Significant differences emerged between individuals within and between occasions, with the greatest differences existing between those of contrasting cultural traditions. Respondents themselves gave an opinion as to the expected determinants of their scores, including their personality, their mood, the season and the weather. In a broad sense these were in conformity with the differences actually detected between individuals and successive applications, but the power of the actual landscape viewed was generally underrated. Season and weather had surprisingly minor significance. These results in themselves are not really economic ones, but they provide a framework within which economic valuations might efficiently be organised.

Historically tastes have changed, but studies of causal factors are rather thin on the ground. There is scope for research here, in particular on whether there are stable factors underlying taste, or whether taste can – or should – be subject to a certain amount of manipulation in the name of expediency.

Landscape may also provide benefits indirectly, via effects on health (Grahn and Stigsdotter, 2003) and on civil behaviour (Sullivan, 2001). It influences local taxation, and so may induce strategic behaviour of municipalities (Choumert and Salanié, 2008). The valuation methods detailed below may be used to address such effects. However, some differentials of neighbourhood and municipality value may as easily reflect degrading of impoverished localities as they may the enrichment of wealthy ones (Price, 1995).

## *Valuation*

It is in relation to evaluative procedures that the predispositions and approaches of landscape architecture and of economics are most likely to be seen as being in opposition. It also seems to be where most economic research relating to landscape is concentrated, so the conflict is not trivial. Economics – according to the official story – is not only client-centred, but client-obsessed: the consumer is monarch, and landscape valuation processes try not only to find what consumers prefer, but to quantify and monetise those preferences. Now there is an apparent parallel within the landscape arena, in that clients are prepared to pay for a design. But the thought-mode of landscape architects is that clients *pay for* aesthetic expertise, and not for a defined product in such a way as to *displace* aesthetic expertise.

As already noted, methods for valuing non-market benefits and costs generally fall under eight headings. All have been used in monetary valuation of landscape (see Price, 1994a for a review). Although they should be aware of these methods, and prepared to collaborate with those who practise them, landscape architects should also furnish themselves with an informed scepticism about their accuracy and applicability. Numbers exert a certain persuasiveness, especially if they have a currency sign in front of them. But the routes by which the numbers are achieved involve assumptions, sometimes implausible ones. The reservations expressed by members of other disciplines may have justification, and the limitations and weaknesses of the evaluative techniques, as well as their powers, should be understood, lest an unwarranted enthusiasm from their proponents should prove infectious. The library of critical material is too extensive for even representative reference. A few hints only about the major debates are given below.

### Future financial costs are saved, or imposed, or voluntarily undertaken as a consequence of current acts or failures to act

For example, the cost of replacing amenity trees might be avoided, or at least delayed, by better care and maintenance of the existing tree stock. This seems to be the underlying philosophy of the Council of Tree and Landscape Appraisers' approach to tree valuation (Council of Tree and Landscape Appraisers, 1983 and frequent revisions). And the required cost of restoring mineral extraction sites evaluates part of what is lost by exploitation.

In a human context, the physical and mental health benefits of recreation in tranquil landscapes can, it is claimed, be measured by reduced costs of morbidity or early mortality. This rather skeletal assessment of value can be criticised even in its own terms. Healthy minds in healthy bodies do not offer immortality, and it is not logically clear that dying of degenerative disease at the end of a long, otherwise healthy life imposes lighter costs on health and social services, than does premature death attributable to poor life-style. Expressing such a viewpoint tends to evoke derision and opprobrium. But you cannot claim the benefits of costs avoided, and then dismiss, as heartless accountancy, the long-term costs concomitantly imposed.

## Costs of *past* decisions made to favour environmental gain (or avoid environmental loss) are taken to measure the gain (or avoided loss)

The argument is rather different from that above. An example is the expenditure that was made on trying to protect elm trees in the English landscape against the fungal pathogen known in the UK as 'Dutch elm disease' (Jones, 1979). Bearing in mind that such treatment involved repeated expenditure, and was not assured of efficacy, the implied benefit of assuredly maintaining each tree in the landscape through its normal life span would be around €2000. The argument is, that such decisions were made by rational people, and so would not have been taken, had the benefits not been at least equal to the costs.

Although this approach has been much used, it begs the question of how past decision makers could have *known* that the benefits were at least equal to costs, given that it was neither their money being spent nor their profit forgone, and it was not they alone who were the beneficiaries.

## Marketable benefits are created or lost elsewhere in the economy

The health benefits of an attractive environment are not confined to cost savings. Early recovery from surgery (Ulrich, 1984) or other illness, resulting from such an environment, also increases the productive output of the economy. If people are not aware that these effects can benefit them, or do not care sufficiently about them (so the argument goes), individual choice provides them insufficiently, and the case is the same for providing good environment as a public good. Nowadays trees are seen as particularly therapeutic, and many European forestry organisations are promoting the health benefits of woodlands (Nilsson *et al.*, 2011).

A case often made in consultants' assessments is that good landscape attracts inward investment and so promotes location-specific economic activity, with benefits to regional income. While this belief is manifest in promotional materials aimed at influencing commercial and industrial location, it is not so clear what *overall* social benefit arises from persuading businesses to relocate to town X rather than city Y.

These approaches value only the financial consequences of aesthetic pleasure's *existing*, not aesthetic pleasure *per se*. This is what the following approaches aim to establish.

## Market parallels

While landscape is mostly experienced free of charge, some aesthetic resources which can readily be fenced off – waterfalls, cliff-tops, historic gardens – do charge an admission price. Resources judged to be of equivalent aesthetic value might then be considered also to have equivalent monetary value. Econometric approaches might in theory be applied, to identify and price components of the aesthetic experience. For example, entry charges paid for Forestry Commission arboreta, broadleaved and coniferous ones, in summer and in winter, suggest that *leafiness* is the valued characteristic (Price, 2007).

## Voluntary subscriptions are made to organisations or campaigns dedicated to protecting or enhancing aesthetic value

The English and Welsh National Trust for Places of Historic Interest or Natural Beauty has a membership of about 3.6 million, and raises about £120 million annually from subscriptions. The problem is that, while evincing a certain willingness to pay for beautiful landscapes and/or historical buildings, such figures tell us nothing about the value of particular properties or characteristics, nor of one form of landscape design rather than another. A more targeted value arises from specific campaigns. A sum of £6.5 million was raised in a few weeks when structural weakness threatened the spire of Salisbury Cathedral, centrepiece of one of Britain's most distinctive cityscapes.

Even then, however, voluntary contributions do not reveal full willingness to pay, since beneficiaries from open-access resources can free-ride, on the basis that the contributions of others would suffice to maintain the resource in its desired state. Until quite recent times, Lincoln Cathedral – argued by some to be the finest Gothic building in Europe – received a mean voluntary donation of 3.5 pence per visitor. Yet, now the entry charge is £5, the visitors continue to come, for whatever combination of spiritual, historical and aesthetic experience they derive from it. Its external appearance, as the overwhelmingly dominant element of an urban landscape, remains free, with no obvious mechanism for those wishing to contribute to the upkeep of this role, to do so.

Thus this approach to valuation, offering no more than a vague, general, and lower-bound valuation, has not formed the basis for substantial research, and is unlikely to do so (Price, 1994b).

## People are asked what they would be willing to pay: contingent valuation and the landscape designer

Of all methods of environmental valuation, the contingent valuation method (CVM) and its variants have become by far the most popular, and they have the advantage of directness. One seeks what price people would be willing to pay for the preservation of a beautiful landscape: and so one asks them just that: 'what price would you be willing to pay for . . .?' At a stroke this approach addresses two problems seen as inherent in leaving judgement to landscape professionals: lack of representativeness, and absence of a cash equivalent value for the preferred form of landscape (so that landscape issues can be weighed against other ones).

The method is equally applicable to degradation (Randall *et al.*, 1974) or enhancement (Tyrväinen, 1997) of landscape. When an environmental valuation problem arises, the instinct nowadays seems to be to 'reach for the CVM manual'.

It has even been (incorrectly) claimed that for certain categories of environmental effect CVM is the *only* method capable of delivering a cash value. CVM generally compares one real situation (the status quo) with one or more hypothetical alternatives (Randall, 1994). In this sense it is effectively presenting 'an artist's impression' – or more likely nowadays a computer visualisation – of the proposal, which may involve enhancement or degradation of landscape.

Contingent referenda have a different format, focusing on political and collective preferences, rather than financial and individual ones. A measure is proposed, say to restore the appearance of a land tract subjected to mineral extraction, and citizens are asked whether they would support the measure if it was accompanied by some specified increase of taxation.

Choice experiments attempt to avoid headlining what may be seen as an inappropriate choice between money and environment, by asking respondents to choose among options which differ in several characteristics, perhaps a number of aesthetic and leisure experiences. A sum of money is included in all options, and statistical analysis maps out the trade-offs made between all the components, including the monetary one.

While some formats of questionnaire are less prone to bias than others, a number of unresolved problems affect them all. For example:

- What *objects* are being valued? What components of a landscape with lakes and rocks and woods and hedgerows elicit the willingness to pay?
- In *what context* are they valued? Yes, the effect of trees seen through a window is that hospital beds are vacated more quickly, and people earn money again sooner. But before we can transfer benefits valued in this context to another context, we need to understand how value arises. Do trees have a therapeutic effect on people in hospital beds? Or do people hate trees and discharge themselves early, to avoid having to look at them any longer than necessary?
- Are we valuing *acts* (as with intervention to save threatened trees), or *states* (as of continuing woodland cover of steep gorges)? Contingent valuation usually seeks to elicit a difference in willingness to pay, between two contrasted states, without reference to the means of moving between those states. But, in the real world, we are usually trying to evaluate acts. Valuation should focus on *what changes* as a result of acts, and over what time-span.
- Who values? Some systems are explicitly expert-based. But the accepted past cost approach, too, implicitly – and quite unreasonably – assumes that past experts had made reasoned professional judgements, that the benefits of action were at least equal to the costs. By contrast, contingent valuation makes a virtue of the representativeness of respondents to questionnaires.
- From what perspective? Individuals have multiple perspectives on value, and it is widely held that the values expressed as consumers differ from those held as (public-spirited) citizens (Ovaskainen and Kniivilä, 2005). On the other hand it has also been argued that questions asked of people as citizens might elicit more accurate representations of respondents' valuations as consumers (Price, 2006).

- Can subjectivity be avoided? Even when performance against a criterion is objectively measurable, as in the case of tree crown size, judging *which are the appropriate criteria* is subjective; and more so when qualities such as suitability to setting must be judged: for example, are cabbage palms (*Cordyline australis* Hook.) on the western seaboard of the UK an affront to ecological purism, or an emblem of cultural tradition?
- How is value scaled? When asked what they are willing to pay for 'trees in the landscape', how large a tract of landscape do respondents envisage? Or do they respond in a symbolic manner, treating the questionnaire as a chance to express their wider concern about the environment (Blamey, 1996)? There are preferences for states of a particular landscape, and preferences for portfolios of experience. Research has too often sought preference between one state of landscape and another, as though that state would then be rolled out across the entire landscape.
- How is value partitioned? Amenity does not generally come as a distinct entity, but forms part of a more-or-less complex package. Subscribers to environmental organisations make payments for provision of a mixture of amenity, access and nature conservation, and very probably for the warm glow that comes with supporting the public provision of these services. Within the amenity category, numerous features contribute to value. The hedonic pricing method aims to separate these by statistical methods, but the key problem is that features are *not* separable in their contribution to valued landscape: steep topography, varied land use and presence of water features all contribute to *good* landscape, but it is their composition that makes landscape *excellent*. So far, and for the foreseeable future, no formula has been devised that allows features to be aggregated into an overall landscape value. And thus landscape value, if assessed at all, needs to be assessed holistically.

Given the popularity of this collection of techniques and the frequency with which their results are introduced to debates about land use change, it is incumbent on all who take part in such debates to be conversant with their weaknesses no less than their merits.

## Experts define a systematic relationship between measurable characteristics and deemed money values

In what may be the earliest published monetisation of aesthetic values, Helliwell (1967) describes how an urban tree might be valued in terms of seven physical, locational and social characteristics, and a value per point scored under a set of rules for aggregating these characteristics. Economists would be instinctively suspicious of such a procedure, which makes no direct reference to consumers' willingness to pay. That said, the method often seems to deliver 'reasonable' values, and, being subject to repeated scrutiny when used in court cases, could be argued to represent in general terms an expert consensus. At present a fierce discussion is raging between the proponents of the Council of Tree and Landscape Appraisers (CTLA) and similar approaches, and the advocates of the Helliwell approach and its variants.

The former argue that costs represent hard financial data and thus are a securer base for calculating value than are judgements of benefit. The latter might respond that it is actually a measure of benefit that is sought; that cost-based approaches require their own judgement, viz. that replacement costs are *worth* incurring – indicating an implicit judgement of benefit; and that the objectivity of costing is in any case compromised by a number of adjustment factors in the CTLA process whose relevance and importance are again a matter of judgement.

## Costs are willingly incurred to gain access to good landscape: the travel cost method

Popular up to the early 1990s, this method analysed how the cost of travel affected the proportion of population who visited a destination, usually a countryside recreation site (Clawson and Knetsch, 1966). By comparing sites of different landscape quality, it is possible to infer a cash value in terms of the extra distance travelled to enjoy superior quality of recreational background (Bergin and Price, 1994). The method had problems such as: how to treat time costs in travel? How to attribute cost across the multiple sites visited on many countryside trips? How to assess substitute sites' effect both on demand for a particular site, and on the consequences of that site's removal from the system?

Though at present eclipsed by CVM and allied methods, this approach appeals to economists, in that it is based on actual rather than hypothetical willingness to pay. It may be expected to reappear when fashion, or the perception of fatal weaknesses, takes the spotlight off CVM.

## Costs are willingly incurred to gain access to good landscape: the hedonic house price method

Views may be free, but houses that have views are not. The price of a house is determined by its physical characteristics, its location in relation to desired amenities and its environmental quality of which landscape is a major component (Figure 9.5). Studies exist from the early 1970s up to the present, which show how trees (Payne and Strom, 1973; Tyrväinen and Väänänen, 1998), water bodies (Luttik, 2000) and urban greenspace (Choumert and Travers, 2010) affect house price. The analytical procedures required are the regression techniques customary in the valuation of the characteristics of other market goods.

The approach of the hedonic pricing method (HPM) is an atomistic one, breaking landscape down into measurable components. It is thus in conflict with landscape architects' innate focus upon the overall design. It bypasses the landscape architect in a profounder way than CVM, which at least generally seeks to value views or environmental conditions or intended design plans as a whole. HPM seems to be designing by numbers. Arrangement of elements is of no consequence, and thus that most subtle aspect of aesthetics, composition, is consigned to the wastepaper basket of residual variation.

Here the differences between economist and aesthete come into sharpest focus. It is here that the members of different professional guilds are in gravest danger of closing ranks, and denying that members of other guilds have

**Figure 9.5**
An urban view that, according to an estate agent in 1990, put £30,000 on the price of a house commanding it. But how much for Durham Cathedral, how much for the trees, how much for the topography . . . ? (Photo Colin Price)

meaningful access to truth or could make helpful input to the design process. For aesthetes, the decompositional approach of economists evinces incomprehension of aesthetic value, while economists might charge aesthetes with isolationism and elitism. In the meantime, decisions wait to be made.

### Access costs: ethical premia

In recent years the term 'payment for environmental services' has become a popular topic in publications and in conferences. It entails compensating the provider of environmental benefits in some way, often through a mark-up on price that environmentally conscious persons are willing to pay for an assured product. The consumer thus purchases a *mode of production* as well as a product. Certified timber, bird-friendly coffee and locally produced foodstuffs are oft-quoted examples.

Although such premia have most frequently been explained as demand for ecosystem-friendly and sustainable management, in truth there is little to prove that there is not an element in them of aesthetic preference for the landscapes in which are grown the products of traditional practice. The assurance provided by certification is generally cast in vague terms of environmental sustainability, legal procurement and social justice, rather than defining the actual difference of states between the conditions of production, and those of uncertified production. To the extent that the product is visible but the process is scarcely defined, a more productive understanding is that such premia represent 'purchase of moral satisfaction' (Kahneman and Knetsch, 1992) or of warm glows (Price *et al.*, 2008). They should be interpreted as a vote in favour of pursuing cost–benefit analysis, or some other more direct means of environmental valuation.

## Valuation on a 'landscape scale': operations research/modelling approaches

The owners of Europe's great landed estates were once able to enact, by autocratic processes, comprehensive schemes for large landscapes. Their demise leaves a public role in devising overarching landscape plans which, by contrast, explicitly consider the interests of a broad stakeholder group.

The broad sweep of such decisions has led to complex computer modelling, with large algorithms defining an iterative process in which a small change in each element is evaluated with respect to the current state of all other elements (contributions in Helles *et al.*, 1999). But, because of subtle aesthetic and other environmental interactions between spatially adjacent landscape elements, an 'ideal landscape' cannot be constructed piecemeal from elements valued by simple and stable equations. Landscape architects need at least to be assured that complexity and opacity are not necessarily guarantors of subtlety and relevance.

## Integrated approaches

A contrast has been noted above between the leaning of economics towards an atomistic view – each landscape element having, apparently, a marginal utility – and the penchant of landscape architects for holistic evaluation – the aesthetic design is exactly that, both a composition of elements, and an *intention* that the elements should be so combined. Given the lack of monetisation, and the charge of unrepresentativeness that limits acceptance of aesthetes' judgements by economists, it might be apposite to consider whether the approaches could be creatively combined.

Price (1978), Abelson (1979), and Henry (1994) have all suggested that the role of economic valuation is to monetise design quality, assessed by whoever is deemed competent. Such an approach could maintain the design and even the aesthetic evaluation elements of an overall valuation process. Or landscape architects could present alternative designs, each of high aesthetic quality, for inspection and client evaluation.

Taste is conditioned by one's customary background. And a tradition based on what one is accustomed to can certainly become stultifying. But still an economist would assert that the safest basis for social arrangements lies in respecting consumers' preferences, both among aesthetic possibilities, and between high-cost aesthetic experiences and other costly acts of consumption. How one resolves this fundamental difference of perspective lies in the professional aesthete offering the public some good alternatives to choose among. This combines the conservative public taste recorded by economists and other social scientists, with the innovative design offered by landscape architects and other design professionals.

Landscape assessment is a straddling discipline, potentially allowing many disciplines an appropriate input. Geographers and planners have long toyed with landscape evaluation in aesthetic terms. Their processes might be applied to aesthetic professionals' designs. The role of economists is then to assess what it is worth in cash terms to advance up the aesthetic scale (Price and Thomas, 2001).

Protected areas such as the UK's national parks provide something of a special case. While their status as IUCN category V protected areas affords no absolute protection against degradation, their designation implies a prior evaluation which neither should be ignored, nor can be completely accommodated by any normal form of cost–benefit analysis. Suggestions have been offered (Price, 1977) on what might be an appropriate reconfiguration of economic assessment in such circumstances, which takes due account of specific designation, yet does no certain violence to general economic principles. This is an area for potential compromise with which, so far, none of the disciplines involved has engaged.

## Intervention

Despite the preferences of some market-orientated economists, regulation through planning and legal processes appears to be the preferred mode of intervention against adverse environmental impact. As for beneficial services, the current consensus among economists and politicians appears to be that payments should be made for the services themselves, and not for farm products associated only loosely with those services.

In forestry and in relation to regionally certified foodstuffs, the argument has taken a different turn, consumers being invited to 'intervene' themselves, to the extent that they favour and will pay a premium for a particular mode or location of production. But, as has already been argued, such premia may be symbolic of a general concern for environment and justice and sustainability, rather than for the modification of packages of environmental services that is concomitant on one more unit of certified production.

There remain economists who are convinced that, with sufficient deregulation and with sufficient incentives for the creation of markets, the market mechanism will assure the 'right' amount of landscape. However, the absence of environmental markets is not accidental, but structural, a consequence of the nature of landscape and the processes by which it is enhanced or degraded.

## The sustainability debate

'Sustainable' is a word that has almost been stripped of its real meaning – 'capable of continuing indefinitely' – by use in politics and the media as a synonym for 'everything I approve of'. Yet it remains a powerful word, emblematic of those long-term benevolent sentiments that everyone in environmental management would like to be thought of as possessing.

In particular, formal economic evaluation of persistent change in environment generally entails some application of discounting procedures. The disquiet that these have raised – particularly in light of sustainability pronouncements – has to an extent modified governments' own approach to discounting (HM Treasury, undated). Yet the process itself still has disturbing overtones, typically reducing the value that would be attributed to future effects, to a tiny fraction of what would be ascribed, if the benefit or cost were delivered today. In the preface to an extensive critique of conventional discounting (Price, 1993: xv), I commented:

> Authors are often asked for whom their books are written. My answer is, for everyone concerned, which I now think means *everyone*. But four specific groups of people have been in my mind while I wrote:

(a) those who apply or implement the results of discounted cash flow (including politicians);
[. . .]
[. . .]
(d) any citizens concerned about the world around them, and about what economists and politicians are recommending should be done with it, on those citizens' behalf.

Landscape architects may, knowingly or not, fall into (a), and should certainly see themselves as part of (d). I would argue that familiarity with the critiques of discounting should be prerequisite for anyone qualifying as a landscape architect, and certainly for those who join economists in collaborative research.

In particular, a frail justification has been offered for discounting, through the concept of 'weak sustainability', which requires only that the sum of natural and fabricated capital should not be reduced. Thus, experience of particular wild and cultural landscapes can be replaced by generalised theme park landscapes, or by virtual landscape experience, or ultimately by any flow of aesthetic stimuli (Price, 2007). Read behind the conceptual stuff, and see what in reality is entailed by such feeble requirements!

## Big questions for research

It is hardly appropriate for one individual to define what should be the collaborative research agenda for landscape architects and economists. My suggestion is that a joint programme with environmental psychologists is desirable, to compare responses to and values of landscapes across Europe; then to seek to explain differences in terms of landscapes' characteristics in themselves, in individual psyches and common cultural backgrounds, in familiarity with types of landscape and individual landscapes, in the search for diversity of landscape experience. The days of international jet-set landscape architecture, interpreted crudely, are, thankfully, over, but we still lack understanding of the factors that relate designs to desires of particular consumers in particular places, and how those designs might be valued. It is important too that methods be developed which can attribute landscape values to subtle causes of value, such as familiarity with a particular landscape in a particular state.

There is also a valuable research task in *defining* a framework of research into which individual research groups can fit as a voluntary act, such that gaps in knowledge are filled, and unproductive overlap of effort is avoided. Because of the variability of landscape and of the people who experience it, there will never be any *wholly* wasteful duplication. But *deliberated* duplication always offers rich possibilities for comparative studies.

## Conclusions

Concerned with similar issues, yet from very different perspectives, landscape architects and economists have great scope for disagreement. Yet in practice different approaches may lead towards common conclusions. For example, in a project to redesign forest boundaries within the Snowdonia National Park in Wales, the modifications proposed by the landscape designers were considered by the forest economists to increase rather than curtail financial profitability.

There is, moreover, a new willingness among economists to broaden the definition of well-being, and to lengthen the customary short-term perspectives to ones that would have been familiar to the constructors of the Neolithic ritual landscapes and the naturalistic landscape gardeners.

This does not mean that there will be no conflict of perspectives, nor of prescriptions. It does mean that failure to understand and to collaborate will miss creative opportunities. One of my periodic reflections on integrated approaches to valuation (Price and Thomas, 2001: 202) runs as follows:

> Taken individually, none of the various monetary and non-monetary approaches to valuing landscape can make all the necessary links between a landscape design, and its value in comparison with cost or other kinds of benefit. The economist's approach in theory comes nearest to it, but in reality its results have been 'disappointing', as even its own practitioners acknowledge (Hanley and Ruffell, 1993).
>
> Taken together, however, each fills in weaknesses in the capabilities of other approaches, and so makes possible a logical, stepwise progression from design to monetary value. The result will not please purists. Neoclassical economists will be affronted by its reliance on subjective judgement (as though values generally were *not* based on such judgements): aesthetes will no doubt revile the taint of money impinging upon matters of the soul. Pragmatists, however, may welcome the possibility of bringing a range of relevant expertise to bear on real problems, which require implementable solutions.

Forty years ago I began my exploration of landscape economics with high hopes that rationality would always prevail over sectoral interest. Ultimately perhaps, it will. Better sooner, however, than later.

## Bibliography

Abelson, P. (1979). *Cost Benefit Analysis and Environmental Problems*. Saxon House, Farnborough, Hants, UK.

Alonso, W. (1964). *Location and Land Use*. Harvard University Press, Cambridge, Mass.

Aristotle (c4 BC). *Politica*. Trans. by Jowett, B. Oxford University Press, 1921.

Bergin, J. and Price, C. (1994). 'The travel cost method and landscape quality'. *Landscape Research*, 19 (1), 21–3.

Blamey, R.K. (1996). 'Citizens, consumers and contingent valuation: clarification and the expression of citizen values and issue-opinions'. In Adamowicz, W.L., Boxall, P.C., Luckert, M.K., Phillips, W.E. and White, W.A. (eds). *Forestry, Economics and the Environment*. CAB International, Wallingford, Berks, pp.103-33.

Bowers, J.K. and Cheshire, P.C. (1983). *Agriculture, the Countryside and Land Use: An Economic Critique*. Methuen, London.

Choumert, J. and Salanié, J., (2008). 'Provision of urban green spaces: some insights from economics'. *Landscape Research*, 33, 331–345.

Choumert, J. and Travers, M. (2010). 'La capitalisation immobilière des espaces verts dans la ville d'Angers: une approche hédoniste'. *Revue Economique,* 61, 821–36.
Christaller, W. (1966). *Central Places in Southern Germany.* Trans. by Baskin, C., Prentice Hall, Englewood Cliffs.
Clawson, M. and Knetsch, J.L. (1966). *Economics of Outdoor Recreation.* Johns Hopkins University Press, Baltimore.
Council of Tree and Landscape Appraisers (1983, later revisions). *Guide for Establishing Values of Trees and Other Plants.* International Society of Arboriculture, CTLA, Urbana, II.
Evans, A.W. (1973). *The Economics of Residential Location.* Macmillan, London.
Friedman, M. (1962). *Capitalism and Freedom.* University of Chicago Press.
Grahn, P. and Stigsdotter, U.A. (2003). 'Landscape planning and stress'. *Urban Forestry and Urban Greening,* 2, 1–18.
Hampson, S. F. (1972). 'Highland forestry: an evaluation'. *Journal of Agricultural Economics,* 23, 49–57.
Hanley, N.D. and Ruffell, R. (1993). 'The contingent valuation of forest characteristics'. *Journal of Agricultural Economics,* 44, 218–29.
Hawkins, C. J. and Pearce, D. W. (1971). *Capital Investment Appraisal.* Macmillan, London.
Helles, F., Holten-Andersen, P. and Wichmann, L. (eds) (1999). *Multiple Use of Forests and Other Natural Resources.* Kluwer, Dordrecht.
Helliwell, D.R. (1967). 'The amenity value of trees and woodlands'. *Arboricultural Association Journal,* 1, 128–31.
Henry, M.S. (1994). 'The contribution of landscaping to the price of single family houses: a study of home sales in Greenville, South Carolina'. *Journal of Environmental Horticulture,* 12 (2), 65–70.
Hiley, W. E. (1956). *Economics of Plantations.* Faber, London.
HM Treasury (undated). *The Green Book: Appraisal and Evaluation in Central Government.* The Stationery Office, London.
Isard, W. (1956). *Location and Space Economy.* Wiley, New York.
Jevons, W.S. (1871). *The Theory of Political Economy.* Macmillan, London.
Jones, P. (1979). 'A gap in nature – Dutch elm disease in Britain'. In Clouston, B. and Stansfield, K. (eds). *After the Elm.* Heinemann, London, UK, Pp. 80–98.
Kahneman, D. and Knetsch, J.L. (1992). 'Valuing public goods: the purchase of moral satisfaction'. *Journal of Environmental Economics and Management,* 22, 57–70.
Keynes, J.M. (1936). *The General Theory of Employment, Interest and Money.* Macmillan, London.
Krutilla, J.V. and Fisher, A.C. (1972). *The Economics of Natural Environments.* Johns Hopkins University Press, Baltimore.
Langdon, D. (2009). *Spon's External Works and Landscape Price Book.* Taylor and Francis, London.
Lifran R. and Oueslati W. (2007). 'Eléments d'économie du paysage', *Economie Rurale,* 297–298, 85–98.
Lipsey, R.G. (1989). *An Introduction to Positive Economics,* 7th edn. Weidenfeld and Nicolson, London.
Luttik, J. (2000). 'The value of trees, water and open space as reflected by house prices in the Netherlands'. *Landscape and Urban Planning,* 48, 161–7.
Malthus, T.R. (1798). *Essay on the Principle of Population as it Affects the Future Improvement of Society.* Ward Lock, London.
Marshall, A. (1890). *Principles of Economics,* 8th edn. Macmillan, London, 1920.
Marx, K. (1867, 1885, 1894). *Capital.* Translated by Moore, S. and Aveling, E. Allen and Unwin, London, 1946.
Mishan, E. J. (1975). *Cost-Benefit Analysis* (2nd edn). Allen and Unwin, London.
Nilsson, K., Sangster, M., Gallis, C., Hartig, T., de Vries, S., Seeland, K. and Schipperijn, J. (2011). *Forests, Trees and Human Health.* Springer, Berlin.
Ovaskainen, V. and Kniivilä, M., (2005). 'Consumer versus citizen preferences: evidence on the role of question framing'. *Australian Journal of Agricultural and Resource Economics,* 49, 379–94.
Parsons, R. and Daniel, T.C. (2002). 'Good looking: in defense of scenic landscape aesthetics'. *Landscape and Urban Planning,* 60, 43–56.

Payne, B.R. and Strom, S. (1973). 'The contribution of trees to the appraised value of unimproved residential land'. *Valuation*, 22 (2), 36–45.
Pearce, D.W. and Turner, R.K. (1990). *The Economics of Natural Resources and the Environment*. Harvester Wheatsheaf, London.
Plato (c4 BC). *Protagoras*. Trans. Taylor, C.C.W. Oxford University Press, 1976.
Price, C. (1966). *Forestry in the Dartmoor National Park*. Youth Hostels Association, St. Albans.
Price, C. (1971a). *Social Benefit from Forestry in the UK*. Department of Forestry, Oxford University.
Price, C. (1971b). 'The effect of tax concessions on social benefit from afforestation'. *Forestry*, 44, 87–94.
Price, C. (1977). 'Cost–benefit analysis, national parks and the pursuit of geographically segregated objectives'. *Journal of Environmental Management*, 5, 87–97.
Price, C. (1978). *Landscape Economics*. Macmillan, London.
Price, C. (1982). 'Residential density and spatial externalities'. *Urban Studies*, 19, 293–302.
Price, C. (1984). 'Project appraisal and planning for over-developed countries'. *Environmental Management*, 8, 221–242.
Price, C. (1988). Does social cost–benefit analysis measure overall utility change? *Economics Letters*, 26, 357–61.
Price, C. (1989). *The Theory and Application of Forest Economics*. Blackwell, Oxford.
Price, C. 1990. 'Pecuniary externalities and project appraisal'. *Project Appraisal*, 5, 39–48.
Price, C. (1993). *Time, Discounting and Value*. Blackwell, Oxford.
Price, C. (1994a). 'Literature review [of landscape valuation]'. *Landscape Research*, 19 (1): 38–55.
Price, C. (1994b). Donations, charges and willingness to pay. *Landscape Research*, 19 (1), 9–12.
Price, C. (1995). 'Pros and cons of alternative evaluation methods'. In Willis, K.G. and Corkindale, J. (eds). *Environmental Valuation: New Directions*, CAB International, Wallingford, Berks, pp.160–77.
Price, C. (2006). 'Superficial citizens and sophisticated consumers: what questions do respondents to stated preference surveys really answer?' *Scandinavian Forest Economics*, 41, 285–96.
Price, C. (2007a). 'Putting a value on trees: an economist's perspective'. *Arboricultural Journal*, 30, 7–19.
Price, C. (2007b). 'The landscape of sustainable economics'. In Benson, J.F. and Roe, M. (eds). *Landscape and Sustainability* (2nd ed.). E & F Spon, Cheltenham, pp. 37–57.
Price, C. (2011). 'Subjectivity and objectivity in landscape evaluation: an old topic revisited'. In Heijman, W. and van der Heide, M. (eds). *The Economic Value of Landscapes*. Routledge, Abingdon.
Price, C., Cooper, R.J. and Taylor, R.C. (2008). 'Further thoughts on certification and markets'. *Scandinavian Forest Economics*, 42, 66–74.
Price, C. and A.Ll. Thomas (2001). 'Evaluating the impact of farm woodland on the landscape: a case of blending perspectives'. In Sievanen, T., Konijnendijk, C.C., Langner, L. and Nilsson, K. (eds). 'Forest and Social Services – the Role of Research'. *Research Paper* 815 of Finnish Forest Research Institute, Vantaa, Finland, pp. 191–203.
Randall, A. (1994). 'Contingent valuation: an introduction'. *Landscape Research*, 19 (1), 12–14.
Randall, A., Ives, B. and Eastman, C. (1974). 'Bidding games for evaluation of aesthetic environmental improvements'. *Journal of Environmental Economics and Management*, 1, 132–49.
Ricardo, D. (1817). *On the Principles of Political Economy and Taxation*. John Murray.
Robinson, J. (1956). *The Accumulation of* Capital. Macmillan, London.
Samuelson, P. A. (1980). *Economics*, 11th edn. McGraw-Hill, New York.
Sheppard, S.R.J. and Harshaw, H.W. (eds). (2001). *Forests and Landscapes: Linking Ecology, Sustainability and Aesthetics*. CABI Publishing, Wallingford, Berks, pp. 149–72.
Smith, A. (1776). *An Inquiry into the Nature and Causes of the Wealth of Nations*. Strahan and Cadell, London.
Stigler, G. and Becker, G. (1977). *De gustibus non est disputandum*. *American Economic Review*, 67, 76–90.
Sullivan, W.C. (2001). 'Neighborhood landscapes, democracy, and community'. *Landscape Journal*, 20, 198–201.

Tyrväinen, L. (1997). 'The amenity value of the urban forest: an application of the hedonic pricing method'. *Landscape and Urban Planning*, 37, 211–22.

Tyrväinen, L. and Väänänen, H. (1998). 'The economic value of urban forest amenities: an application of the contingent valuation method'. *Landscape and Urban Planning*, 43, 105–18.

Ulrich, R.S. (1984). 'View through a window may influence recovery from surgery'. *Science*, 224, 420–1.

Von Thünen, J. H. (1875). *Der Isolierte Staat in Beziehung auf Landwirtschaft und Nationalökonomie.* Schumaucher Zarchlin, Berlin.

Whitby. M. (ed.) (1994). *Incentives for Countryside Management.* CAB International, Wallingford, Berks.

Wu, J. and Irwin, E. (2008). 'Optimal land development with endogenous environmental amenities'. *American Journal of Agricultural Economics*, 90, 232–48.

Chapter 10

# Space, place, site and locality

## The study of landscape in cultural anthropology

*Robert Rotenberg*

**Introduction**

The discipline of cultural anthropology deals directly with questions about the groups of people landscape architects serve. Cultural anthropology's methodological base in ethnography provides deep or 'thick' descriptions of the everyday lives of people, providing a rich source of information about patterns of behaviour, common meanings and associations people attach to places, and their values and aspirations for the future that can inform and guide the landscape architect. Anthropological studies often produce unexpected findings. They may reveal order where disorder is anticipated, power where marginality is assumed, negative practical outcomes from contradictions embedded in design ideologies, and unintended consequences resulting from the best laid plans. Cultural anthropology not only provides cultural information but, at its best, a critique of landscape design. It serves to enlighten those who seek to impose a particular vision on the landscape of the hazards involved in such actions. Anthropology has come to its own self-critique as the postcolonial discipline *par excellence*. For this reason, it lays the foundation for a truly reflexive and ethical regime for assessing how better to respond to functional and aesthetic needs through the transformation of the landscape.

All design involves two simultaneous goals: effective function and evocative aesthetics. These are held in tension and must be balanced during the design process. For landscape architects, functional concerns include site limitations such as topography, drainage, climate and sustainability issues related to soil, water and habitats, in the context of human behaviour activities. Aesthetic concerns include the plant materials, such as palette, size and scale, and composition, and hardscape features that accommodate human uses. Both goals entail people as agents of activities or as perpetrators of uses that must be accommodated. This is where the cultural anthropologist has a contribution to make to the landscape architect.

**Robert Rotenberg**

Anthropology is a discipline that has a rich and important history. The discipline was born in the heady, confusing, nineteenth century as Europeans tried to make sense of the human experiences they encountered while incorporating distant lands into market and empire. The demarcation of 'race' categories defined the nineteenth-century science's boundaries. In the twentieth century, anthropologists incorporated other kinds of difference, including categories of experience that were already studied by other disciplines: class, nation, region, occupation, gender, religion and ability. This mixing of foci in research has led to a confusion of the boundaries between anthropology, sociology and geography by scholars outside the disciplines. For practitioners, however, the differences between these disciplines are very clear, both in theory and in practice.

The nineteenth-century academic enterprise spawned several academic traditions. In most of Europe, anthropology most often refers to human biology and even more specifically to human palaeontology. One also finds the related disciplines of ethnology, national ethnography and social anthropology in European universities. The academic traditions in North America, Latin America, Africa, China, Japan and India only complicate the matter even further. It would seem that the anthropological enterprise is a canvas onto which intellectuals project their concern for the role of some basic human 'nature' in the origins or outcomes of contemporary issues. This often has resonance with political concerns, such as immigration, multiculturalism, national identity, dialect preservation or official folklore.

I write from the tradition of North American cultural anthropology. This tradition dates from the late nineteenth century and can be traced to the work of a single scholar, Franz Boas. Trained as physicist and geographer, he became interested in the lives of Arctic peoples living in Greenland and in British Columbia. His great insight was that race, language and culture were the products of separate human experiences and developed according to different influences and processes. Compared to the racial thinking of the nineteenth century, this was a radical idea. It took some years before Boas could find an academic post. Eventually he taught anthropology at Columbia University (1896). He trained many anthropologists who then established the first anthropology departments in the other universities of the United States, Canada and Mexico.

The work of Boas and his students is known as the Boasian School. This academic tradition insists that an anthropologist should be equally knowledgeable in human biology, human palaeontology, descriptive, historical and comparative linguistics, pre-historic and historic archaeology, and ethnology, also known as cultural anthropology. This last field, ethnology, is not the same as the one with the same name in Europe. For Boas, ethnology is about the distribution of traits, artefacts and practices in space, regardless of the political, linguistic or environmental features of the people who possess them. Cultural anthropology incorporates both ethnography and ethnology to understanding how culture shapes the human experience. This integration of several different disciplinary traditions within one academic department sets North American anthropology apart from the European tendency to separate these disciplines. Contemporary anthropologists have these multiple fields as the core of their academic training, but specialise in one of them. Two of these fields, archaeology and cultural anthropology, have a bearing on landscape architecture. Here I discuss landscape design in the context of cultural anthropology.

Cultural anthropology deepens our understanding of how culture shapes the human experience. For anthropologists culture is something that must be explained. One can never assume it exists as an independent feature of human experience. 'How can this be?' you ask, 'Do not all humans possess culture, just as all humans possess a biology?' Of course, but if we were to stop at such a statement there would be very little need for further inquiry. The real questions are what are the traits, artefacts and practices every human possesses and how did these come to be in the possession of a specific person. There are many ways of going about answering these questions. Each way constitutes a distinct theory of culture. I wish to focus on three of them that I believe have the greatest relevance for landscape architects. These theories are *mediation*, *interpretation* and *distribution*. These refer to specific sets of ideas to understand culture. The words do not mean what you might commonly assume that they mean. They are a short-hand way for anthropologists to talk to each other.

I will begin with mediation (and with apologies to Viollet-le-Duc; the similarity between what follows and his work *Discourses on Architecture* (1875) are purely coincidental). We can assume that everything that is beyond the immediate control of human beings can be lumped together under the term 'nature'. Because we lack control over it, nature is continually surprising us with its variability; weather, famine, drought, plague, predators and pollution increase and decrease threats in our lives seemingly without pattern. Humans are cultural beings because we can protect ourselves from these variations in nature. It rains. We can stand naked in the rain and get wet, then wait for it to stop raining, and wait again for the wind and sun to dry our bodies. Or, we can walk to a tree and seek shelter under its leaves. We could take an animal hide, dried in the sun, and hold it over our heads to ward off the rain, while we stay in place. Finally, we could fashion a frame and tie the hide to it, holding the frame with a single hand while we go about our business in the rain with the other hand. In the first case, we are facing nature directly and we get wet. In the second case, we alter our behaviour, seeking shelter under the tree and we stay dry. In the third case, we have created a dry barrier between nature and ourselves. That barrier, however, requires us to maintain it (hold it up; repair any holes) using our energy and distracting our attention. In the final case, we have created a barrier that requires less effort and attention. The first two cases are examples of unmediated behaviours, much as you would find among animals. The third and fourth cases are mediated behaviours found among primates and humans.

In that fourth case, if you stitch several hides together and cover a frame that is well anchored in the ground, you have a nice, dry hut, one of the first buildings. Landscape design mediates between the variability of nature and human action. Through design humans extend control to a world that was previously natural. In this way, landscape architecture produces and maintains the boundary between culture and nature. The elements that construct a landscape design are inorganic and organic features of nature. The rearrangement simulates a version of nature in which variability has been brought under control. Sometimes it uses elements that are nearby and native. At other times it uses elements from distant places, creating a more fanciful design, or even one that never existed in nature. Yet all designed landscapes mediate between people and nature. Mediation theories of culture are limited because they see design as a rational solution to the problem of insulating humans from harm. This leads to a 'form follows function' view of design that is restrictive at best.

Let's turn now to interpretative approaches to culture. If you have a 'bag of culture' in your hand, the contents of the bag will consist of various ideas, behaviours and practices through which you create your everyday experience. Is the bag of culture what we mean when we say that someone 'possesses' a culture? No, not really. That bag may contain all sorts of things that you would never use because the practice is old fashioned or because there are several options to solve the same problem and you habitually choose some behaviours over others. In other words, you know more about the possibilities for acting, thinking or believing than you actually use. If we were to assume that everything in the bag is relevant to the way culture affects people's lives, we would have to accept all sorts of archaic and extraneous information. This error led previous generations of anthropologists to make inappropriate generalisations about what a group of people 'believed' about the world. The interpretation perspective helps us avoid the fallacy that humans are constrained to act out pre-determined 'cultural' performances in all situations, even when their better judgement warns them against it. The interpretation perspective instead reminds us that all individuals are masters of their 'bags of culture,' picking and choosing the ideas, behaviours and practices that make the most sense for the situations they find themselves in. People can even invent new practices that are not in the bag. We are more or less conscious of these choices. We can usually explain them if someone, like a visiting anthropologist, were to ask us why we did what we did.

Landscape designers also make decisions about what to include or exclude in the production of landscapes. In so doing the designer selects from the bag of cultural possibilities. The landscape architect produces material possibilities for others through these choices. Many of the possibilities for finding meaning in space, interacting with the material qualities of space and developing habits of visiting or use of specific spaces exist for people because of the work of landscape designers. The profession is a significant generator of culture.

We live in routines. The situations we find ourselves in vary less and less over time. We have made the same choices so often we hardly think about the alternatives anymore. Visiting a new place can stimulate new choices. In fact, the creative side of landscape architecture asks people to break from their routines, encounter new possibilities and invent adaptations that can then be added to their bag of culture. Designed landscapes are particularly conducive to exploration and invention by the people who visit them. Take two city parks, for example. One of them is an early eighteenth-century garden attached to a palace and restored to a form of historical accuracy. The requirements of maintaining the park confine visitors to stroll only on the walks. The second park is a late nineteenth-century functional design with large grass beds, curving walkways lined with benches, tree groups that create 'walls' around the several 'rooms' for the citizens to occupy. People cluster around water elements, buildings and playgrounds. Every square meter of the park is open and available for use. Now, let's imagine two business people intent upon having a serious conversation outside the office. The office is equally distant from both parks. They decide to hold their conversation while walking in a park, but which one? They must decide which space is the appropriate one for this conversation. There is no single right answer here. Different people would make different choices, taking into consideration the topic of conversation, the relationship between the two people, the time of day, the weather, and the amount of time they wanted to spend in the park. In other words, every social act requires interpretation. A choice is potentially a novel behaviour that could become part of a routine over time. Or, it may remain a singular event, never to be repeated. Culture has determined nothing. People have chosen how they want to act and think in that situation. They continue to do so once they get to the park and interact with the space the landscape architect has designed for them.

Culture is not merely complicated because intellectuals like to complicate ideas. Rather, it is complicated because people are complicated. Investigating culture forces us to embrace people in all their complexity. Nowhere is this better illustrated than in the final perspective I want to discuss, the distributive quality of culture. So far, our discussion has tended to focus on the individual and the locations for meaningful activity. Now I want to focus on the traits, artefacts and practices that could be shared by a group.

If you and I were to empty our respective bags of culture for each other to see, what are the odds that the contents will be exactly the same? Given the way those bags came to be filled in the first place, the following sequence of events would have had to occur: we were raised in the same home by the same parents in the same neighbourhood. We went to the same schools from early childhood through university. We participated in the same kinds of activities, clubs, religious organisations and sports teams. We worked in the same organisations under the same managers with the same co-workers during approximately the same time in the organisations' development. We shared the same intimate relationships with the same people during the same period in their lives. The same state policies, market influences and social movements influenced us. In other words, it is practically impossible for two people to have identical cultural possibilities to draw from when living their lives. Instead, a few elements of culture are distributed widely across a great number of people while the overwhelming number of elements is more narrowly distributed or unique to the individual.

An important insight of Franz Boas was that any sense of unity that the concept of culture implicitly predicts for a group is really a subjective unity, one that is constituted only in the mind of the observer, such as a politician, a market strategist, an urban planner, an artist or a social scientist. Boas did not mean to undervalue the observations of these actors. They are responsible for creating any sense of community we possess. For ordinary people, however, the unity inherent in the cultural possibilities is an abstraction, an imagined unity. Edward Sapir, a student of Boas, elaborated this further (1924), saying that ordinary people perceive a commonality of culture through relations of mutual comprehension rather than an actual sameness or identity. People need to predict each other's actions and reaction, if only partially and imperfectly. The commonness of culture reduces those moments that we are surprised or shocked by people. Given these insights, it is better to describe what people have in common as cultural proximity rather than a cultural unity.

Boas argued that culture can never be fully integrated. Integration is at best an ongoing process that cannot be completed. It was best found in styles of art and architecture, in patterns of symbols and motivation, in selective perception and valuation, and in efforts to distil distinctive character qualities from a group's historical experience. In this way, the designed landscape can be understood not only as a mediation with nature and as an interpretative canvas upon which people can invent practices, but also as an opportunity to realise an integration of cultural elements, common sense meanings and shared historical experience. An artefact as large and as important in people's lives as a green belt embodies a pattern of symbols, motivations, perceptions, valuations and distinctions that contrast with the qualities of other green belts.

Culture is not an integrated system, a text, or an aggregation of traits or behaviours. It is a population of meanings. These meanings have material forms, such as landscapes. The meanings may be expressed in speech and other forms of action, or transmitted in writing and other artefacts, but they are always things in the world, rather than abstractions (Schwartz 1978 p. 423; Sperber 1996 pp. 77–78).

There are two contradictory trends in the development of culture. On the one hand, people have unique experiences that endow them with knowledge they alone possess. On the other, states, markets and social movements impose ideas, behaviours and practices on vast numbers of people. For example, states attempt to produce a uniform understanding of the 'state-person' through residence registration, licenses, military and civil service, the census and taxation. Through advertising and displays, the market distributes images of alternative lives that products or services can make possible. Designed landscapes are part of market displays. Social movements of various kinds revolutionise the way people see the world and to reset their behaviours and practices. Social movements affect everyone, regardless of the acceptability of the ideas. Because of states, markets and movements, individuals never quite succeed in constructing separate worlds for themselves. So, too, the totally conformist state is the stuff of dystopian fiction. Most people can readily resist the demands of states, markets and social movements when those demands clash with their experiences.

The outcome of these contradictory processes is an unequal distribution of knowledge. Some people know a great deal about their world, anticipating changes and acting proactively, while others always seem to be surprised by changes. In specific areas of knowledge, we can speak of differences between experts, novices and the uninformed. These are not merely indications of differences in education. They are also differences in social power. Those who know more about a situation can command the actions of those who know less. The phrase 'knowledge is power' may be a cliché, but it is also a social reality. The social distribution of knowledge, therefore, is not merely an artefact of a process of balancing the contradictory processes of individuation and integration. It is a product of the ability of socially powerful people to hold on to their privileges. If access to the knowledge is restricted and controlled, their privilege of that knowledge is protected. This feature of culture is found throughout the world. An unequal distribution of the powerful is directly related to the uneven distribution of knowledge. In societies where the distribution of knowledge is relatively equal across all categories of persons, so, too, is the distribution of power. In contrast, where the distribution of knowledge across all categories of persons is unequal, as in our own society, the distribution of power is also unequal.

The designed environment can embody these differences in knowledge and power in society. The most obvious power feature in landscape design is accessibility. If some people can move through the space more easily than others, the design sends the message that it regards those people as more desirable patrons. Differential access is often hidden within the design under other, seemingly more desirable design outcomes. The feature of sustainability, for example, may restrict public access to some sections of a landscape during certain times of the year, but this restriction does not apply to the caretaker, the caretaker's supervisor or the visiting landscape designer from another city. Historical reconstructions restrict access in the name of preserving the details of the design. In addition to restricted accessibility, landscapes can embody expert knowledge in the form of hybrid botanicals, historical references in the land and bed forms, or simulations of specific ecologies. Without signage or human guides to instruct the visitor what to look for, the expert's efforts are often hidden from the public. If the designer's work is not transparent to all, then for whom is the work intended? Finally, differences in knowledge can lead to contests between different people in defining the role of a landscape in their lives. Such contests are particularly acute in situations where different sets of life experiences share the same landscape, as in the ethnic diversity of large cities.

I focused this discussion on mediation, interpretation and distribution because I have found these to be the most relevant perspectives for my own work in understanding the role of landscape in the cultural lives of people. I have tried to find examples that would speak to landscape designers. These ideas prepare you to understand the areas of basic knowledge in the cultural anthropology of landscape that I will now discuss.

## Areas of basic knowledge in cultural anthropology

> What anthropology can contribute to the study of landscape is first and foremost the unpacking of the Western landscape concept, but also a theorising of landscape as a cultural process that is dynamic, multi-sensual and constantly oscillating between a 'foreground' of everyday experience and a 'background' of social potential.
>
> (Hirsch and O'Hanlon 1995 p. 3)

The basic building blocks of the cultural analysis of landscape are bound up in four concepts: space, place, site and locality. In a classic article about how residents of New York City describe their apartments, Linde and Labov (1975) discovered that all the descriptions fell into only two types, The first type is some variation of the following: 'The bedroom is next to the kitchen'; the second type sounds like this: 'You turn right and come into the living room.' These are labelled, respectively, the 'map' and the 'tour.' In this particular study, only three per cent of the people interviewed chose to describe their apartment using the 'map' style. All the rest chose the 'tour' style. This study inspires anthropologists to consider all the ways that people experience landscape through language. When the experience of a landscape is put into words, people reveal the meaningful elements with clarity and precision. To hear this, however, one has to know what to listen for. The areas of basic knowledge of cultural anthropologists with respect to landscape consist of a series of general statements about what to listen for.

These two types of descriptions, the map and tour, illustrate a long-standing and critical difference in how people in the Western tradition understand our environment: seeing vs. going, presenting a tableau vs. organising someone's movements. These ways of describing an environment coincide with the distinction between the opposed terms 'place' (*lieu*, *Ort*,) and 'space' (*espace*, *Raum*). The terms are opposed to each other because they do not co-exist in experience. One is either attentive to place or one is moving through space. Place is static, the being-there of something dead and unchanging. Space is dynamic, the process of eventually arriving at a destination (a place) by a living person. Space cannot be separated from movement and place never moves. There are as many spaces as there are distinct paths people can take to attain a place. Places, however, are finite. They become defined by memory and imbued with meanings, both mundane and symbolic.

When spaces and places bear a coherent relationship with each other, such that spaces lead to places and a series of places define a space, we can speak of a 'site' (*site*, *Anlage*) of human action. A landscape is a site. Sites have several features that are worth noting. Descriptions of sites, like the description of an apartment, assume a relationship between the spaces and places much like the 'map' type of description. While this map may remain un-spoken when the site is described, the resulting itinerary could not exist without it. The description of the site includes effects ('you will see . . .'), limits ('there is a wall'), possibilities ('there is a door'), and directives ('look to your left'). This chain of spatial descriptions produces a representation of the spaces and places that people can narrate to each other, bringing the site into social existence.

When a site comes into focus in people's lives it simultaneously creates a 'locality' (*endroit*, *Ortschaft*). The manner in which people narrate the features of a landscape to each other is the landscape's locality. The term describes the marking out of elements that separate this site from other, especially contiguous sites. Locality is a social distinction, a way of evaluating one site as distinct from others. It is not dependent on the un-spoken map, although people often describe localities as the sum of their constituting places. This 'story' of the locality is a narrative that integrates the stories of the separate places and establishes them as a single spatial entity. In the example of the apartment description, the apartment becomes a locality of our private life because it is comprised of the bedroom where we sleep, the bathroom where we wash, the kitchen where we prepare and eat our food, and the living room where we bring guests into our private lives. Localities can scale down to the very small, like the apartment or the café, or up to the very large, like the metropolis or the nation. Localities can serve as both a container for human actions and as a license for action, permitting or requiring some behaviour while forbidding or sanctioning others.

When anthropologists use categories such as place, space, site and locality to frame their analysis of landscape, it enables us to focus on the creative forces that integrate them into a single social experience. This is only the first frame of what a cultural analysis of landscape makes possible, the phenomenological. With this arrangement of basic parts before us, we can now explore three additional frames of analysis: spatial discourse, social production of space and spatial practices.

## *Spatial discourse*

Place is a location of elements that we find meaningful. It might be an address, a park, a battlefield, an office building where we work, or a beach we go to in our minds when we want a little peace and quiet. Place does not have to be real. The most satisfying places combine elements of real locations with imaginary ones. Place is difficult to produce. It lies at the intersection of discourses and productive processes. It is the stuff of history, memory and mythology. One experiences place through memory, narrative and monument. One becomes attached to places emotionally or intellectually through associations that one builds in the mind between memories, narrative and monuments.

Place enters all mutual understandings of meaning. Like time, identity and event, it becomes a dominating site of symbolic production (Sahlins 1978 p. 211). That is, the qualities of a site can generate new meanings in addition to serving as a repository for established meanings. To the extent that a person is paying attention to the environment, the 'I' that is moving from place to place reinterprets that awareness through categories of memory, history, civility, spirituality, practicality, and so forth. These categories are not unique to the individual, but commonly known among local residents. By participating in this act of reinterpreting place within a commonly known category of meaning, the person is adding to the category. It is almost as if there were a silent conversation between people where each contributes a bit of meaning to the topic, and in turn receives the interpretations of others. We name the ongoing conversation between people that elaborates upon this mutual understanding of the social experience of place a discourse.

The discourse on place applies to both the most modest and domestic of sites and the most grandiose and ambitious. The homeowner considers how others will judge the condition of the property. As the social standing of the family changes, so does the thinking and investment in the condition of the property, always with a view to how the changes will be perceived by others. Politicians produce elaborate and complete representations of their vision of the metropolis, believing that they are responding to the values of the people who elected them. As politicians succeed each other in power, they appropriate a specific set of public landscape design possibilities to represent their vision. The previous group's forms continue to exist along with the new models. The newer forms borrow design ideas from the old, sometimes in polite emulation of them, sometimes to invert and transform them. To accomplish this, homeowners, designers, nurserymen and politicians must develop a common language of design. They do so by borrowing from the existing, ongoing discourse on social space: what is the boundary between the private and public in metropolitan life? How do family, community, the municipal agencies, health and safety, or the market understand this boundary differently? What is the best way for actors to mark their boundaries? This conversation connects the spatial forms with the vision of metropolitan life the ensuing landscape will represent. There are as many voices of design as there are visions of what urban life can be.

Among the writers on historic preservation practice, there is a saying that every centimetre of pavement has a history, but not every history is worth preserving. This is a good example of the general principle that places can be created through the spatial discourse, but also through non-discursive actions. To understand this distinction, consider that in the course of an ordinary day there are moments when you are aware of your thoughts, actions and habits in relation to others. There are other moments when your thoughts are within yourself, private moments when you are alone or even in public when you are lost in your own thoughts. These moments are not part of the ongoing discourses that connect you to others through a system of mutual comprehension. You are living in a non-discursive moment. You don't care if others comprehend what you are thinking or not. It is enough that you comprehend it. Place-making also has its non-discursive modes. Place-making is about seeing. The discursive and non-discursive modes of seeing refer to our understandings of place as part of some common narrative or as a personal, unshared memory or insight. Thus, I have my favourite table at the coffee shop, or a preferred parking place at work. In the course of a day, our encounter with places varies between these two modes.

There are several areas in which uneven distributions of knowledge influence the direction of spatial discourses. Among experts, design regimes can form. These are a set of rules through which experts over a particular period of time impose and enforce design standards. This can occur in all areas of design and planning, including scientific research, election campaigns, zoning, or landscape design. The effect is to shape the discourse around such design and planning. It becomes increasingly difficult to legitimately introduce topics or support ideas that run counter to the design regime. With diminished diversity of ideas, the regime becomes increasingly dominant in people's minds. Patronage and legislation follow the common sense. Everyone wants their place to conform to rules. Eventually, place becomes unthinkable unless it is couched in terms of the regime's design rules.

Access is another way in which uneven distributions of knowledge shape discourse. Keith Thomas has documented a movement in England in the eighteenth century to collect and catalogue the plant knowledge of English villagers (Thomas 1996). This effort followed in the wake of the publication of Linnaeus' *Systema Naturae*. These plants often bore local, colourful names, alluding to local stories or events, or to side effects if eaten. The same plant could have different names in villages a few kilometres apart. The naturalists quickly renamed the species without bothering to inform the people who had gladly assisted in the collection and identification of the plants. Within a short time, the local names for the plants were competing with the official names. This would not have mattered to the isolated farmers, but their world was quickly changing and their contacts with outsiders increased. Cosmopolitans educated outside the district, such as clergy, doctors and other professionals who commanded respect, would call the plant by its official name, often to utter confusion of the locals. The experience of locality itself was undermined. The community could no longer identify its members through the names on local plant varieties. Finally, the knowledge of proper names was locked up in universities and research centres where rural folk were unwelcome, preventing them accessing the very knowledge that they had helped to create.

Lastly, uneven distributions of knowledge can result in contestation, open conflict and resistance within a discourse. More than mere disagreement about the meaning of a place, knowledge distribution issues can lead to counter discourses that can unseat design regimes and restricted access. They can even result in a complete re-evaluation of the meaning of a place. A memorial square dedicated to the victims of fascism, a nudist beach, the re-zoning of a derelict cemetery for a housing project, the banning of skate boarding from a public park are all examples of discourses on place that have led to contests between members of the community who support the action and those who are opposed to any form of the action. Differing sets of experience leading to different knowledge sets creates the imbalance. This can split the community, leading to destructive actions. Such conflicts are thorny issues for designers because they never occur at convenient times in the project cycle. Yet, time is the critical variable in the effort of the community to rebalance the discourse.

The spatial discourse produces places through an interpretation of sensory impressions within existing categories of interpretation, design regimes, systems of access and conflicting understanding. The products of this conversation are a set of conventional understandings that describe the commonalities and differences between sites. These are meaningful to analyse because they contrast with those actions that actually move earth in the production of new space. In everyday experience, we do not distinguish between constructing places in earth and sky, and constructing them in our imaginations. Teasing apart this difference is one of the contributions of cultural anthropology to the study of landscape. The next section, however, will focus on constructing places in actual landscapes.

## *Social production of space*

Societies with professional landscape architects have one thing in common, as societies: differences in social power between individuals enter in all human relationships. This is commonly understood as the social structuring features of race, class, gender, expertise and physical ability. The social production of space is a research focus that concerns itself with the production of spatial objects that privilege and reinforce society's distinctions. Landscape architects are among the producers of social space. You are implicated in the question of how do we as a society acquire locations that are identified with specific classes, races, genders, expertise or abilities?

The French philosopher Henri Lefebvre is most closely associated with this question. He, in turn, influenced two contemporary researchers, Edward Soja (1989; 1996) and David Harvey (1989a; 1989b; 2001). Lefebvre's great insight is that 'space is a social product – the space produced in a certain manner serves as a tool of thought and action. It is not only a means of production but also a means of control, and hence of domination/power' (Lefebvre 1992 p. 26). Take, for example, a baroque palace garden. Lefebvre would argue that the look of this garden style is neither accidental nor separate from the model of society in the mind of the patron who paid to build it. Instead, everything about this style is consistent with that model of society: the regulation of social orders in the geometrical layout, the control of nature in the topiary, the grandeur of the nobility in the scale of the garden and the aristocracy's rule through surveillance of the lower orders revealed through the vistas of palace and garden. A specific designer produced this palace and garden.

Lefebvre argued that every society, which he understood through the Marxist concept of mode of production, produces space that mirrors the view of the dominant class, race and gender. He gives the example of the city in the ancient world. It was not a mere agglomeration of people and things in space. Its arrangements of parts in space required a specific way of moving about the city, the congregating and dispersing of groupings of *paterfamilii*, slaves, women, religious workers, soldiers, citizens and strangers. The social space produced through the filter of power simultaneously produces behavioural practices and intellectual outcomes that reinforce the existing social order. The intellectual climate of the city in the ancient world arose in spaces designed to cultivate abstract conversation. Those who congregated together could converse, while others would be left out of the conversation. Civic space was privileged space.

Furthermore, Lefebvre argued that a social movement aspiring to power, but not producing its own space, would remain an abstraction that will never escape its ideological paralysis. He criticized the Soviet urban planners of his day for failing to replace the modernist model of urban design with a space wholly defined by socialist arrangements and practices.

Lefebvre's vision of the social production of space operates below our consciousness because, before his analysis, there was no conversation about the ways that the differences in power in society were made concrete in the planted and built environment. Lefebvre's work is an example of how hegemony can be exposed through analysis. Hegemony is the common sense, everyday practices and shared beliefs that provide the foundation for domination by the powerful (Gramsci 1992 pp. 233–38). Hegemony operates below people's consciousness. The thousands of little decisions we make every day, such as what shoes to buy, what means of transportation to use, what events to pay attention to, comprise the hegemony of contemporary life. We believe we have freedom of choice, when, in fact, our choices have been circumscribed for us and we actually choose from a predetermined set of options that represent the most desirable outcomes for the system as a whole. It is this system that maintains the differences in power. In this way, the dominant class, race and gender shape spaces by limiting the range of choices in which designers can work. Lefebvre demonstrates that reducing the complexity of space limits choices, directing designers to focus on some aspect of spaces, but not others. In its full complexity, we can see three distinct aspects of space.

The first to be produced is the registry (*cadastre, Kataster*) of surveyed parcels, which Lefebvre calls the *absolute aspect of space*. This is the ground plan on which all further acts of production will unfold. The parcels can be zoned for different uses, filled with roads and services, owned or transferred by and between private or public interests, or bounded in ways that inhibit or expedite further production. Absolute space is the landscape architect's drawing of the ground plan, the space of design and planning, and the space of governmental registration, and surveillance.

The second aspect of space is the everyday experience of the space and the behaviours of the people who inhabit it, which Lefebvre calls *lived space*. This includes the places that the spatial discourses produce out of memory, history, civility, spirituality, practicality, and so forth. It is the habitual paths we take between routine destinations as we move through our days. It is the street where we live, our favourite pub, the park our children play in and the cemetery where our loved ones are buried.

The third aspect of space is comprised of structures that channel design and planning, on the one hand, and lived experience, on the other, toward specific socially defined ends, which Lefebvre calls *representational space*. He sees these structures as distortions from some hypothesized ideal that sets out to grant privileges of access, use and disposal of specific spaces to some people, while simultaneously denying this privilege to others. Every space, he observes, includes a set of rules for containing a limited set of activities and a set of rules for permitting those activities. When challenged, the authorities who help to enforce these structures deflect criticism by alluding to the requirements of absolute space ('It's not zoned for that'), or the custom of the anonymous, local people ('That sort of thing is not tolerated here'). As a result, the insistence of a dominant group to maintain its privileges is made invisible, and thereby, hegemonic.

This political analysis of space is pertinent to the study of landscape by cultural anthropologists because it begins to answer the question 'For whom is the landscape being built?' The question is double-edged because it can refer to both the owner of the space and the user of the space. Landscape architects ask this question with every project. Much of the programme the designer follows is concerned with user needs. The idea that there is a category of person that we can call a 'user' or an 'owner,' and a set of behaviours that we can label 'needs' is an example of the hidden forces that shape design. The political analysis of space is pertinent to the landscape designer, if only to make visible the forces shaping the design.

The social production of space directs our attention to the ways that differences in power in society distort our actions in spaces, both public and private. An example of this distortion in private spaces came to my attention while doing research on domestic gardens in a suburb. The sustainable gardening movement was in its early stages. One enthusiast had decided to tear out the early twentieth-century house garden beside his house along with its fruit trees and well-kept grass lawn lined with flower beds. All of the neighbouring houses kept up gardens of this style. As representational space, house gardens privileged private property ownership. The landscape consistently reproduces planted property markers that enclose an outdoor living space. In place of these features, our sustainable garden enthusiast installed a small pond with no natural water source and plants from the neighbouring hills. Then, instead of tending to the growth of these features, he let the garden develop in whatever way 'it chose.' As representational space, his garden privileged the subordination of property to the processes of nature and the trans-species ethical community in which dandelion, nettle and mosquito had their place in the balanced order of the world. The resulting conflict of representational spaces was swift, dramatic and catastrophic. The government happily sided with the property-oriented neighbours and a park department backhoe made short work of this experiment in sustainability.

An even more extreme example of producing spaces is found in the construction of emptiness. Empty lots may be devoid of certain recognizable constructions, but are often filled with images and practices. As described by Gary McDonogh (1993), there is a particular anonymity available for people in spaces labelled as empty. The emptiness can be nostalgic, a place where a personal landmark once stood. It can be a deviant place 'used only by dogs, drug addicts and malingerers'. It can be a boundary zone between the acceptable and unacceptable behaviours, a 'no man's land' where upright citizens do not go. It can be intentionally fallow, promising, 'a future of speculation and development', a street of 'burned out or boarded up houses in a slum neighbourhood'. The phrases in quotes are references to discourses on urban life that are widely experienced (Ford 2003). The same social forces that produced other spaces produce empty places.

The construction of landscapes is never politically neutral. Each movement of earth, placement of beds and walks, and even the choice of vegetation result in some people maintaining privilege while restricting the actions of others. It is against this background of produced spaces that we turn to our final area of basic knowledge: the practices that help us differentiate between the ordinary and the extraordinary in our understanding of place.

## *Spatial practices*

Practice, practical sense and practical consciousness all refer to going about our everyday business. The focus on practices reveals how our bodies are transformed by our contact with different kinds of places. Places are one of the channels through which this transformation occurs. This might be as ordinary as holding an umbrella as we walk down a street, or as singular as wearing a wet suit and breathing apparatus to explore a sea bottom. Spatial practices are what we do when we are in a particular place. For the cultural anthropologist, the focus on landscape practices answers the question 'What meanings do people and designers give to a specific site?'

Practices are about doing what is expected and avoiding what is unexpected. Of course, 'what is expected' varies as we move from place to place and through time. Walking down a street holding an umbrella while the sun is shining will attract more attention than doing so when it is raining. It is often easier to grasp a practice by referring to actions that shift from the ordinary (sunshine) to the extraordinary (rain), or vice versa. The cultural anthropologist's interest in spatial practices lies in understanding how place affects our activities and, hence, our way of being ourselves in places. The landscape architect should be interested in spatial practices. They represent our best guess of how the design will evolve over time.

For example, in my city, Chicago, there is a 22-kilometre long strip of park along the shore of Lake Michigan, known as Lincoln Park. My fellow citizens use the park very often throughout the year, but each uses it a different way: strolling alone, in pairs or in groups, jogging, cycling, roller-blading, walking dogs, skate-boarding, scootering, sitting on benches, lying on the grass, picnicking, playing Frisbee with other people or dogs, playing volleyball, badminton, swimming, playing on the beach, fishing off a pier, kayaking, sculling, canoeing and, in the case of the grounds crew and police, driving vehicles and working. Which activities people choose to do in the park are, first, particular to the possibilities the place contains; second, particular to skills and inclinations of the people involved; and third, restricted to those possibilities that are appropriate to the park and the people who are around the activity at any given point in time. To illustrate this point, consider the following: drinking alcoholic beverages is officially prohibited in the park. Yet, anyone enjoying a day at the beach or a picnic under the trees is likely to be drinking beer or wine. It is understood by the visitors and the police alike, that drinking is tolerated as long as no one complains and no one is too conspicuous.

When the activities chosen are particular to skills and inclinations of the people involved, the implication is that such skills and inclinations are not even distributed across a population. Different groupings of people are more likely to be interested in, say, jogging, while others finding jogging a senseless pursuit and are more involved in dog-walking. The French sociologist Bourdieu has written extensively on the class basis of everyday practices. He would argue that there is really less choice in these activities than anyone suspects. Instead, the activities we enact in places are narrowed by the qualities of age, gender, class and education. While exceptions are certainly possible, he demonstrated in several studies that these qualities predict our actions (1998 pp. 1–13).

Practices are also limited by convention. The place's designer seeks a mutual understanding of possibilities for action with users, but cannot anticipate all the understandings users may bring. Cultural anthropologists use the Greek word *topos*, place, to describe various combinations of real and imaginary places that represent fundamental differences in these mutual understandings. *Utopia*, literally 'no place', is a literary genre for imagining a society whose practices strike the writer as more satisfying. *Dystopia*, on the other hand, is a 'sick place' where people behave in a far less satisfying way. An ordinary place can be described by the term orthotopia, while a place that has something truly extraordinary about it is a heterotopia. Places that have no inherent meaning at all are *atopia*, or non-places. Finally, it is possible for us to create our own places, *autotopia*, where governmental regulation is ignored. Each of these places engenders different possibilities for action.

## Orthotopos

Ordinary places develop when people relate to others in public with as little friction as possible (Gehl 2001; Whyte 1980). In ordinary places, we read the possibilities for action by observing the people who are already in the space. Examples might include a street, a café, a bank lobby, or a classroom. Even strangers passing by, whether they indicate each other's presence or not, read each other and form a silent, momentary relationship. Ordinary places make the practices of the locality visible.

Ordinary places tend toward the invisible, but never really disappear, such as the street we walk down to get from a bus stop to our office. That street has all the qualities of a place. At another time and circumstance, it could be a destination, perhaps the ideal place to participate in a public demonstration, or the meeting place for an intimate rendezvous. Short of such circumstance, it remains partially invisible to us as a place.

Ordinary places contain the things of everyday experience. They gather these things. Using the example of the street between the parking place and the office again, we can see that the following things are contained there: pavements, cars, debris, dog faeces, beggars, signs and pedestrians walking towards us, with us and entering from doorways and from between parked cars. We are paying attention to all of these things. We must do so to avoid collisions. They bring about actions on our part that make the movement in the place carefree: turning our bodies to pass by three people in group who are talking to each other and taking up more than the usual space on the pavement, shifting direction to avoid someone entering from a doorway on the right, or slowing down to avoid stepping on the heel of the person walking in front of us. The actions are perfectly suited to this place, as indeed all orthotopia engender the most appropriate action responses from us. These actions are conventional. We learn them as children and practise them without thinking all our lives. These actions reduce conflict by making everyone's trajectory predictable to everyone else. Imagine the chaos that ensues when, say, a drunken man stumbles out of pub onto a busy pavement and is too slow to make the kinds of quick adjustments that allow sober people to walk down a pavement. The hallmark of an ordinary place is that it constantly reminds us that we are embedded in a social fabric in which who we are matters less than how we enact the conventions that reduce conflict. This is the primary characteristic of orthotopic spatial practice.

Orthotopia, like all unmarked features of our experience are most useful for what they tell us about non-ordinary places. Our flats and houses, offices, classrooms, dining halls, parking facilities, neighbourhood food shops and the paths we take to get back and forth between them are all ordinary places, while highways, shopping centres, parks, football stadia, theatres and airports are not.

## Heterotopos

What exactly makes a place extraordinary? De Sousa Santos has proposed that something becomes extraordinary when it results in a radical displacement within the same place, such as the movement (actual or imagined) from the centre to the margin, that allows us to view the centre from afar, and thus begin to understand what the centre cannot or will not contain (1995 p. 481). The extraordinary is bound up with the place where we experience it. Something happens to us when we are in such a place that makes us see things differently and thereby, act differently. Heterotopia are extraordinary places. They concentrate the practices of the locality intensely, permitting us to become conscious of these practices for the first time.

Extraordinary places must be contiguous with ordinary ones. They are separated from the ordinary, marked in significant ways, as if the perceptions they permit would be slightly dangerous, or at least provocative, if allowed to leak out into ordinary spaces. Heterotopic sites reflect everyday experience, but do so in a way that is highly selective. This selection marks these sites. Ordinary sites have minimal specification and demarcation. We know where we are, but it is not particularly noteworthy. Ordinary places may not even have a name. Even though they may gather important personal and social meanings, such places retain their ordinariness. Heterotopic sites are the 'other' places that exist within the landscapes of our daily lives. We enter them or not, freely or under duress, and exit them again to go about our business. But when we are in them, the shift in focus is palpable and transformative. The possibilities for action are singular and potentially subversive of social order. As you might imagine, cultural anthropologists have a particular keenness for exploring heterotopias whenever we encounter them.

Foucault defines heterotopias as 'real places – places that do exist and that are formed at the very founding of society – which are something like counter-sites, a kind of effectively enacted utopia in which real sites, all the other real sites that can be found within cultures, are simultaneously represented, contested, and inverted' (1986 p. 22). He identifies six features that separate an extraordinary place from an ordinary one. He describes these features in a lecture given in 1967 called 'Of other spaces' (1986). These include (1) how the people project their understanding of nature in these places, (2) how they express the fulfilment of some utopian ideal in these places, (3) how people refer to unresolved social issues in these places, (4) how they transform time in these places, (5) how people create boundaries to separate the place from ordinary places, and (6) how they close off, camouflage or mystify everyday experience so that the experience of the place can exist apart. These sites must be seen as absolutely different from all the sites that they reflect and speak about. This contradiction between the need to be different but linked to the ordinary gives the experience of heterotopias their appeal, their teaching quality. They are neither utopic nor abstract. They are fully formed, real places that are designed to illustrate an ideal. That ideal is the key to the extraordinary meaning of the site and the spatial practices of people when they occupy heterotopia.

All such sites have a quality of social universality. The ideal they are trying to illustrate is one that is believed by the people who built the site to be a common experience of all people. The site should be a common place, in spite of its special qualities. Unique or temporary sites do not qualify, unless their uniqueness or temporariness is intended to project a universal ideal. There are two ways that this universality can be realized. They can be 'places of crisis', such as funeral homes or hospitals, or 'places of deviation,' such as asylums and homeless shelters. These are Foucault's name for the universal qualities. The ideals they project are those of shared life-cycle crises and the containment of deviance. There are probably as many such 'places' as there are ideals that communities have identified as worthy of projecting.

An urban park, for example, is heterotopic because it attempts to illustrate an ideal of nature in the city. Nature is a universal experience, a commonsense category that describes all of the aspects of reality that humans feel are beyond their control. Thus, the most exquisite of human artifices, the built form of the city, is contrasted with the world of plants, animals and climatic forces. The power of humans to build is contrasted with nature's power to grow.

Such sites have identifiable functions. Foucault suggests that the cemetery best illustrates this heterotopic practice. In periods of stronger religious belief this site was centrally located. Concern for the integrity of the physical remains was absent. Cemeteries could be small and internally undifferentiated. Under conditions of weaker religious belief, the growing concern for the integrity of the remains requires larger areas, systems of streets and hierarchies of neighbourhoods. The identifiable function is seen in the way the design decisions reflect the concerns and practices of the community.

An urban park could reflect this second feature in a variety of ways. An old palace garden could be converted to a historically accurate public park to reflect the community's need to connect to its history, perhaps as a reflection of its sense of grandeur as its prominence is waning. Or, the park could be designed to emphasize its accessibility, thus embodying ideals of pluralism, diversity and democracy, even as prejudice and disenfranchisement increase.

Such sites resist being reduced to a single meaning. They are multi-vocal landscapes that convey different things to different people at different times in the same community. Foucault offers the example of the Persian garden reduced to a design on a carpet that can be carried to the Mosque for prayer, but still exemplifies the geography of heaven. The carpet is simultaneously a carpet, a model of a garden, the garden itself, a model of heaven and heaven itself.

An urban park is simultaneously a place to walk in peace and quiet in the middle of the busy city, a playground for children, a rendezvous for lovers, a private place to hold a business meeting, a gallery for flower enthusiasts, a laboratory for urban landscape practices, a model of gardening for home gardeners, a place to experience nature and nature itself.

Such sites are *heterochronic*. Just as space can be orthotopic or heterotopic, so time can be ordinary or extraordinary. Heterotopias break the continuity of ordinary time, as well as that of space. This is achieved through the accumulation of meanings over time. The contemporary meaning of the place and the aggregate of its past meanings are indistinguishable. The museum and memorial square become heterotopias through their ability to suspend the passage of time. The temporal break also can be achieved through the creation of the fleeting, the transitory or the precarious. An example of this is the circus that appears overnight in an open field and disappears again a few days later. In domestic gardens this heterochrony is served by the contrast between the annual life cycle of botanicals and the social conventions of metropolitan time schedules.

Parks gather memories of communal events, celebrations and crises that are remembered differently by different groupings within the community. Some remember an event as the community's greatest triumph, and others remember the event as its greatest shame. While this event could just as easily have transformed any ordinary streetscape, it may have been specifically sited in the park because of its heterotopic character. Parks tend to gather extraordinary events over time, preserving threads of different experiences, both personal and communal. Like museums, they freeze time as all memories are remembered as equally contemporary.

Such sites are neither completely inaccessible, nor are they completely open. Instead, entry is either compulsory, as with the army barracks or the prison, or it is available only through permission from some kind of authority. Foucault identifies 'places of purification' as heterotopias that achieve their extraordinariness primarily through the manner of their control of access, such as the Moslem hammam, the Jewish mikva, or the Finnish sauna, along with places of sexual intimacy, rooms marked 'Authorized Personnel Only' and drug houses. Most domestic gardens have a fence and a gate. Opening can refer to sight as well as site. Some landscapes can only be seen from the inside outward, while others are open to viewing by passersby.

The urban park has its own system of opening and closing, beginning with the signage at its gates stipulating whether the visiting hours are limited. Such parks have gates, even if these are merely cuts in a hedge wall. Streets, pavements and sometimes fences bound them. More importantly, we see them from either all vantage points or from only specific vantage points.

Finally, such sites link to the ordinary places in society. The nature of the link can be as complex and multi-vocal as the sites themselves. The link creates an illusion that the site is not what it appears to be. The same aspects of everyday experience that seem to be closed off, shut out, mystified or camouflaged by the site are precisely the ones a person is most aware of. They are conspicuous in their absence. To be effective fantasies of a society reduced to its universal qualities, these sites must encourage visitors to suspend disbelief, as in a theatrical performance. They do so by excluding those social realities that contradict the idealized view enshrined in their design. A theme park on the scale of Disneyland is a prime example of exclusionary linking. Visitors to such sites can choose to accept the camouflage, agreeing to suspend disbelief that an ideal world coexists with the real one they occupied before entering the park. They exchange these worlds for a satisfying, momentarily ordered meditation on the contrast between the ideal with the real community. Such linkages are immediate and self-evident to the visitor. They are an integral part of the experience of the site.

An urban park closes off access to the people that are deemed upsetting to the decorum of a public place: the rowdy, the homeless, the derelict and the deviant. To the greatest extent possible, it shuts out the sound and sights of the surrounding city, as if to preserve the illusion of an all-embracing nature. In doing so, it mystifies the relationship between the rural and the urban, the condition of nature in the city and nature in nature, and the construction of nature by people and the unintended growth and distribution of plants. Finally, the urban park camouflages its teaching function by never directly referencing the ideals it was designed to project. These can only be glimpsed indirectly, strengthening the power of the design to communicate these ideals without contradiction or contradistinction.

The difference between the ordinary and extraordinary is not one that community members themselves will easily make. They, too, have to be shown the features that distinguish one site from another (Rotenberg 1995). A single site need not emphasize all of these features in order to qualify as a heterotopia. The task belongs to the analyst to demonstrate that a site qualifies through the practices of the people who visit it. This is most often the case when the analyst wants to reinforce the teaching quality of the landscape for the community.

## Atopos

Webber (1964) first described what he called a non-place in the mid-1960s as 'a sprawling, polycentric landscape characterized by the steady erasure of locality by the generic forms of a diversified yet ultimately homogenizing market culture' (Rutheiser 1997). Sorkin (1992) and Zukin (1991) have also described several efforts at creating these non-places. More recently, Marc Augé has described these atopia as two complementary but distinct realities: spaces formed in relation to certain specific urban activities, usually transport, transit, commerce, and leisure, and the relations that individuals have with these spaces (Augé 1995 p. 94).

A public bus is not an ordinary place, but neither is it extraordinary. One bus is very much like another. Something meaningful can happen to a person on a bus that might be the basis for place-making, but that particular bus, its number, its peculiarities among other buses, will not be part of the memory. Rather, the event took place on 'a' bus. As for the other people on the bus, their relationship to each other is the same as their relationship to the activity they are engaged in: solitary and anonymous. The bus is an atopos, a non-place.

A bus has the characteristics of a space. One moves on a bus, even as the bus moves through the streets from bus stop to bus stop. In his analysis of this movement, Augé notes that the stops of the Paris metro inevitably reference monuments and historic districts of one sort or another, in other words, places. This is one of the features of atopias that make them interesting to think about. They are in the same position as ordinary places even though they are devoid of the memory of relations to the people and things. We do not become emotionally attached to a bus. They are non-places because the only relationship possible is a contractual one, represented by the ticket and the authority of the driver. Unlike the conventionality of the street, the contract of a bus ride is negotiable. A range of behaviour is possible, as determined by the driver and the other passengers. We all have stories of improbable behaviour that was tolerated on a bus, and that would never have been tolerated on a street.

The bus ticket is a contract between the transit authority and a single rider, not a group or community. You are truly alone on a bus. Can one undertake a more solitary activity in public? Even though someone may be sitting next to you, no interaction is expected. You can have as much space to yourself as the design of the seat and the girth of the passenger next you will allow.

Augé sees the spatial practices of atopia increasing in our cities. Non-places are closely associated with what scholars call the global neoliberal regime (Brenner and Theodore 2002; Hackworth 2007; Swyngedouw *et al.* 2002). Investors, planners and regional governments enact this regime by transforming large, and from their perspective, under-utilized sections of urban centres to spaces suitable for investment and profit-taking. What follows is an onslaught of commoditisation, hyper-gentrification, cultural deracination, corporate takeover of municipal services and spiralling costs. The specific targets are transport sites (airports, train stations, inter-city bus terminals), transit sites (taxis, cars, buses, subways, escalators), commercial sites (of the chain store, franchise restaurant, mall outlet variety) and leisure sites (the theme park, urban attraction, 'must see' vista, or staged festival). These have their parallels in ordinary spaces: the shared ride using the personal cars of each rider in rotation, the corner 'mom and pop' grocery where names and greetings are exchanged with each transaction, and the regular Saturday morning chess game in the park with the same three people for the last five years, weather permitting. Not only is the former list contractual and solitary, while the latter are consensual and social, the scales of the non-places are large enough to accommodate many more people.

The creation of large, open spaces in city contexts generates a marked contrast with the local tolerance of crowding. Such spaces are produced according to formulas, such as the faux nostalgia of neo-urban landscape design, or the adaptive reuse of historically preserved/conserved landscapes. They are meant to generate income. The people who move through them eventually become numb to such places, responding increasingly like programmed robots; they act only according to expectations. Atopia represents the intrusive presence of regimentation and aesthetic domination (Herzfeld 2006).

The quintessential atopos is the shopping centre. From the moment one enters the parking lot to the moment one leaves again, almost all of the relations are solitary and contractual. There are ordinary places mixed in, such as the walkways between shops and the dining sites. These are all the more invisible because of the overwhelming difference with the atopic parking lot and commercial sites. Selecting a parking space involves a set of spatial practices almost too complex to describe here. Each space seems to have a particular value attached to it, the spaces closer to an entrance having a higher value than spaces farther away. Spaces where the adjacent spaces are empty have a higher value than those where the adjacent ones are occupied. The value that one achieves by parking the car gives one a moment of self-knowledge: ordinarily it is something on the order of 'achievement of one's goals often involves compromise'. What is important about the games we play with ourselves over parking spaces is the solitary, exclusionary, anti-social moment that parking engenders. There is a parking contract: one cannot park in two spaces at once; one must park fully within the space and not permit the car to stick out into the driving lane; and one must open the doors so as not to dent the car in the adjacent space. More could be said about parking and the negotiation of actions with drivers in other cars, all of which is different from, but analogous to the process that takes place on streets. However, it is time to enter the centre and do some shopping. Here, too, the solitude that began in the parking lot continues. We are preoccupied with our own person, our body, how our body looks, how our body is reflected in the bodies of others, and how the bodies of others reflect on the value of shopping at this particular centre for our body.

There is a contract to shopping in these stores. It varies slightly in different communities, but the clerk and especially the manager is in the position of the bus driver, interpreting the relationship of the store to the customer to permit a flexibility of actions than exists in the street. This is true of all stores, but shopping in such centres is unique because they represent a concentration and variety of stores that would rarely be found on a single street. Their design is closer to that of an entire town or neighbourhood. The flexibility concerns practices that reduce the risk to the shopper, like trying out or trying on a product, comparing prices between stores, negotiating alterations and negotiating price. One leaves the shopping centre with one's purchases having confirmed one's sense of self. This over-arching valuing of the experience of place as an experience of self is the primary characteristic of atopic spatial practices.

## Autotopos

The most recent development in understanding spatial practices emphasizes the role of non-expert, ordinary residents in the construction of places. This is slightly different from the architectural historian's category of vernacular design. Autotopic places are most often constructed in opposition to some sort of governmental regime, such as zoning, district covenants, lease agreements, building codes, and official 'taste.' The most concentrated form of the autotopia is the squatter settlement. Using whatever materials are at hand and the technical ingenuity born of necessity, the autotopic place is slightly dangerous, exciting and democratic. Autotopia are not confined to impoverished populations. In any community where there is an extension of voting rights with restricted access to property rights, rights of residence and/or limited economic mobility, there is the potential for people to take places into their own hands and appropriate them to their own ends. Holston calls this insurgent citizenship (Holston 2008). This is an involving area of research as anthropologists attribute places to these autonomous spatial practices.

## Research approaches in cultural anthropology

Cultural anthropologists have two different styles of research: ethnography and ethnology. Ethnography is the set of research practices that culminate in a description of the lives of people. The description can be targeted to a specific condition, problem, region or period. It can involve a single site or multiple sites. Ethnology is the analysis of the distributions and patterns that emerge when the lives of people are compared in different conditions, through different problems, or across different periods or regions. It always involves multiple sites.

### *Ethnography*

The primary research practice in ethnography is long-term fieldwork. This involves living with the group of people, learning their language and adjusting one's behaviour so that it is predictable to the community you are living in. All of this is accomplished with a high degree of self-consciousness, note taking and question asking. By long term, most anthropologists would agree that multiple years of commitment to a community are necessary, though this is often interrupted with trips home. Fieldwork has a strong linguistic focus. It tends to give priority to forms of local knowledge and to localised forms of expressing that knowledge. Other tools include the formal interview with a consistent set of questions asked to community members, photographic documentation of the sites and archival research in specialised libraries and collections to recover past experiences with sites and published expert commentary.

Unlike survey research, in which an ideal sample size can be known ahead of time, the lack of consistent and evenly distributed knowledge in a community requires the ethnographer to ask similar questions of a variety of people. The questioning continues until the researcher understands why most answers are the same and why some answers are different. This can take quite a long time, but it will happen eventually. Underlying the uneven distribution of knowledge is a process of mutual comprehension that makes community life possible. That is, even though two people may have differing knowledge of a phenomenon in their locality, they understand when such differences are crucial to predicting how someone will act and when the differences are inconsequential. The ethnographic sample is complete when the researcher is sufficiently familiar with this underlying process of mutual understanding that questioning is no longer necessary.

The fieldwork describes the ways people encounter places, perceive them and invest them with significance. Your disciplinary training in the culture theory, previous research experience and conversations with others engaged in similar research combine to produce a competent and convincing description. Having community members read and criticise it validates this narrative.

The ethnography of landscape describes specific ways in which places naturalise different ways of making sense of the world (Feld and Basso 1996 p. 8). That is, we see the reasonableness of an arrangement of a specific community life represented in the landscape. To this end, ethnographers collect verbal descriptions of sites and localities and detailed spoken narratives of places. However, the advantage of being present in the community is that we can put our own bodies in these communities and these landscapes, observing the actions of the people around us, but also reacting to the spaces as a 'community member in training', learning the hard way which behaviours are permitted and which are not. It is from this direct involvement in the sites we seek to analyse that cultural anthropologists make their greatest contribution to the study of landscape.

## *Ethnology*

Beyond the landscapes of particular communities lies the theoretical problem of whether aspects of the experience of landscape are common to all people. The weighing of evidence from different ethnographies in an effort to answer this question is known to cultural anthropologists in the Boasian tradition as the science of ethnology. I realise that the term has a different meaning in many European universities where it refers to research that describes the cultural coherence of a region of the world. In the United States, the European usage of the term was transformed at the turn of the century to put less emphasis on cultural coherence and more emphasis on the historical processes through which common understandings come in being across localities.

For example, in European ethnology it would be appropriate to describe the persistence of a French vernacular landscape style as distinct from, say, a Dutch vernacular. In American ethnology it would be appropriate to ask how the form of garden colonies spread through Europe following the publication of Ebenezer Howard's 'To-Morrow: A Peaceful Path to Real Reform' (1898) interacting with the already established pedagogical gardens of D. G. M. Schreber and E. I. Hauschild (1864) to create the allotment garden movement. The former ethnology is an analysis of static qualities. The latter focuses on flow and movement.

This form of ethnology is only as good as the ethnography on which it is based. That is, we first have to know that the allotment garden movement was indeed a movement. That it had the potential to reshape the form of European cities. That it was a durable and persistent force throughout the twentieth century. That people knowledgeable of urban land use policies can reasonably disagree on the ultimate value of the movement. Allotment gardens are a particular form of land use with specific, constantly changing legal, political and social features. The ethnology that emerges from these ethnographic observations is critical. It is not merely history in the service of ethnography. It documents the changes in landscape meanings through time, in much the same way that archaeology documents the changes in material culture through time.

## Concluding thoughts

I have tried in these few pages to summarize fifty years of research on a complex facet of the human experience: the meaning we derive from our experience with specific spaces and places. A landscape architect may rightly ask, 'When I am designing a site, how much of my design is the product of my local and my professional communities, and how much of it is my creative innovation?' Culture is not a straightjacket. It is like a set of grooves in our lives. We can easily move within the grooves, or we can choose to step out of the grooves and walk beside them. The greater our awareness of where the grooves lie, the broader our range of choice. In other words, the landscape designer decides how much of the design responds to the issues and concerns in the professional community or the community of users, and how much derives from creativity.

It is helpful to have someone around who can describe those grooves and explain why they have come to exist. Collaboration is possible between the landscape architect and cultural anthropologist. When landscape architects take the time to engage in ethnographic research themselves, their designs become more deeply rooted to the locality. It is a tool for discovering how a community will interact with a design. It is a process for evaluating a design after it is built. It is a path to self-knowledge for the designer who is open to discovering the spatial discourses and practices that have shaped the work.

Designers have effectively teamed with social scientists in the past. In North America, a professional association known as the Environmental Design Research Association (EDRA) and in Europe as the International Association for People-Environment Studies (IAPS) has brought together landscape architects, architects, regional planners, preservationists, environmental psychologists, geographers and cultural anthropologists to share research ideas and techniques since 1969. I have taken part in four EDRA conferences and found the conversations with designers, planners and fellow researchers highly stimulating. Several scientific journals are also devoted to this collaboration including *Environment and Behavior*, *Journal of Architectural and Planning Research*, *Journal of Environmental Planning and Management*, *Journal of Planning Literature*, *Landscape Journal*, *Places* and *Research Design Connections*. I urge all landscape architects to take advantage of the potential for such collaboration. Our communities can only benefit.

## Bibliography

Augé, M. 1995, *Non-places: Introduction to an Anthropology of Supermodernity*. J. Howe, transl. Verso, New York.

Bourdieu, P. 1998, *Practical Reason: On the Theory of Action*. Stanford University Press, Stanford, CA.

Brenner, N., and N. Theodore, 2002, 'Cities and the geographies of "actually existing neoliberalism".' *Antipode* 34 (3): 349–79.

de Sousa Santos, B., 1995, *Toward a New Common Sense: Law, Science and Politics in the Paradigmatic Transition*. Routledge, London.

Feld, S. and K. Basso, eds, 1996, *Senses of Place*. School of American Research Press, Santa Fe, NM.

Ford, L., 2003, *America's New Downtowns*. Johns Hopkins University Press, Baltimore.

Foucault, M., 1986, 'Of other spaces.' *Diacritics* 16 (1): 22–27.

Gehl, J., 2001, *Life between Buildings: Using Public Space*. The Danish Architectural Press, Kopenhagen.

Gramsci, A., 1992, *Prison Notebooks*. Columbia University Press New York.

Hackworth, J., 2007, *The Neoliberal City: Governance, Ideology, and Development in American Urbanism*. Cornell University Press, Ithaca, NY.

Harvey, D., 1989a, *The Condition of Postmodernity: An Enquiry into the Origins of Cultural Change*. Wiley-Blackwell, New York.

Harvey, D., 1989b, *The Urban Experience*. The Johns Hopkins University Press, Baltimore.

Harvey, D., 2001, *Spaces of Capital: Towards a Critical Geography*. Routledge, London.

Herzfeld, M., 2006, 'Spatial cleansing: monumental vacuity and the idea of the West.' *Journal of Material Culture* 11 (1-2): 127–51.

Hirsch, E. and M. O'Hanlon, eds., 1995, *The Anthropology of Landscape: Perspectives from Space and Place*. Clarendon Press, Oxford.

Holston, J., 2008, *Insurgent Citizenship*. Princeton University Press, Princeton, NJ.

Lefebvre, H., 1992, *The Production of Space*. D. Nicolson-Smith, transl. Wiley-Blackwell New York.

Linde, C., & W. Labov, 1975, 'Spatial networks as a site for the study of language and thought.' *Language* 51: 924–939.

McDonogh, G. W., 1993, 'The geography of emptiness.' in *The Cultural Meaning of Urban Space*. R. Rotenberg and G.W. McDonogh, eds. Pp. 3–15. Contemporary Urban Studies. Bergin & Garvey, Westport, CT.

Rotenberg, R., 1995, *Landscape and Power in Vienna*. Johns Hopkins University Press, Baltimore.

Rutheiser, C., 1997, 'Making public space in a nonplace urban realm: "re-vitalizing" Olympia Atlanta'. *Urban Anthropology and Studies of Cultural Systems and World Economic Development* 26(1).

Sahlins, M., 1978, *Culture and Practical Reason*. University of Chicago Press, Chicago.

Sapir, E., 1924, 'Culture, genuine and spurious'. *American Journal of Sociology*, 29 (4): 401–429.
Schwartz, T., 1978, 'Where is the culture? Personality as the distributive locus of culture'. In *The Making of Psychological Anthropology*. G.D. Spindler, ed. Pp. 419–441. University of California Press, Berkeley.
Soja, E., 1989, *Postmodern Geographies: The Reassertion of Space in Critical Social Theory*. Verso Press, London.
Soja, E., 1996, *Thirdspace: Journeys to Los Angeles and Other Real-and-Imagined Places*. Wiley-Blackwell, New York.
Sorkin, M., ed., 1992, *Variation on a Theme Park: The New American City and the Death of Public Space*. Hill and Wang, New York.
Sperber, D., 1996, *Explaining Culture: A Naturalistic Approach*. Blackwell, Oxford.
Swyngedouw, E., F. Moulaert and A. Rodriguez, 2002, 'Neoliberal urbanization in Europe: large-scale urban development projects and the new urban policy'. *Antipode* 34(3): 542–577.
Thomas, K, 1996, *Man and the Natural World: Changing Attitudes in England 1500–1800*. Oxford University Press, New York.
Viollet-le-Duc, E. E., 1875, *Discourses on Architecture*. B. Bucknall, transl. Grove Press (1959), New York.
Webber, M., 1964, 'The urban place and the nonplace urban realm.' In *Explorations into Urban Structure*. M. Webber, ed. Pp. 79-153. University of Pennsylvania Press, Philadelphia.
Whyte, W. H., 1980, *The Social Life of Small Urban Spaces*. Conservation Foundation Washington, D.C.
Zukin, S., 1991, *Landscapes of Power: From Detroit to Disney World*. University of California Press, Berkeley.

Chapter 11

# Greening planning

## Regional planning and landscape architecture

*Marco Venturi*

**Introduction**

Until today, the relationships between the world of spatial planners and that of landscape architects have been extremely dialectic. On the one hand has been the tentative 'residual' use of landscape experts (what should be done to prevent extreme events? What measures should be taken with a particularly valuable pocket?). On the other hand there has been a preventive war against any new use of open areas, without paying particular attention to possible alternatives.

At a moment in which the urban population of the world has exceeded 50 per cent, and in Europe new forms of widespread urbanization cover 25 per cent of the land area with infrastructures (but much more in terms of the percentage of new building, population and work places), the tendential inversion of the relationship between open and built-up spaces is creating a series of dichotomies that are central for the future of physical planning at different scales.

Historically (but the dichotomies overlap in both space and time, leading to unheard of complexities in the landscape and in interventions to transform it) one goes from the contraposition between actions of landscape protection and those to maintain political consensus, to a conflict between the collective wealth that sites and their private use represent, to a comparison between the costs of maintaining the territories and those of their decay (environmental decay restricts the efficiency of the economic system, while the inefficiency of the productive system exacerbates environmental decay).

Instead of being aware of the potential and collective wealth that is represented by the biodiversity of Europe, at a territorial level environmental policies are conditioned by the fear of ecological disasters, in town planning terms by the mediation of conflicting interests, and in economic terms by the ignorance of the resources that have been consolidated in landscapes over the ages.

## The frame

Now, what we have is neither a new phase in old cycles of development nor the simple addition of new parts to the traditional city, but rather an indication of a structural change in the urban entity itself. Yet, while hope is being placed in 'strong', compelling and univocal interpretations of what is happening, the changes that are actually taking place seem to be the outcome of many small-scale – and often discordant – behaviours rather than the result of one single thrust operating in one clearly definable direction. Thus, there is awareness of the changes taking place but no ability to read the pattern of which they are a part. Instead, these very changes are seen as annoyances and irregularities to be corrected – in order to make the unforeseen and the misunderstood fit in with the known and established. This explains the tendency towards the definition of outlying city areas, of increasing the density in suburban sectors with geometrically regular patterns – relying on axes that often are only to be found in the two-dimensional world of technical plans and not in the multi-dimensional perception of everyday life – or by focusing attention on those urban sites that are the historical core of the idea of the city (the public square, the public park, the main shopping street, city gateways).

Very often our landscape is compared to a palimpsest, the signs of which are more the fruit of accumulation than substitution. Today, an understanding of what was conserved and what was denied, destroyed or transformed in each age can help us in innovating an overall project of a particular territory, within which it is possible to outline the specific responsibilities of territorial planning and landscape architecture.

There are two great tendencies that appear to generate the clearest transformations: on the one hand homogenization – the diffusion of infrastructures everywhere, productive techniques and uniform settlement typologies – while on the other, the differentiation of districts, and the concentration of the environmental impact on limited areas. A meeting of these two tendencies and their overlapping on territories that have already been structured by successive anthropologization and greatly varying geological characteristics leads to fragments that are so diverse that it seems difficult to lead them back to great interpretative paradigm.

Hence arises the myth of incomprehensibility: everything is too complex, too global and – at one and the same time – too fragmentary, to admit analysis. However, I would argue that this is an exceptionally propitious time for territorial analysis – a moment when such analysis could put forward interpretative schema that might well be of use to disciplines that are normally our creditors rather than debtors.

The main problem lies in changing the temporal and spatial focus of analysis; shifting away from the long-term and the national to the more immediate and the regional. This more segmented focus could produce a sort of pointillist composite, illustrating forms of development that are imperceptible if viewed at a different scale of magnification. In such cases it is worth following the advice of the old French proverb 'reculer pour mieux sauter' and starting again from basics – from a simply inventory of the changes that are taking place. First of all, we have to accept that the instruments at our disposal are not really adequate to this descriptive task, and that we will have to resort to an interdisciplinary approach, borrowing anew from others. Statistics, censuses, cartographic representations, illustrated reports, frameworks of regulation and various other elaborations of data have all been used in our discipline, but their focus has been on the long term, on changes that take place over a certain period of time. We do not seem to have the tools for rapidly identifying the meaning and significance of tumultuous change, nor for intervening in these processes of change in 'real time'.

One only needs to cite the example of what happened in the countries of Eastern Europe after 1989: just when it was most important to be able to exercise control over the situation and the changes that were taking place, urban planning offices were cast adrift, 'playing it by ear' until the situation had settled down enough for them to start re-applying the old, tried-and-trusted tools of the trade. Undoubtedly, the very complexity of problems – the presence of such phenomena as globalization and increasing mobility – make it more difficult to interpret what is happening in this apparently fluid and rapidly evolving situation.

First of all, a quantitative observation: the crucial diminishing aspect in Europe is the specific weight of the centres within a widespread urbanization compared to the concentration prevailing in the rest of the world. Furthermore, what were once the paradigmatic functions of urbanity, i.e. housing and production, are no longer to be found in urban centres. In the cities there is also a general refusal, if not abandonment, of more 'planned' areas and quarters, of the symbols of the modern movement urbanism and welfare that had been achieved in Europe by the 1970s. It is very likely that this had to be the case, since for more than a generation we have seen a total reduction in investments in fixed social capital, and a restructuring towards forms that have scarce visibility or are not specifically urban.

Public investments seem to be moving from what is real and concrete to what is monetary and imaginary, thus losing their capacity for symbolic representation. Corresponding to the fall of policies of support for urban development is decreased development, or development in different sectors that is seen in more individual characteristics and advantages that are not as widely divided. From the point of view of planning, up to now the main problem has been of recognizing existing resources and dividing them between places and social groups. Now, however, there is a tendency that does not yet have theoretical foundations and that no longer regards planning as a way of dividing resources but as creating them, mainly through the formation of a system of attractiveness and enhancement of places. The main idea is that the result must be greater than the sum of parts and that the division of costs will be less painful in times of general growth, even if the benefits will stay unequally distributed. In reality, it is the latest attempt to quantify the relationship between space and society, to reduce places to the functions imposed on them.

There is no doubt that European cities have been penalized by uncertainties regarding strategies, and above all, a ruinous welfare policy, one that is not only perceived by those who are paying the price as too expensive and badly managed but also perceived as one that does not correspond to the expectations of those using it, since it is offered in a standardized manner, while the ideas of well-being of different groups generally tend to be diverse. However, it is possible, when seen in perspective, that the most innovative phenomena can be observed in our regions.

The supporters of globalization emphasize the decentralization of production but forget that not only are services and management concentrated in a city but also, and above all, so is the production of the city itself. From this point of view, which includes not only the already accumulated fixed capital but also the increasing location values being carried out, it is these that appear to create greatly differing positions of European regions in comparison to apparently more fascinating areas.

## The forms of transformations

For the first time a change of this size and importance cannot be expressed in the usual terms of growth but in terms of transformation and redevelopment. Eighty per cent of building permits are not for new constructions but for redevelopment or re-structuring. The EU forecasts that after the completion of the present phase of building work intended to bring underprovided areas up to standard, the percentage of new buildings will decline even further. This is something which requires us to review the nature of those professions engaged in the design and exploitation of urban space. In fact, urban planning legislation is predicated on expansion, so is ill-equipped to deal with redevelopment (which has generally been seen as a rather special case, concerned more with preservation and restoration than with the actual modification of the urban fabric). However, the scale of such redevelopment in both real and percentage terms means that there will have to be a radical re-thinking of approach – given that the projects being carried out could lead to a total transformation of cities over the next thirty years. What is more, every year disused industrial areas account for 5–10 per cent of land available, so they naturally lead to a steady increase in the number of redevelopment projects.

This would be an opportunity for their strategic adaptation to a new *forma urbis* – if it were not also true that present development potential is too low to cope with the redevelopment of suburban and outer city areas (and the creation of an entirely redeveloped general framework). For the first time there is an overall crisis throughout the urban infrastructural system. Whilst the emergence of the industrial city involved enormous investment in technological renewal and the creation of new public services, the last fifty years have been characterised by over-exploitation of these services (without any substantial renewal) which has ultimately resulted in the break-down of the system inherited from the pre-war years. Railway stations and trams, water towers and purifiers, drains, sewers and water pipes, ports and canals, power stations and slaughterhouses – there is an entire body of technological facilities that is now obsolete and must be replaced (Figure 11.1). At the same time, all of those public buildings and structures that were part of the industrial city – hospitals and barracks, universities and places of entertainment, sports and social centres – all seem to be undergoing re-location and reorganization.

If it is true that city population figures are in decline, it is also true that this drop is not accounted for by moves to small or medium-sized towns or to suburban areas (as was claimed by traditional 'urban-centric' explanatory models). What is happening is that people are moving to new, largely autonomous areas,

**Greening planning**

**Figure 11.1 a and b**
Koper (Capodistria), Slovenia: Proposals for the expansion of the port with environmental and energy arrangements. The port is vast area, dwarfing the old Venetian town. Planning is a vital task. Authors: Venturi, Azman.

following what might be called a 'self-referential' logic of settlement. The new parts of the city do not, therefore, seem to be parts of a city; they arise outside the city, using the entire territorial system until they eventually undermine the traditional centre of gravity of that system. Greater emphasis is now placed on areas of medium-low density which do not have a single centre but a number of centres. But if there is no centre there are also no suburbs, and the dichotomies city/country and culture/nature change in meaning. Thence arise new paradigms for interpretation.

    The metaphor of the network seems to be particularly successful; it manages to take into account not only the layout of the individual 'links' (each with its internal logic) but also the 'old' urban centres, including both in a schema which in some way transcends them. What emerges as a result is not a new type of city or a non-city, but rather many types of coexistent cities. The great phases of technological innovation also upset the space-time relations within the city. The speed of movement of goods, people and information, once similar in the various sectors, is gradually being differentiated: people are relatively stable compared to the increase in information transmitted. This raises new problems in renewing infrastructures and of the perception and appropriation of spaces, as well as the nature of bonds in communities or social and political groupings.

    Social spatial organization now undermines the very possibility of belonging to a single community, or sharing an urban ethos or a set of customs. The end of a city morality implies, however, the end of the *forma urbis*: the traditional city is seen as an obstacle to the various interests of the individual social groupings as well as a hindrance to the most important form of freedom recognized today – freedom of movement.

    The city is no longer perceived as a device for maximizing social interaction, but as an obstacle to general mobility, which is expressed in various ways in the surrounding area. The city is thus socially and physically fragmented. People identify first with their quarter (neighbourhood) and then with the city – more with a social group than with a place. Often whole groups move and streets and places are only of secondary importance in the criteria for residential options. The whole system of powerful 'social dampers' disappears through this fragmentation, since they were linked to places.

## *Landscape patterns*

The phenomenon of new urban expansions has led to their being standardized according to endogenous criteria, linked to features in buildings rather than the specific need of the site on which they are constructed. It becomes difficult to identify cities on a map without their historical centres: the logic of the building industry obeys autonomous criteria, which refer to the object and not the site, to quantity and not to quality.

    The criterion of value between quality and quantity thus begins to be reversed: the relation which was directly proportional for centuries (in traditional planning 'big is beautiful') is reversed and quality becomes the attribute of rarity or the exception and therefore once more linked to geographic specificity. The task of planners is split in two: on one hand they must optimize conditions to invent and spread more and more new quantities, and on the other, they need to differentiate similar quantities in the search for quality, or possibly uniqueness. Both ways lead to the loss of some of the traditional features of historic cities. Those had in common the continuity of the urban fabric, the mix of functions inside the blocks and the variety of architectural expressions in similar economic and social contexts. All these features remain in the case of unplanned settlements, while planning rules seem to bring about the contrary: discontinuity in the fabric, monofunctionality of the buildings and homogeneous and repetitive typologies for similar areas or purposes.

However, the most important dynamic feature of the city-as-form (its great innovative capacity compared to any other form of coexistence) is linked to its ability to provide the conditions for a non-planned synthesis of different cultures and experiences. The success and the visibility of a city will, therefore, not depend on its uniformity, but its diversity, not on its 'rational' subdivision but on the fruitful reciprocal accessibility of its parts. Of great interest is the fact that these new structures tend to prefer the older stretches of the roads rather than face the new motorways or railways. Not only is it difficult for these to be integrated in the local infrastructures, but everywhere the presence of the old urbanization structures such as sewers, waterworks, power-lines and even old cadastral divisions seem to play a decisive role for new settlements.

Historically established settlements are therefore 'ready-for-use' whereas new structures require longer to take root and to generate the secondary structures necessary if they are to be fully exploited – a factor that is incompatible in times of upheaval. The roads therefore remain permanent while their surroundings undergo transformation – the old stretches seem to be the only fixed point in an ever-changing, non-isotropic landscape. Their current primary function therefore appears to be that of re-weaving the many intersecting links where the junctions also belong to other superimposed networks.

The sudden success of certain stretches that had remained relatively undeveloped until recently could be explained by the very fact that these were the stretches that were needed for the connection with other networks. Persevering with the old stretches of continuity – which, it is important to point out, was not planned – thus had the aim of making a mobile system more fluid, a system that had to be able to rearrange itself as a whole, questioning the old criteria of location with new forms of accessibility that tend to be ubiquitous. Thus, the 'difference' no longer lies so much in the architecture or the model, no matter what the variation, but in the accesses, the interstitial spaces, and the street edge – one could almost say in the pauses, with what is left.

Faced with modes of use in tumultuous transformation, this means not only recording the 'differences', the various ways in which the areas in question actually reacted to analogous inputs, but also their diverse inclination to change. Indeed, this is the first possible observation – that of the different speeds of transformation, with places that have been stable for long periods of time showing much more inertia and resilience, while others appear to have the tendency to be more open to innovation, 'asking' for future adjustment. The specificity of roads is to maintain urban characteristics all over the place, in a manner of speaking, to bring the city to the countryside. In the very same way that a city is not just made of buildings but also of the relationships between its characteristic elements, the study has to concentrate on the discontinuities, on the 'differences'.

This is particularly true in more specifically urban areas. Residential areas traditionally offered an extremely homogeneous mixture of functions – at least in similar settlements. Today, however, the junctions of the reticular continuum are beginning to diversify: their attractiveness depends on their ability to offer something that another place does not have. Together with the different degrees of inertia, this diverse reaction to change becomes a method of highlighting identity and specificity on which to build new projects (Figure 11.2).

**Figure 11.2 a and b**
Stanezice, Slovenia: A new city centre for 3000 dwellings to be planned within the region and to take advantage of new communication infrastructure. Authors: Venturi, Azman.

At the moment consolidated cities appear to have been reduced to objects for consumption, while the production of ideas and commodities has moved to the links between the centres. The phenomenon of the differentiation in the speed of transformations is evident. There is a clear inclination towards change in some parts of the city and it is on this that new functions automatically concentrate while elsewhere consolidation is also manifested as the public adhesion to the physical configuration that has been achieved with time and is therefore resistant to these very changes. However, if urban policies do not face the new scale of problems, at least at a district or regional dimension, the inertia regarding the stabilized centres risks being translated into their progressive deterioration. The longer the attempt to defend gravitational models and traditional hierarchies persists, the more painful it will be to identify the operations needed for the networking of the polycentric structures required by the new scale economies.

It is more a case of facilitating the processes and guiding individual interests towards convincing scenes rather than modifying the hierarchies with 'profound' interventions. The logic seems to be that of eliminating the obstacles, whether physical or procedural, to the connection of all the elements with a tendency towards change. In reality, new intervention tends to be reduced to the modest, albeit strategically significant completion of the pieces missing for the networking of the territory.

At the same time this induces a necessary modification of the knots of this re-stitching: in the old centripetal models, competition could be seen in the imitation and repetition of the same mix of successful functions. Administrative policies tend towards a 'me too' attitude – if the cities I compare myself to nearby attract investments due to the services they offer, I have to have them, too. This results in a superabundance of museums, stadiums, technological parks, fairs, shopping malls or integrated stations.

However, in a city where mobility allows one to take advantage of the whole region, while avoiding commuting and crossing the centres, the success of these knots lies in their specialization, in their capacity to offer what is lacking elsewhere and not what can be found elsewhere. The attractiveness of a knot lies in its complementarity and not in its competitiveness with its neighbours.

## *The features of innovation*

So, there is innovation in Europe – even if it does require a change in outlook in order to be perceived. It is just like a picture in a pointillist painting, a multitude of projects without a plan reveal a complex design – although only visible to the long-sighted. The transformation that is taking place here does not seem to compare itself to other models that have been absorbed, but rather to our very past. The attempt appears to be that of overturning the inherited city system. The emerging city does not limit itself to contradicting or competing with what preceded it but to swallowing it up and metabolizing it by changing both its position and role.

Some characteristics can already be seen even if they concern more the transformation processes than their forms.

## Densification-dilution
From the very beginning, the difference in density has always been at the heart of urban history, increasing in the centres and decreasing towards the countryside beyond. For the very first time, density (and everything that is interrelated, above all the intensity of social relationships) is diminishing in most European centres, whereas the connections between them are increasing.

## Concentration-deconcentration
The difference in potential between centres and suburbs is overturned in favour of a tendency for indifference in location: what currently matters is the total critical mass, not its internal articulation.

## Continuity-discontinuity
New urbanized territory is no longer isotropic as in modern urbanisms' dreams. There is no longer any continuity, neither physical nor social. The new city is made of fragments, specific solutions, the search for individual well-being that expresses itself in the isolation of one building from another and in the definition of territorial limits at the expense of shared spaces.

## Centripetal-centrifugal
What is new is to be found in the new parts of the city, leaving the representative centres 'behind'. The life of new urban areas is determined by increasingly unforeseeable flows that no longer commute between the centre and the outskirts. Those who move do so in between the cities, with a lifestyle that is the very opposite of that of our parents.

## Symmetrical-asymmetrical
In practice if not in policy, the search for balance is replaced by the acceptance and emphasis of asymmetry, which is regarded as the engine of both growth and innovation.

## Innovation-conservation
The production of innovation appears to be reduced in the traditional centres, which are characterized by traditional know-how. The capacity for innovation seems to be increasingly linked to moveable goods and not fixed capital, to software not hardware.

## Dot-like-network-like
The capacity of the nucleus' attraction seems to have been replaced by that of the interconnected areas. Urban projects are destined to make citizens who abandoned the city for inter-urban areas to return there – not as inhabitants but as consumers.

## Competitive-complementary
Where the city's ranking order once depended on its ability to out-do its competitors in the same fields, success now depends on it offering something that is lacking elsewhere, on the complementary nature of the urban system networks.

## Stability-change

Different speeds of transformation are no novelty – what is new is the attempt to institutionalize them, to recognize areas with a strong tendency to change and those that are more inert. Differentiated intervention systems, increasing the fluidity and flexibility of the former, can all lead to the preservation of the latter.

The list could go on with opposing pairs in different fields.

Amidst the consolidated past and new aggregations that are held together by technological networks in which service quality and quantity are already superior to those of the past, there is no lack in plans and proposals that allude to possible futures for our urban systems. But at present, only one thing appears to be certain. It will be a series of different cities with one thing in common – cities opposing instead of developing the tradition of the European city. Since innovation is situated in the interstices between two kinds of phenomena – i.e. temporal and spatial – it provides a yardstick for them. Innovation is a break – a rift in an otherwise supposedly linear development. The 'linear' model made up of many small innovations – together they form a tradition, a long-lasting internal order – is measured by the eruption of the new – by change.

It would seem that planning, or at least the tradition of planning, is only able to cope with linear development processes. At times of tumultuous innovation, the shortcomings of our toolbox are self-evident, not only in its poor capacities for predicting phenomena, but above all in its inability to record and represent them.

The inability of the traditional toolbox to support operative decision-making is particularly clear at present. Transformations of this kind and size are not compatible with the usual statistical, graphic and administrative methods. Just when we need to know in real time what is happening in the city and how, the reaction times of the traditional apparatus turn out to be totally inadequate. This is so much the case that the major cities have opted for voluntary choices, which eschew any claims at planning rationality.

Planning programmes seem to be measured in the long term, therefore, while the innovative processes are moving increasingly swiftly. This could be a 'structural' explanation of the current predilection for 'big events' and 'grand projects' that fall outside traditional planning logic. The same kind of attempt to respond to deep changes in urban policies with special tools and programmes was already tried in 1848 in Europe and in the whole of the West after 1929.

Radical innovations in planning theories can be seen in this direction. No matter how much it is dwindling, the old policy of plans as the division and allocation of existing resources appears to leave the field to planning as the creation of new resources. What is more, this is by means of processes that enhance the localities and create systems of attractiveness in which the image and brand both play a key role.

The other innovative element is the transfer of impacts between those of the planning of space to those of time – in the past urbanization processes coincided with those of the concentration of human and economic resources, thus creating the basis for the acceleration of cultural innovation. Today however, we are living in a period of great transformation characterized by low density, where what actually matters is the range and quality of exchanges rather than their intensity and frequency.

We are therefore experiencing a new phenomenon – innovation processes now coincide with dispersion and not with concentration and high density. While traditional town planning used precise densification to provoke temporal acceleration, attention is now paid to the planning of deadlines, anniversaries (big events, jubilees, world championships, universal exhibitions, etc.), so increasing the urban velocity in order to attract a concentration of resources on pre-selected spatial knots.

## Subjects for landscape architecture education

Students of landscape architecture should therefore be encouraged more towards a change in viewpoint than in subject: struggling with the causes more than with the effects and foreseeing the spatial repercussions of the decisions taken in other sectors. It means thinking 'long' rather than 'big', starting with operations of preventive interventions of the ground, thus conditioning any possible future building rather than intervening afterwards in the spaces resulting from development operations, waiting for years before any green can be seen growing. 'Erst begrünen, dann bebauen' (first make it green, then build) is a motto that should circulate outside professional offices.

Landscape professions are, by definition, professions that require patience, and are long term. This means planning early and being ready to adapt the initial project gradually, without losing sight of its overall coherence. It is difficult to convince someone to pay today for the results that might be enjoyed by others, but it is in this very sector that many techniques of regional planning could once again be of use if they were to be taught in landscape schools. I am not just thinking of provisional techniques but also those regarding negotiation and mediation, in the sectors of communication and management.

So a special preparation, missing up until now in landscape architecture schools, should focus on the ability to foresee the effects of many small, single interventions that are cumulatively able to cause an evolution in the structure of whole areas: rows of trees, agricultural fields, rules regarding the limits and verges, water runoff, the treatment of gradients, etc.

Another line of research that should play a considerable role in the schools is that of the representation of the territory. Until now, representation systems have not been coherent with the complexity of the problems, and in general are much more behind the times in comparison with the novelty of phenomena. However, this is one of the aspects with the greatest potential: the need to represent three-dimensional phenomena, flows and variations in time has the potential not merely to result in an innovative use of computer tools, but could also contribute to the positioning of landscape disciplines in regard to neighbouring faculties.

In the past century, while the art of gardens and the forest sciences were still anchored in academic styles of representation (the beaux arts), architecture and town planning were linked to the renewal of artistic languages, breaking with systems of spatial continuity. Recently, this relationship appears to have been interrupted. While the visual arts are exploring three-dimensionality, new materials, the interaction with users, flows and fields, architecture and town planning seem to have remained locked in the bi-dimensionality of a scene on which objects are to be arranged. On the contrary, the very nature of territory requires the study and representation of what is above and below the surface, the flows, and of changes that are both cyclic and in constant evolution.

For an even greater reason, the same can be said for the treatment of information concerning territorial transformations: gas, water and electricity companies, telephone and computer connections, etc. are all equipped with computerised maps that, if linked to those of the soil, satellite photos and climatic data, can supply an outline of the evolution of the use of the ground in real time, day by day. An understanding that does not lie in a map, but in the 'range' of the open spaces could help us change the nature of the clients' questions: no longer technical networks following construction, but a flexible treatment of the territory that overcomes the engineering of technological restoration to re-acquire the treatment of water and its recycling – also for building and other materials.

In the field of regional and landscape transformations, one handicap has always been that of the times and costs of experimentation, which is in itself extremely complicated: the engineers' solution following that of estate agents can now be overcome by simulation techniques that are not only possible, but also economically acceptable, and that result in completely innovative solutions. Therefore, a representation taking advantage of new computer technology could contribute to the rejuvenation of the relationship of the disciplines of spatial transformation with those of cultural change.

I therefore consider the teaching of anthropology indispensable if landscape is to be linked to culture and cultivation. Often, for example in the South Tyrol and Tuscany in Italy and in the mountain regions in Austria, the policies aimed at defending the local identity, which appeared to have been successful, actually risk leading to its progressive falsification, to a hypostatization of homogenized interpretations, while a more correct anthropologically-based understanding would highlight the differences and allow more positive developments in the landscape.

A society that marginalizes the phatic diversity is also in trouble in protecting biodiversity, and the students of today, who will probably work in regions other than where they grew up, will therefore need to manage sophisticated skills if they want to understand the relationships between a society and its landscape. This appears to be particularly important in one sector: that of the production of landscapes and alternative models for the single house. A site dictates its own specific rules, contrary to the self-referential nature of the peripheral architecture, and could justify the use of convincing and innovative models.

Continuing with the list of subjects to be anticipated in education, before they become the prerogative of other faculties or professional interests, that of the place of free time in people's lives is essential. This is becoming central in both the theory and practice of planning, after a century of being limited to just the first two parts of the great tripartition of modernist theory – *habiter* and *travailler*: the *loisir* was the result of interventions in the first two fields and was often expressed in terms of square metres of the green calculated, but not planned.

Today, however, free time has become a decisive factor in planning: only stimulating cities can attract flows of human and material resources, and the structuring of leisure services, that are both routine and exceptional, that satisfy the needs of citizens is becoming a priority for local administrations. The participation and development of entertainment engineering techniques (amusement parks, acoustics, light and music, computerized scenography, artificial realities, etc.) is therefore fundamental. The new leisure landscapes will probably be expressed by unheard of modelling capacities using extremely innovative mixed natural-artificial technologies.

It is clear that the tourism and leisure time may place a burden on the protection, recovery and development of areas that are attractive for other uses. It is our ability to propose convincing alternative models of landscape to ensure that they are not destined to succumb to the pressure of strong economic interests.

An example of the success of this type of behaviour took place in the expansion of landscape projects to areas that had previously been neglected such as old industrial areas and brownfields: moving from the initial demand for removing polluting materials and renewal interventions to more general regeneration projects was a fundamental step in landscape architecture, and one that, in my opinion, was mainly due to the courage of several colleagues in rediscovering the entertainment value of degraded areas and the unexpressed possibilities of industrial plants.

The next step will be that of rediscovering *all* public infrastructures as elements of environmental planning: advancing from the engineer's culture of fear, of the need to isolate artefacts from each other and to avert the risk of improper use, to a culture of urban projects in which the infrastructures and facilities are once again a reason for civic pride, not hidden but integrated and used as part of the urban landscape of our times – this is one of the challenges that, at least partially, the alliance between town planners and landscape architects is successfully overcoming today (Figure 11.3).

## Research fields

In the current crisis of the status of the professional and of the social credibility of planners (second only to dentists in a recent survey of the most hated professions), landscape planning is central to the renovation of tools and objectives of physical design. As is always the case, a mechanical transfer of experiences and techniques (best practices) from one discipline to another is not possible, but at the very least the attempt should be made to avoid mistakes that others have already made in the transfer of social behaviour to a territorial dimension.

The first example concerns the recent transformation in the composition of families: their fragmentation alone, especially in urbanized areas, leads to an increase in numbers of households without any demographic increases or internal migration. The absence of a suitable response by the housing market and of research into these new and differentiated life styles, with proposals for the subdivision and restructuration of one-family houses and large apartments, runs the risk of placing unsustainable pressure on green areas, which could be avoided if opportune and timely measures were taken.

**Figure 11.3 a and b**
Rovigo, Italy: Castle, civic park and system of squares, showing planning and design in the context of an old urban fabric. Authors: Venturi, Bolner, Ferrarese, Navarrini

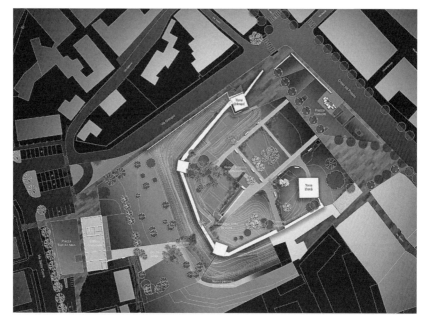

Likewise, in the field of mobility, if long-distance commuting is taken into consideration, the transformations of the employment market are almost ignored: on average one changes address ten times in one's working life, not to mention phenomena such as productive de-localization abroad and the consequences for individual or family strategies as regards the position, dimensions and furnishing in a house (for example, I do not believe any research has been carried out on the growing phenomena where 'home' becomes or remains a small villa in the country, while small rented apartments in urban centres follow the development of a professional career).

Another emblematic case is that of the treatment of the by-products of urban and agricultural metabolism, brownfield sites in particular. The anxiety to decontaminate and clean these areas to make them available on the market often has the opposite effect of keeping them off the market: after costly interventions, the areas become too expensive and of little interest precisely because they are bald, 'neutral' places.

However, many city councils manage to avoid pressure on green areas by demanding that before anything can be built on agricultural or 'greenfield' sites, 'recycled' areas must be shown to be available. This does not mean levelling them, but an attempt to develop their historic and morphological characteristics. A paradigmatic example of 'overplanning' in which the landscape professions can benefit from the experiences of town planning is that of the forecast of consequences in other sectors of building, hygiene and other regulations.

A study carried out in the Charlottenburg area of Berlin, one of the city's most attractive quarters, shows that all the current rules (distance from the roads, from other buildings and cross roads, standards concerning public and private parking, schools and social services, the distancing of handicraft and productive activities, the height of the eaves, regulations regarding the health services and fire access, etc.) would not only lead to a strong reduction in the amount of buildable cubic metres on the same areas, but to the construction of buildings that are unconnected to each other and their context, in other words, the loss of an urban landscape that was a valuable asset of that very quarter.

## Concluding thoughts

The European Landscape Convention finds the universities at a time in which they are undergoing generalized restructuring and one in which it is both difficult and probably premature to establish obligatory courses beyond what has already been indicated by EFLA, IFLA, and ECLAS.

Landscape schools are currently very diversified (from professional courses lasting between one and two to eight years, including practical training, to specialization courses after a degree in other specific educational subjects) that it is possible to believe that the most important thing is experimentation of alternative paths, paths that still link the treatment of the physical dimension through the modification of what exists towards sustainability and compatibility of interventions and the energy optimization of the processes involved. The constant education of technicians and the recycling of the skills of professionals in neighbouring fields in the logic and practice of a common landscape project remains a priority today.

Specific national interests may contribute to common progress: for example, I am thinking of the relationship between the landscape and historic and archaeological pre-existence in Italy or of interventions on coastal and insular systems. Other potential research fields depend greatly on the personality of the university teachers and local demand: it suffices to think of the proposals for the regeneration of residential spaces, especially the possible merging of individual gardens in a neighbourhood to form community spaces, to projects for mobility and car parks, the arrangement for workspaces and logistics, in particular on and in buildings, the internal treatment of large warehouses and their surrounding areas, the proposals of alternative use for abandoned agricultural areas and temporary installations or land art.

In the transition phase we have to reconsider the relationship between man and the earth: the geometry of the landscape reproduces the geometry of its social complexity and forces us to give integrated answers. Until now, the prevalent point of view has been that environmental politics started from the city; urban centres cover 2 per cent of the area but absorb 75 per cent of the resources, thus also functionally absorbing natural areas.

If town planners and landscape architects are to work together, they must change perspective: designing the city 'as a garden', working not just on the interstices but also on the buildings as part of an overall project, highlighting the internal structure and potential, in accordance with the teachings of Michelangelo, 'not by addition, but by subtraction'.

## Bibliography

Baldeschi, P. (2002). *Dalla Razionalità all'Identità*, Alinea.
Bulgarelli, V. (2004). *Città e Ambiente*. Angeli.
Caravaggi, L. (2004). *Paesaggi di Paesaggi*. Meltemi.
Donadieu, P. (2002). *La Société Paysagiste*. Actes Sud.
Huet, B. (1995). 'Park design and urban continuity'. In: *Modern Park Design*, THOTH: 23–24.
Lanzani, A. (2004). *I paesaggi Italiani*. Meltemi.
Le Dantec, J.P. (2002). *Le Sauvage et le Regulier*. Moniteur.
Magnaghi, A. (2000). *Il Progetto Locale*. Bollati Boringhieri.
Masboungi, A. (2002). *Penser la Ville par le Paysage*. Villette.
Vigano, P. (ed.) (2001) *Territories of a New Modernity* Electa Napoli

Chapter 12

# The place of landscape
Conversing with cultural geography

*Stephen Daniels*

## Introduction

Cultural geography is presently one of the most vibrant and influential fields of human geography, particularly in the English-speaking world, and has increasing levels of recognition and influence beyond the discipline. Cultural geography has expanded rapidly in the last twenty years, in connection with cognate subjects such as historical, economic and political geography, enlarging, enriching and reshaping these fields of enquiry. The concept of culture in cultural geography bridges humanities and social science traditions in human geography, focusing on the importance of values, imagination, identity, power relations, creativity and critical interpretation, in the making and meaning of places, both in the mind and on the ground. Cultural geography also exerts an influence beyond geography, in exchanges with disciplines which have an established focus on place and space, landscape and environment, like archaeology and anthropology, and others like literature, music and cultural history in which a landscape focus is emergent. Cultural geography connects with practice-led subjects like urban design, planning and heritage management as well as landscape architecture, if some of these connections are potential rather than actual, yet to be conducted as disciplinary exchanges.

Geography, like architecture, has always been multi-disciplinary, encompassing social, physical and technical perspectives and practices in relation to the built environment. In this chapter I want to set out some of the creative meeting points between cultural geography and landscape architecture, mainly with that field of landscape architecture concerned with history and theory, ideas and representation, in areas of research and scholarship, post-graduate training and undergraduate teaching. As an arena of enquiry and practice, landscape itself is a meeting place, a forum for dialogue between these subjects as part of wider multi-disciplinary and inter-disciplinary conversations. Much of this chapter is informed by my own encounters (in person and through their writings) with both cultural geographers and landscape architects who have a knowledge of each discipline and their wider domains, by my recent experience directing a multi-disciplinary research programme in landscape and environment, and by a career developing cultural geography to address questions of landscape representation, design and management.

The conversational nature of cultural geography is signified by the title of its main journal, *cultural geographies* (plural and lower case), designed to accommodate articles from a range of scholars addressing issues of space and place, landscape and environment.

## Landscape and culture

Landscape is a key, foreground concept for cultural geography. Landscape is deployed in an expanded and enriched way, co-ordinating geographies at different scales, mediating representation as well as reality, and intersecting with other key terms such as place, space, nature and environment.

Landscape was central to the initial establishment of cultural geography, as the core of Anglophone geography generally, in the United States in the mid-twentieth century, when the concept owed a good deal to inter-disciplinary (and international) connections with other field sciences, in geology, botany, archaeology and anthropology, and decisively to a Germanic genealogy of the term as *Landschaft*, the material (physical and cultural) shaping of territory, and with it a focus on vernacular structures and rural ways of life. Cultural geography lapsed as a designated field of enquiry in the Anglophone world until the 1980s (although its former concerns were sustained under other names by some human geographers) as geography underwent a positivist restructuring as a consciously modern and forward looking, spatial and behavioural science (and one which had – and in places still has – a profound influence on practical disciplines such as urban and regional planning). Cultural geography was revived, and to a degree re-invented, in the 1980s in relation to an expansion and reformulation of the term 'culture' more widely, in the humanities and social sciences, to describe a modern, cosmopolitan world of cities, mass media, artworks and popular and radical lifestyles – in the past as well as the present. The term landscape was central to cultural geography, as it once again expanded and colonized the discipline, but with an emphasis on the term's more Dutch genealogy as *landskip*, a view, an image, as in a traditional landscape painting or a modern virtual reality projection. Landscape's re-definition in cultural geography landscape, as representation as well as reality, a way of seeing as well as a way of life, owed much to its currency in the interpretative scholarship of art and architectural history, critical theory and cultural studies, for examples through the book *The Iconography of landscape* (Cosgrove and Daniels 1988) (Figure 12.1).

Landscape has continued to be reshaped as a concept in cultural geography, in part to recover and renovate its meaning as a physical world which is lived in, worked on and moved through as well as looked at. Landscape is recognised as a manifold material and cultural medium. One approach to landscape is *vertical*, probing landscape as a site of human experience, collective memory and imagination, of touch, smell and sound as well as sight. Landscape as a way of seeing the world is still significant, but it is more than a matter of *looking at* landscape, but of *seeing with* landscape, the way a place shapes people's outlook on the world. Here landscape is substance, *terra firma*. The other, complementary approach to landscape is

**Figure 12.1**
*Fu Kei No Zuzogaku* (2001) Japanese edition of *The Iconography of Landscape* (CUP 1988) Tokyo, Chijin Shobo. (Courtesy Stephen Daniels)

*horizontal*, more a matter of *routes* than *roots*, surface not depth, a *terra infirma* of fluency and process. In this perspective landscape is articulated by wide ranging flows and networks, including the circulation of capital, people and information (Wylie 2007).

While many of these ideas of landscape remain a conceptual, even conjectural, part of a cultural geography's profile as subject of scholarly interpretation, including a strain of personal reflection, there are also signs of more material, practice-led developments in the field, in various activities from heritage site interpretation to digital mapping. One of the liveliest sections of the journal *cultural geographies* is devoted to 'cultural geographies in practice' and includes shorter, more informal pieces on a variety of practical projects, including architecture and design, film making, dance, curatorship, installation art and urban walking, and projects which bring together academic, artistic and activist work. Inspired by the performative turn in the arts, humanities and social sciences, cultural geographers are now exploring conversational exchanges between critical and creative writing, action and reflection, life stories and social histories, the processes of ruin and abandonment and construction and restoration, in landscapes well beyond canonical scenic landscapes (Pearson, 2006). They are offering new perspectives on the place of landscape as a cultural medium and professional practice in wider forms and processes of living in the material world.

## Conversations with landscape architecture

The reformulations of landscape in cultural geography have drawn on wider conversation with the literature of architecture generally, and of landscape architecture in particular. My own landscape studies were influenced by writings on buildings, design and landscapes by Reyner Banham (1971), Mark Girouard (1978) and John Summerson (1945) as much for the writing style of these authors as for what they were writing about, whether Banham's pop style texts or Summerson's more measured, classical prose. The sensibility here was not just literary, or scholarly, for their writing appeared in wonderfully well-designed books and journals, with due attention to typography and image layout. As much as studio work and site visits in architecture, or field trips and map making in geography, this was practice as well as thought, part of the art of landscape as a practical accomplishment.

This architectural history tradition shaped my book *Humphry Repton: Landscape Gardening and the Geography of Georgian England* published by Yale University Press, a press chosen as much for its fine tradition of book making as for their publication list of fine scholarship (Daniels 1999). Moreover the emphasis on representation was entirely suited to its subject, Repton's designs, which were created as much to be seen on the page, articulated in his Red Books of watercolours and text, as to be executed on the ground; indeed such was Repton's experience of seeing his plans ignored, botched, mismanaged or decayed that he reckoned, rightly enough, that his art of landscape would survive more in books than in parks and gardens (Figure 12.2). Repton's art of landscaping combined down-to-earth matters of architecture, horticulture, arboriculture and hydraulics but it was its appearance on the page which proved so highly influential in shaping the aesthetics of landscape for the profession of landscape architecture when it became established in the United States in the mid-nineteenth century. It was as much a matter of words and images as of roads and trees, the product of conversations with clients on site, and the complex exchanges of representation and reality, between texts read in the libraries and places experienced in parks and pleasure grounds, pathways and turnpikes.

Writings on cultural geography engage widely with the history and practice of landscape design and management. A number take a broad view of the making and meaning of landscape and the role of professional expertise in shaping it, situating plans and designs in wider worlds of landscape representation and of material worlds of land tenure, land use, infrastructure, management, movement and circulation. Books include *The Palladian Landscape* (Cosgrove 1993) where the idea of landscape is taken beyond Palladio's building projects to frame relations of country and city, land and water in the Veneto region; *Landscape and Englishness* (Matless 1998) which situates Town and Country planning and New Town design in nationalistic geographies of conservation and reconstruction, alongside other schemes such as ordnance survey mapping and soil conservation; and *Concrete and*

Figure 12.2 Humphry Repton's Trade Card. (Private collection, permission of owner)

*Clay* (Gandy 2002) which charts the cultural and material structures and processes of 'metropolitan nature' in New York City as they shape design and planning projects, including Central Park.

Alongside the study of pre-meditated schemes of landscape planning, design and management, these volumes contribute a concern with on-going, piecemeal modes of place making, improvised and unconscious perceptions, practices and customs, including routine uses, forms of work and livelihood, collective memories, emotional attachments, as well as small scale, material adjustments to the fabric of places, matters of mend and make do. While focused on the ways culture shapes, and reshapes, the material world, they are also conscious that the physical environment is not an inert or neutral setting but works on, and places limits upon, designs and developments. The geographies of landscape, the sites and spaces implicated in landscape, range in scale from regions and nations to smaller scale places and sites, including the internal spaces, the micro-geographies, of houses, gardens, factories and offices. Key concepts in cultural geography range across these spaces, for example a recent focus on 'home' moves in and through the material sites of houses or residential districts to consider the affective and political role of domestic space, from the actual or proverbial kitchen sink to real or imagined homelands, and their place in wider regional and global relations of migration and settlement, estrangement and belonging (Duncan and Lambert, 2004).

There are connections here with the conceptual frameworks of critical histories of architecture, like the concept of 'cultural landscape' which Dell Upton deploys to displace the heroic 'master narrative' of the human creation of landscape in design education, the architect and monument-centred nature of many histories which serve to justify the claims of the profession. 'The act of architecture' he asserts 'is one gesture in an endlessly recursive articulation of the individual and the landscape'. Here landscape is uncoupled from its professional enclosure as part of design and planning and released into wider processes of nature and culture:

> once introduced into the landscape, the identity of a building and the intentions of its makers are dissolved within confusing patterns of human perception, imagination and use. Consequently the meaning of a building is determined primarily by its viewers and users. The process of creation goes on long after the crew leaves the site; it never stops.
>
> (Upton, 1991, p. 195)

It is worth noting the edited collections and issues of multi-disciplinary journals in which cultural geography and landscape architecture are included, often arising from conferences and more lasting institutional exchanges in the field of landscape, for example *Representing Landscape Architecture* (Treib, 2008), *Sites Unseen* (Harris and Ruggles, 2007) and *Recovering Landscape* (Corner, 2007). 'Geography' as an emergent field of enquiry rather than an established discipline has been a framework for recent work in landscape and environment, particularly in the United States. Thus the journal of Harvard's Graduate School of Design is entitled *New Geographies* and includes contributions from a range of scholars and practitioners, including landscape architects and cultural geographers. Many articles in *Garden History* are based on research beyond the academy to professional projects of landscape restoration and conservation and by people keen to learn about the role of designed spaces in the places where they live, that connection of locality, landscape and lay scholarship which is so powerful in Britain. The range of such research, with a strong academic representation, is to be found in the special issue of *Garden History* I co-edited on 'The Cultural and Historical Geographies of the Arboretum', addressing the role of communities of both plants and people in arboreta throughout the world and their place as a model for other designed landscapes like public parks and cemeteries (Elliott et al., 2007).

A special mention should be made of the publications of the Landscape Research Group, both its journal, *Landscape Research* which has many thematic issues, and its books. An early book *Landscape, Meanings and Values* (Penning-Rowsell and Lowenthal, 1986) is among its most ambitious volumes, bringing together geographers David Lowenthal, Brian Goodey and Jay Appleton with landscape architect Hal Moggridge, environmental psychologist Kenneth Craik, evolutionary biologist Gordon H. Orians and landscape historian J. B. Jackson. As an audacious exercise in bridge building between the practice of academic scholarship and the ideas of professionals, the implications of this gathering remain to be fully developed and tested. The challenges should not be underestimated. The running commentaries in the book, interventions from those on the floor of the conference, were often sceptical of some of the social and physical scientific claims of the main speakers, and included interjections from cultural geographers who were developing a highly contextualized, interpretative perspective on the meaning and value of landscape.

## Stephen Daniels

### Living landscapes

Landscape is a meeting place, but it can seem less common ground than a field parcelled up into disciplinary enclosures, even a contested discursive terrain in which landscape seems less a world out there, on which we might have different but complementary perspectives, than a projection of particular forms of knowledge, in which cultural construction of the world matters more than the material world itself. A way forward, in matters of history and theory as well as practice and policy, is collaborative research and teaching arising from it.

The multi-disciplinary Arts and Humanities Research Council Landscape and Environment programme, which I have directed since 2005, includes fifty projects, focused on places both within and beyond the UK. A central theme of the programme is exploring the range of values, including aesthetic, commercial, spiritual, scientific, social, historical and ethical values, which are expressed in the way landscape is seen, designed, made and managed (Daniels and Cowell, 2010).

The programme reflects a resurgence of public interest and scholarly research, both nationally and internationally, in the ways in which landscape is valued and managed (ESF/COST 2010). Places may have their price, as creators of tourist revenue as well as the costs of conservation, but the value of Stonehenge to the English landscape, no less than the value of Shakespeare to the English language, is a matter of public good that cannot be measured adequately in monetary terms and which arts and humanities research helps us identify and enhance. Such valuation in terms of landscape takes us beyond the protection of certain sites to the wider, changeful world of their historical and geographical significance, including overlooked places and periods, and the views of those who live and work around them as well as visit them from afar. This is a growing feature of conservation and heritage management policy internationally.

The Landscape and Environment programme has created a collaborative community within and beyond the academy, including multi-disciplinary project teams in partnership with a range of people and organisations with a practical stake in landscape matters, notably on questions of heritage, conservation and public understanding. This section focuses on two example programme projects on northern England which include National Trust properties, Eskdale and Hadrian's Wall.

One of Britain's best known historic monuments, Hadrian's Wall provides a vantage point for exploring wide-ranging issues of landscape value. The Wall has been opened up to a wider world in the last decade, with sites made more physically accessible and publically understandable along its 117 km, coast-to-coast, length from the Solway to the Tyne, as it runs across upland pasture to post-industrial towns. As a linear insertion in the landscape, the Wall is a complex and challenging monument to maintain and manage, with over 50 organizations and 700 private individuals owning particular sections. It is a requirement of the Wall's designation as a World Heritage Site that it is both managed as a single entity along its length, and a ten-mile zone to either side, and that its historical significance is framed by a broad field of cultural values, accommodating the range of interests of those who conserve, use and enjoy the site including the million or so people who live or work within the Wall's regional sphere of influence.

**Figure 12.3**
Housesteads Roman fort on Hadrian's Wall, Northumbria, UK, now owned and managed by the National Trust. (Photo courtesy National Trust)

The scholarly concern to explore the wider world of Hadrian's Wall for its three-century Roman history, including its dynamic relations with the land and life of the territories it passed through, is beginning to be presented to a wider public, in on-site interpretation and associated guide books. What has been much less understood, and scarcely presented, is the Wall's long, post-Roman history, and its impact on the changing fabric of the Wall and how these express changing ideas of Roman Britain (Figure 12.3). The Landscape and Environment project 'Tales of the Frontier' has pioneered this form of understanding, by researching key periods in the Wall's post-Roman history, including that of the present, to uncover contrasting and competing perspectives on its Roman past, and their material effects in developing the surroundings of the Wall as well as the structure itself.

The multi-disciplinary team of archaeologists and geographers from Durham University have excavated a rich range of sources, including the rich seam of factual and fictional writings on the Wall, as well as new ethnographic and observational field work, to examine how the meaning of the Wall has been made and remade, materially and imaginatively, and variously represented and experienced, written on the page, performed in re-enactments, encoded in visitor conduct, inscribed in the earth in excavations and set in stone in restorations and reconstructions. They have examined what these findings tell us about wider questions of national, imperial and post-imperial identity. The project has explored the landscape narratives, the framing codes of time and space, which both shape, and are shaped by, the Wall as it transects a range of rural and urban sites and episodes of the region's longer history (Witcher *et al.*, 2010).

Eskdale in Cumbria is a prime example of upland common land (Figure 12.4). 'Contested Common Land' is a project focused on upland commons as a continuing collective resource with multiple, and sometimes conflicting, valuations

**Figure 12.4**
Eskdale in the Lake District National Park, Cumbria, UK, owned and managed by the National Trust. (Photo courtesy National Trust)

and uses. The project brings together trained historians, geographers, anthropologists and legal scholars from both sides of the Pennines, from the Universities of Newcastle and Lancaster, to address the issue of the sustainable use and management of common land. Archival and field work was combined with contemporary ethnographic research, interviews, group discussions and conversations, with farmers, landowners and land managers. Both forms of research helped reconstruct the collective memory of common land, as expressed in documents, embodied in custom, acted upon in practice, and manifested on the ground, in the traces of present and past activity. As well as informing and improving the governance of commons the project has enhanced wider scholarly understanding and public awareness (Rodgers *et al.*, 2010).

The projects at Hadrian's Wall and Eskdale explore a variety of different concepts of landscape and environment, both in the present day and in an historical context. They do so through multi-disciplinary approaches, encompassing legal, archaeological, anthropological and archival research. At Hadrian's Wall, issues of monumentality are addressed, locating the physical structure of the boundary within the fluid movement of people over time and place, from Roman imperial forces to the shipyards of the early twentieth century. At Eskdale, the community of farmers defined their relationship with the landscape through structures of governance centred on the manorial court, and later the commoners' association. In both examples, the National Trust is one of the organisations that have stepped in to protect and conserve the significance of these landscapes in the twenty-first century. Landscapes are composed of multiple layers of value, accreted over time and sustained through the interactions between people and place. Understanding those values requires a kaleidoscopic approach to research, one that looks beyond the boundaries of the academy, embracing both past and present, and theory and practice, in equal measure.

## Landscape and learning

This section sets out some groundwork in cultural geography, the areas of basic knowledge to equip students to engage with and contribute to the wider field of learning about landscape. These areas of knowledge are connected to physical sites, including the lecture theatre, library, archive and the field. This is an environment, a landscape of learning, with its own cultural geography, one we encourage students to reflect on in relation to other places for the practice, performance and production of knowledge about landscape, including the drawing office, the studio, the gallery and the laboratory.

Cultural geography is at present largely a scholarly field, of reading and writing, which has produced a large and growing literature in the form of books and journals, including a growing number of texts aimed at the student market, such as readers, dictionaries, companions and handbooks (a select list is included in the bibliography). This literature reflects cultural geography's disposition as a subject of research-led teaching, of critical analysis, theory building, interpretative exposition and historical perspective, if usually engaged with the practical implications and material effects of wider geographical projects concerned with representing and changing the world, whether through social activism or professional intervention, including schemes of conservation and management, planning and design.

### *Landscape in pictures*

Geography is traditionally a visual discipline, a graphic way of learning about the world. Visual imagery, including maps, diagrams, photographs and models, have long been both sources and methods of learning. If some of the practice of image making, such as surveying and field sketching, has either declined or been displaced in computer applications to more technical fields of the discipline, cultural geography has deepened and widened the critical interpretation of visual imagery, as produced and disseminated by the range of interests concerned with representing places and planning the landscape. So most textbooks on cultural geography (e.g. Gold and Revill, 2004) include exercises in image analysis, in a variety of media and genres, from landscape paintings of great estates and major cities to family photos taken in homes and gardens, from science fiction films to advertisements and promotional posters. Such images, including conspicuously documentary ones like maps, tend to be analysed more for the values they project than for the facts they portray, for the role of the symbolic in the representation of reality, for what pictures mean as well as what they show.

This perspective on images in geography was formulated in the 1980s when I was team teaching with Denis Cosgrove as a way of introducing the role of culture generally, and symbolism specifically, to geographers largely trained to look at the world in a narrowly factual, economic or utilitarian way, to encourage them to attend to wider questions of value, including aesthetics and politics. The method of iconography was adapted from art and architectural history. The procedure is best illustrated by examples from the case studies of artworks, maps, garden designs and buildings in our book *The Iconography of Landscape* (Cosgrove and Daniels, 1988) but two broad methodological points can be made here. First it involves an attention to both word and image, as part of a concept of 'cultural texts', and to the way these texts are embedded in particular historical and geography contexts which shape their meaning and give them life. In subsequent developments of the method, attention is placed to the way such texts function in a wider world. These include their effects on spectators and their material use and display in sites and spaces, for example landscape paintings (real or reproduction) on the walls of domestic interiors and maps pinned up in military headquarters or folded in the glove compartments of cars or pockets of anoraks. In this attention to objects as well as images, all artwork is site specific. Such images are not seen as metaphorical windows on the world or mirrors of cultural attitudes, but also as props in performance, material artefacts which are mobilized in the ways landscape is acted upon (Daniels, 2004). From this perspective framed artworks may be considered as fixtures and fittings like real windows and mirrors, and, in turn, real windows considered for their imagery and associations as well as material fabrication, as mediating interior and exterior space, as in recent research by cultural geographers and architectural historians (Isenstadt, 2007; Jacobs *et al.*, 2010).

Our initial teaching method for iconography was classroom based, using the art historical technique of double slide projection to compare and contrast images. Image projection has always been part of geography, going back to popular lectures of explorers and travellers – it is part of the subject's history as a form of both entertainment and instruction – but we used it in a less illustrative, more immersive and participatory way, first to convey the power of landscape imagery and secondly to prompt discussion and encourage conversation about it. Landscape images were to be looked into, not merely looked at, imaginatively entered and inhabited, moved in and around. While this might seem to anticipate forms of digital landscape visualisation, we had in mind the way geographers – and other outdoor types – were trained to look at maps, detecting the lay of the land, and how you might walk over it, from conventional cartographic signs, so not eliding or mistaking the signified for the sign. In tracing the material constitution and effects of pictures, imagistic teaching techniques are supplemented by other methods in other places, including gallery tours and site visits. Here iconography is combined with ethnography, including interviewing of various kinds, whether with artists on the implications of producing artworks of, or in, landscape, or of picture buyers for how they regard certain landscape paintings, including kitschy popular ones in cheap reproductions they have in their homes, how such artefacts are part of their lives (Daniels, 2004).

Maps are central, long-standing and eloquent forms of geographical representation. They have been subject to a good deal of critical interpretation by cultural geographers, and social criticism too, for their selectivity, for projecting the views of powerful elites and overlooking the views and lives of the poor and the dispossessed and, when translated into plans, of imposing the views of the powerful on the landscape itself. Indeed, maps fell out of favour as constructive visual aids or analytical images in cultural geography texts. Cartography has however undergone something of a revival in cultural geography, in part because 'mapping' has become such a powerful metaphorical shorthand for cultural interpretation in the arts and humanities. 'Deep maps', multi-media representations of localities, combining image, text and performance, draw on older cartographic and chorographic traditions (Pearson and Shanks, 2001, 162–185; Pearson, 2006, 15–16). A number of landscape practitioners, including artists as well as landscape architects, are deploying mapping techniques. Such cartography is often consciously creative and counterpointed to official style topographical maps. The 'parish maps' commissioned by the English conservationist group Common Ground are designed in a variety of sometimes picturesque styles and media to express a local sense of place (Crouch and Matless, 1996). The maps of New York-based landscape architect James Corner are collages including drawing and aerial photography, designed to reveal wide-ranging infrastructures and processes barely perceptible from the ground. Corner's maps fit with an idea of measure in the classical culture of landscape architecture, as a matter of ethics as well as of mathematics, of virtue and vision, *ethos* and *topos;* they are as much speculative instruments for thought as practical blueprints for action (Corner, 1996).

## Landscape in theory

Theory is integral to landscape research in cultural geography, part of its appetite for ideas. Theory is designed to do different things. Some is highly speculative, other is more anchored in empirical observation, often an ingredient of history, as when scholars of landscape design are as concerned to understand the precepts of the past as those of the present, in what may be a dialogue between theoretical perspectives. Theory also functions as a meta-discourse, for conversing with those in other disciplines as well as enlarging our understanding of processes on the ground. Much theory in cultural geography is imported, as theory tends to be in all disciplines, customised and translated from fields beyond landscape, in psychology, philosophy, sociology and literary criticism. If theory in landscape geography does not play quite the provocative role it does in landscape archaeology, the 'archaeology with attitude' which antagonises some field workers (Pearson and Shanks, 2001), it does come with a degree of rhetorical relish. There has for centuries been a stubborn empiricism in English landscape study and practice, a scepticism particularly about French theory, which persists in those who want less on Foucault and Irigaray and more on fortification and irrigation.

Much theoretical ground work in cultural geography is in a sense archaeological, excavating some of the conceptual foundations of geography as a field of study, probing the power and genealogy of its basic ideas and root metaphors, their material relations and effects. An essay which is still useful as a exercise primer on the multiple conceptualisation of landscape is by an old school cultural geographer, D.W. Meinig, in which he considers the way that different analogies for landscape – as nature, habitat, artefact, system, wealth, ideology, history, place and aesthetic – configure the ways we interpret a particular scene (Meinig, 1979).

The work of literary historian and cultural theorist Raymond Williams was influential in emphasising the power of concepts in cultural geography, as it sought to clarify the multiple, and sometimes conflicting meanings of keywords like landscape itself, and its cognates like region, space, environment and nature, country and city. Those following in Williams' path have shown how the language of landscape reveals and also conceals perceptions and values, professional languages concerned with designing, planning and managing the land and the built environment. Landscape may connote both substance and illusion, depth and surface, solid environments and superficial scenery. 'Landscapes can be deceptive', John Berger wrote in his photo-essay *A Fortunate Man:* 'Sometimes a landscape seems less a setting for the life of its inhabitants, than a curtain behind which their struggles, achievements and accidents take place' (Berger, 2005), the lie of the land. Landscape can be placed in critical counterpoint to concepts like place and space, one, more concrete, permeating everyday life, the other more abstract and analytical (Anderson, 2003, 227–282). As a speculative instrument the concept of landscape requires careful handling, as Kenneth Olwig once reminded me 'landscape is good to think with, but it is also good to think against'.

There are a number of cultural geography guidebooks to theory, setting out the strengths and limitations of various positions, if usually from preferred perspectives. Most theoretical texts in cultural geography take a relativist perspective of knowledge and put into question proposals to build any total, universal theory of landscape, including ones which conjecture essential, archetypal or scientifically based ideas or values of landscape, which is a challenge for the role of theory in some practical field of landscape, of design, conservation and heritage designation. *Landscape* by John Wylie moves from cultural materialism to phenomenology, and is unusual among cultural geography texts in having no images, other than the cover illustration of a painting by Cezanne, but fits with the wider geographical take on theory as a world of words (Wylie, 2007). *Landscape* sets a challenge for all landscape researchers, to stand their ground and see what difference their experience of working with landscape, as itself a way of thinking about the world, makes to wider currents of theory, to think of the geography of theory, how theory might be as it were landscaped.

Most discussions of landscape in theory have been spatial, although increasing attention is being paid to questions of time and temporal process, history and narrative, landscape as a way of telling as well as a way of seeing. Matthew Johnson's *Ideas of Landscape* (2007) addresses a consciously English tradition of landscape research 'firmly in the grip of the most unreflective empiricism in which "theory" is a dirty word and the only reality worth holding onto is that of muddy boots' – a direct, unmediated encounter with the real world. In the process he recovers the theoretical implications of this tradition, its 'habits of thought'. The book centres on the writings of W.G. Hoskins. Hostile to academic theory of any kind, especially as a badge of international professional advancement (his book *The Making of the English Landscape* he said might have made a greater impact if it had been called *The Morphogenesis of the Cultural Environment*) Hoskins's writings are structured by a romantic, sometime militant, particularism, consciously so when he launches attacks on the powers in the land which destroyed the detailed local livelihood of peasant cultures. Perhaps inevitably, Hoskins' landscape narrative is an elegiac one, indeed there seems scarcely any landscape in England in his book after the nineteenth century.

Narrative interpretation in landscape study is not just a matter of reflecting on the way we were; it may be seen in terms of a broader frame of cultural analysis which includes William Cronon's work on narratives of environmental history. 'We tell stories', notes Cronon 'to explore the alternative choices that might lead to feared or hoped-for futures' (quoted in Daniels, 2008 p.241). There is an affiliation here with the work of Ann Whiston Spirn set out in her book *The Language of Landscape* in which pictures, including her own expressive photographs, are as important as words in telling 'landscape stories' (Spirn 1998). These stories include those of her own student participation projects for reclaiming the social and physical landscape of a poor district of West Philadelphia, Mill Creek, and promoting local learning about its history and function as a river valley.

## *Landscape in the library*

The concluding essay of John Stilgoe's recent collection on the changing American landscape is about the importance of libraries and archives for intensive landscape research, not just reading the books and manuscripts there, but in navigating such places, wandering the stacks, browsing the shelves and the great card catalogues, the library as a landscape of learning. Now in Stilgoe's own research library at Harvard, with so many old books stored off site, stack wandering is a restricted activity, and with computerised catalogues, students and even faculty 'know nothing of the golden-oak file drawers around which everyone once clustered in some railway depot-like way'. As so often with landscape history, the idea of landscape is a nostalgic one, a world we have lost, here a sociable landscape of learning as much displaced by new technology as the sociable suburban railway by individual automobile commuting. Landscape as a category 'confounds the finest reference libraries and cataloguing systems' and what Stilgoe misses is the serendipity of library research for a field which is not well served by systematic organization, the remark of a colleague around the card catalogues, the volume or manuscript come across by chance (Stilgoe, 2005).

Despite Stilgoe's misgiving, there is still much in modernised libraries and archives which is a rich resource for landscape research, including online browsing of data bases. There is great potential for systematic as well as serendipitous research. In the UK local record offices are by definition place specific. No less than the landscape outside, they are part of the cultural memory of a place. Local archives often catalogue their records accordingly, with extensive collections of maps and plans, deeds and directories, which may be linked systematically. Deposits of family estate papers are particularly promising for situating proposals of landscape design within the larger business of estate management, say within processes of agriculture and forestry. Documents of all kinds can be assembled on the library tables, from ledgers of accounts to correspondence between architects, clients and estate stewards.

While there may be comprehensive sets of information, much of the pleasure of archival landscape research is piecing together eclectic and fragmentary sources to build up a picture of a place and its livelihood, rather like a field archaeologist putting together fragments of material. This is the perspective of a seasoned archival researcher, for whom such libraries are a comfort zone. It looks very different starting out. While record offices are keen to promote wide access to archives as a matter of their public funding, it takes a while to find one's way into and around a collection, initially involving a good deal of waiting for document delivery (and often a document which turns out not to be as useful as the catalogue suggested) and this can be frustrating to students and young scholars used to instant access to information. It has prompting some recent reflections among cultural geographers on the cultural barriers to archival awareness and conduct, likening such libraries to labyrinths and fortresses (Lorimer 2009). Like any place worth knowing, a library takes time.

In the historical variant of cultural geography I am experienced in teaching, archival work is reciprocally connected with field work, the trip to the record office with the transect of the countryside. In the mythology of landscape history, the scholar moves between mud of the field and the dust of old documents. It's no accident that one of the presiding metaphors for landscape in historical research is the palimpsest, a text which is continually annotated and overwritten by successive generations, and whose layers of meaning need to be carefully decoded for the landscape to be read. Relations of library and field are complicated. Much landscape fieldwork is urban, not all archival documents are old manuscripts. There is a rich range of archival material, including objects as well as documents, printed matter as well as manuscripts, and increasingly many documents are being scanned and posted on library websites, often to conserve them. Thus, archival research now can be conducted in the comfort of one's home, or anywhere on a portable computer, if perhaps something of the material culture of an archive is lost to researchers when they no longer handle old paper or pore over parchment in situ.

Not every document in a library is relevant to landscape research, and many that are are not collected, curated and conserved. Stilgoe notes that much on paper of interest to the landscape historian, especially of everyday places, is either not collected by research libraries, like mail order catalogues or motoring maps, or cut out of their material as ephemera, such as advertisements from magazines when they are bound as periodicals. So some scholars of the byways of landscape research go in search of documents and artefacts for their own private research collections, memorabilia like old photos and cigarette cards in 'haphazard encounters with rummage sales, flea markets and barn auctions', a new antiquarianism, even a vernacular version of the connoisseurship of cultural historians who once collected landscape paintings and fine prints, manuscripts and first editions.

## Landscape in the field

While cultural geography seems a largely a bookish pursuit it is still shaped, and inspired, by a physical engagement with the world, and of specific places, localities and regions, by ground truths as well as cultural interpretations. Many of its findings draw on various kinds of site-specific study, including field observation and recording of terrain, building types and land use and various kinds of ethnographic work, interviewing and conversing with people who live and work in places, using methods of oral history and participant observation.

Cultural geographical fieldwork, drawing on a rich range of sources, both documentary, and environmental can promote a deeper and wider understanding of a locality. These include familiar places where students live and work, looking at the overlooked, as well as spectacular places they might visit. Effective fieldwork in cultural geography is as much a matter of mentality as a methodology, as much the cultivation of a receptive and empathetic outlook as the acquisition and deployment of a tool kit of systematic techniques. Formal procedures for surveying, field walking, focus group work, diary keeping and interviewing are effective when deployed with an attentive attitude which comes when researchers care about a place and the people who inherit and inhabit it. Reading the landscape is a synthetic skill, and integrative accomplishment, which I will illustrate in terms of undergraduate field projects which include planned and designed landscapes as part of a wider cultural landscape and its local and regional geographies.

Still a good place to start as a student introduction to cultural geography fieldwork is Denis Cosgrove's essay 'Geography is everywhere: culture and symbolism in human landscapes' (Cosgrove, 1988, 2008) for it showed the cultural richness of what seems at first sight an unremarkable place, the shopping precinct of a small town in the English Midlands, and did so as a matter of informal observation as well as formal investigation, landscape appreciation as a matter of educated citizenship as well as professional training, of self-knowledge as well as knowledge about the world. The precinct contained 'an entirely predictable collection of chain stores', even a symptom of the placelessness critics complain about as they yearn for 'real high streets' with locally owned shops, but considered in detail the precinct was revealed to be 'a highly textured place, with multiple layers of meaning'. This was evident from shopping on a Saturday morning, conscious of the way the place is used, an evangelical distributing tracts, punk teenagers hanging around the concrete base of the decorative tree scowling at shoppers, adverts for window panels for house insulation which will 'destroy the visual harmony of my street'.

> The taken-for-granted landscapes of our daily lives are full of meaning. Much of the most interesting geography lies in decoding them... Because geography is everywhere, reproduced daily by each one of us, the recovery of meaning in our ordinary landscapes tells us much about ourselves.
>
> (Cosgrove 1988, 2008, p. 185)

The lessons of the group field excursion we conducted for a number of years in Italy, in Venice and the Veneto (Figure 12.5), were published under the title *Fieldwork as Theatre* (Cosgrove and Daniels, 1989). This now appears prescient given the current cultural geography concern with issues of performance, in the conduct of investigation as well as in the worlds observed. We were mindful of some established connections between landscape and theatre, in the fields of Renaissance and eighteenth-century design which formed much of the classroom subject matter, if

Stephen Daniels

**Figure 12.5**
The author and Denis Cosgrove leading a group of students in Vicenza in the Veneto, Italy, on one of the field tours described in the text. (Courtesy Stephen Daniels)

we were concerned to extend them beyond spectacular scenographic aesthetics to consider the ways that wider land uses and civil engineering schemes, especially concerned with water and water management, were incorporated in the making and meaning of the Venetian landscape, to consider the landscape as a regional material and cultural system, some of which was hidden while other parts displayed. So the palaces and squares of the city were connected to the villas and gardens of the country, and these more ornamental spaces to the functional spaces of canals, drainage ditches and irrigated farmland.

The more recent field excursions I have organized are part of specialist research-led courses on eighteenth-century England landscape, including its designed landscapes in country and city. The courses are more explicitly inter-disciplinary, available to students from beyond geography, and as part of a Nottingham University Masters degree in Landscape and Culture. The field courses relate to primary scholarship conducted by students on historical texts, archival work on maps and plans and other documentation of estate layout and management in the Midlands, and the interpretation of the imaginative literature, including novels of the period, notably Jane Austen's *Mansfield Park*, which explore landscape improvements as part of wider narrations of characters' knowledge and experience of places and spaces, including global geographies of war and trade.

The main field excursion is through the Derwent Valley of Derbyshire. This covers the area designated as a World Heritage Site in 2001, the 15 miles from Matlock to Derby famous for pioneering the factory production of textiles, mostly cotton mills which are no longer functioning but which have been converted to a variety of new uses, including museums, and areas of ancillary infrastructure, such as mill housing and canals. The excursion goes farther to take in other eighteenth- and nineteenth-century landscapes in the valley, notably country houses and parks and towns, as part of the theme of water power, water to power mills, transport materials, to irrigate parks and be made into water features like the great fountain at Chatsworth, water for health, including the supply of spas and baths for tourists of the time. The point of the excursion is to consider the Derwent Valley as a hydraulic region, a liquid landscape, to look at questions of circulation in its cultural ecology as well as form.

## Concluding thoughts

This chapter developed from a report for LE:NOTRE (see the Introduction), and has in the process shifted in shape and substance to reflect the essay form as a genre, one in which the discursive process of conversation is paramount, between different styles of presentation as well as authorial perspectives: in this essay teaching texts, policy documents, biographies, theoretical treatises, and guidebooks. Much writing on landscape interpretation, especially landscape history and aesthetics, takes an essay form, but arguably the practical projects of design and planning, particularly those concerned with issues of care and conservation, both cultural and environmental, involve a conversational procedure. Thus Catherin Bull's book on projects of landscape architecture in contemporary Australia is titled *New Conversations with an Old Landscape* (Bull, 2002) to describe the capacity of plans and designs to develop complex, reciprocal exchanges between peoples and environments, culture and nature. The place of landscape is a meeting place, a forum for the exchange of knowledge and practice, and the art of landscape is arguably connected to the art of conversation. What would enhance the exchange between cultural geography and landscape architecture is more joint participation collaborative projects, in which polite conversation could be sharpened by practical experience of the limits as well as opportunities for interdisciplinary work.

## Acknowledgments

My thanks to the following conversationalists: Catherine Dee, John Gold, Brian Goody, Dianne Harris, Beth Meyer, Finola O'Kane, Kenneth Olwig, George Revill, Maggie Roe and Charles Waldheim.

## Bibliography

Anderson, K. (2003) *The Handbook of Cultural Geography*. Sage, London.
Banham, R. (1971) *Los Angeles: The Architecture of the Four Ecologies*. Allen Lane, London.
Berger J. (2005) *A Fortunate Man: The Story of a Country Doctor*. London: Royal College of General Practitioners.

Blunt, A. (2003) *Cultural Geography in Practice*. Arnold, London.
Bull, C. (2002) *New Conversations with an Old Landscape*. The Image, Mulgrave, Vict.
Clifford, N and Valentine, G. (eds). (2010) *Key Methods in Geography*. Sage, London.
Corner, J. (1996). 'The Agency of Mapping: Speculation, Critique, and Invention'. In *On Landscape Urbanism* (pp. 148-173). Austin TX: Center for American Architecture and Design University of Texas at Austin School of Architecture.
Corner, J. (ed.) (2007) *Recovering Landscape: Essays on Contemporary Landscape Architecture*. Princeton Architectural Press, Princeton.
Cosgrove, D. (1988, 2008) 'Geography is Everywhere: Culture and Symbolism in Human Landscapes', in Derek Gregory and Rex Walford (eds) *Horizons in Human Geography*. London, 118–135, reprinted in Oakes and Lynn, *The Cultural Geography Reader* 176–185.
Cosgrove, D. (1993) *The Palladian Landscape*. Leicester University Press, Leicester.
Cosgrove, D. and Daniels, S. (eds) (1988). *The Iconography of Landscape*. Cambridge University Press, Cambridge.
Cosgrove, D. and Daniels, S. (1989) 'Fieldwork as Theatre: A Week's Performance in Venice and its Region'. *Journal of Geography in Higher Education* 13, 169–183.
Crang, M. (1998) *Cultural Geography*. Routledge, London.
Crouch, D. and Matless, D. (1996) 'Refiguring Geography: Parish Maps of Common Ground'. *Transactions of the Institute of British Geographers* 21, 236–255.
Daniels, S. (1999) *Humphry Repton: Landscape Gardening and the Geography of Georgian England*. Yale University Press, New Haven and London.
Daniels, S. (2004) 'Landscape and Art', in Duncan *et al.*, *A Companion to Cultural Geography* 430–446.
Daniels, S. (2008) 'Landscape and Narrative', in J. Elkins and R. Delue (eds) *Landscape Theory*, Routledge, Abingdon.
Daniels, S. (2011) 'Classics in Human Geography Revisited: The Iconography of Landscape'. *Progress in Human Geography* 35, 266–270.
Daniels, S. and Cowell, B. (2010) 'Living Landscapes' in J. Bate, *The Public Value of the Humanities* pp. 105–107. Bloomsbury, London.
Daniels, S., DeLyser, D., Entrikin, J. N., and Richardson, D. (eds) (2010) *Envisioning Landscapes, Making Worlds*. Routledge, London.
Delyser, D., Herbert, S. and Aikin, S. (2009) *The SAGE Book of Qualitative Geography*. Sage, London,.
Duncan, J., Johnson, N. and Schein, R. (2004) (eds) *A Companion to Cultural Geography*. Blackwell, Oxford.
Duncan, J. and Lambert, D. (2004) 'Landscapes of Home', in Duncan *et al.*, *A Companion to Cultural Geography,* 382–403.
Elliott, P., Watkins, C. and Daniels, S. (eds) (2007), 'Cultural and Historical Geographies of the Arboretum', *Garden History* 35, supplement 2.
ESF/COST (2010) 'Landscape in a Changing World: Bridging Divides, Integrating Disciplines, Serving Society', European Science Foundation/COST Science Policy Briefing Note 41. Available at: http://www.esf.org/publications/science-policy-briefings.html (accessed 7.4.11).
Gandy, M. (2002) *Concrete and Clay: Reworking Nature in New York City*. MIT Press, Cambridge Mass.
Girouard, M. (1978) *Life in the English Country House*. Yale University Press, New Haven and London.
Gold, J. and Revill, G. (2004) *Representing the Environment*. Routledge, London.
Harris, D. and Ruggles, D. F. (eds) (2007) *Sites Unseen: Landscape and Vision*. University of Pittsburgh Press, Pittsbugh.
Hoskins, W. G.(1955) *The Making of the English Landscape*. Penguin Books, London.
Isenstadt, S. (2007) 'Four views, Three of Them through Glass, in Harris and Ruggles', *Sites Unseen*, 213–240.
Jacobs, J., Cairns, S. and Strebel, I. (2010) 'Materialising Vision: Performing a High Rise View', in Daniels *et al.*, *Envisioning Landscapes, Making Worlds,* 256–268.
Johnson, M. (2007) *Ideas of Landscape*. Oxford, Blackwell.

Jones, M. and Olwig, K. (eds) (2008) *Nordic Landscapes*. University of Minnesota Press, Minneapolis.

Lorimer, H. (2009) 'Caught in the Nick of Time: Archives and Fieldwork', in Delyser *et al.*, *The SAGE Book of Qualitative Geography*, 248–273.

Matless, D. (1998) *Landscape and Englishness*. London, Reaktion.

Meinig, D.W. (1979) 'The Beholding Eye, Ten Versions of the Same Scene', in Meinig (ed.) *The Interpretation of Ordinary Landscapes,* 33–48.

Meinig, D.W. (ed.) (1979) *The Interpretation of Ordinary Landscapes: Geographical Essays*. Oxford University Press, Oxford.

Oakes, T. S and Price, P. L (eds) (2008) *The Cultural Geography Reader*. Routledge, London.

Penning-Rowsell, E. and Lowenthal, D. (eds) (1986) *Landscape, Meanings and Values*. Allen and Unwin, London.

Pearson, M. (2006) *'In Comes I' Performance, Memory and Landscape*. University Press Exeter, Exeter.

Pearson, M. and Shanks, M. (2001) *Theatre/Archaeology*. Routledge, London.

Rodgers, C. P., Straughton, E. A., Winchester, A. J. L. and Pieraccini, M. (2010) *Contested Common Land: Environmental Governance Past and Present*. London and Washington, Earthscan.

Shurmer-Smith, P. (2002) *Doing Cultural Geography*. Sage, London.

Spirn, A. Whiston, (1998) *The Language of Landscape*. Yale University Press, New Haven.

Stilgoe, J. (2005) *Landscape and Image*. University of Virginia Press, Charlottesville.

Summerson, J. (1945) *Georgian London*. Pleiades Books, London.

Thrift, N. and Whatmore S. (eds) (2004) *Cultural Geography,* 2 volumes, Routledge, London.

Treib, M. (ed.) (2008) *Representing Landscape Architecture*. Taylor and Francis, London.

Upton, D. (1991) 'Architectural History or Landscape History', *Journal of Architectural Education* 44, 195–199.

Witcher, R. E., Tolia-Kelly, D. and Hingley R. (2010) 'Archaeologies of Landscape: Excavating the Materialities of Hadrian's Wall', *Journal of Material Culture* 15, 1.

Wylie, J. (2007) *Landscape*. Routledge, London.

# Part 4

# Conclusions

Chapter 13

# Crossing the boundaries?

*Maggie Roe*

## Introduction

This chapter aims to pull out from the previous 12 chapters some key messages and to provide overall reflections from the view of landscape architecture. The content of the chapters, the subjects chosen for particular examination and the style of the writing are all taken into account in this analysis and the discussion is set within the context of a growing focus within environmental disciplines on cross-disciplinary working. This has arisen from an acknowledgement that the complexity of environmental issues and the need to address 'real world problems' requires knowledge from many different discipline areas (ESF/Cost, 2010; Marzano *et al.*, 2006; Brewer, 1999; Brewer and Lövgren, 1999), a point now supported in European policy by the European Landscape Convention (ELC). In the professional sphere, landscape architects nearly always work as part of a multi-disciplinary team and commonly pull in the services of other specialists. However there is some considerable difference between working as separate professionals based on understandings and using techniques of that discipline area, and working in an integrated fashion as part of an interdisciplinary or transdisciplinary team (Figure 13.1). While all methods of working tend to be dynamic (Tress *et al.*, 2006; Marzano *et al.*, 2006) and the boundaries between these ways of working commonly cross or overlap, it is useful in the context of this book to consider the benefits of how such knowledge exchange can help achieve more integrated ways of working.

In the academic world, the knowledge cultures traditionally tend to be quite clear with distinctions in research methodologies, theoretical approaches, ways to collect and validate data etc. (Tress *et al.*, 2006). In the professional world there is also considerable discipline separation, even with disciplines that basically work with the same subject matter. Much of this separation seems to be emphasised or even enforced by professional bodies and codes of conduct. However with the emergence of crossover discipline areas such as urban design, landscape archaeology and landscape ecology, these boundaries become much less clear and in academia interdisciplinary working attempts to bring together discourses from the natural sciences, social sciences and humanities as well as qualitative and

| Term | Involvement | Aims | Way of Working |
|---|---|---|---|
| Disciplinary | Working within one discipline area. | Goals determined by creating new disciplinary knowledge or solving project objectives from a single disciplinary approach. | Working within normal discipline boundaries or recognised professional expertise commonly set by professional institutions. |
| Multidisciplinary | Involving several different disciplines to examine one theme or problem but each discipline retaining its own goals. | Participants set goals under one 'thematic umbrella' and exchange knowledge, but do not aim to cross subject boundaries to create new knowledge and theory. | Loose cooperation where work is carried out in parallel either to coalesce or compare without integrating knowledge or results. |
| Interdisciplinary | Several disciplines (literally 'between disciplines'). | Achieve a common goal and create new knowledge and theories | Crossing subject boundaries. |
| Participatory | Academics and experts working with non-academic and non-expert participants. | Achieve a common goal to solve a common problem. | Knowledge is exchanged but not integrated to form new knowledge. |
| Transdisciplinary | Academic and non-academic participants from various disciplines (including the general public). | Achieve a common goal or solve a common problem and create new knowledge and theories. | Crosses disciplinary and scientific/academic/professional boundaries. Combines interdisciplinarity with a participatory approach |

**Figure 13.1** Interdisciplinary ways of working. (Source: Based on Karlqvist, 1999; Tress et al., 2006)

quantitative methods. Over the years there has been much discussion about whether the 'divides' between humanities/arts and science subjects are real or exaggerated and what the significance of such a divide actually is (see Snow, 2008; Gould, 2004). However the need to rejoin cultures is important, as Snow said in 1959:

> In our society (that is, advanced western society) we have lost even the pretence of a common culture. Persons educated with the greatest intensity we know can no longer communicate with each other on the plane of their major intellectual concern. This is serious for our creative, intellectual and, above all, our normal life. It is leading us to interpret the past wrongly, to misjudge the present, and to deny our hopes of the future. It is making it difficult or impossible for us to take good action.
> (Snow, 1959: 60)

Although the benefits of closing the gap between cultures is recognised, there is a perception that cross-disciplinary working can be risky (Hansson, 1999), and it is still difficult to gain funding for interdisciplinary research although funding for establishing networks for cross-disciplinary research is more generally available.

Crossing the boundaries to work in an integrated manner requires flexibility of approach and outlook but the benefits can be large and most knowledge breakthroughs of long-lasting importance have resulted from such working (Hansson, 1999). Examining problems from different viewpoints can provide knowledge that can both extend and change existing knowledge areas (Bhatia, 2002). Knowledge can be seen as overlapping rather than bounded and perhaps is characterised more by the outlook or approach than real difference. Difference in disciplines is not only about the way that problems are viewed, but about diverse skills and language or communication methods used, thus it is sometimes more difficult to explain what is known (knowledge) rather than how it is known (methodological approaches) (Bahtia, 2002; Karlqvist, 1999; Marzano *et al.*, 2006). As suggested by Price (Chapter 9), collaboration may require more than just turning up to a meeting, but also the abandonment of 'cherished preconditions' for working. A good starting point for cross-disciplinary working is to provide opportunities for an improved understanding of the disciplinary discourses, the methods, outlooks and direction of the gaze (Phillipson *et al.*, 2009) – ways of thinking and working – as well as ensuring such work is with those who have the intellectual agility and willingness to undertake it.

In examining the chapters in this book I have carried out both a content analysis and an analysis of tone and style – or a discourse analysis – to see how the view is being communicated. This analysis is necessarily concise, but of course further detail can be gleaned from the chapters themselves. The editors' brief provided me with the basis for the three main research questions for this task:

- Reflect on positions from different cultures
- Identify key messages
- Provide information that could indicate a way forward for landscape architectural teaching and practice.

There is a wealth of information in these chapters which will be of considerable interest to those who are working or intend to work in cross-disciplinary teams. For the purposes of this chapter, what emerges is a number of key themes which are discussed separately and then I have attempted to pull out some particular messages for the profession of landscape architecture.

## Key messages
### Overall view 'from the outside'
Throughout the varying picture of the relationship with landscape architects there is considerable enthusiasm, both for this particular project but also for the way that landscape architects are willing to reach outside the profession. There are considerable overlaps evident with other disciplines, particularly with the areas that have themselves emerged from narrower disciplines in recent years – such as landscape ecology, landscape archaeology, cultural geography and the increasingly broad areas that fine art covers in relation to the wide range of landscape dimensions that can

be identified (see Figure 13.2). For example, Fairclough (Chapter 4) describes how landscape archaeology is similarly concerned with different scales of investigation and entails problem solving thus providing much potential for crossover with landscape architecture. Both disciplines have diverse theoretical bases. Archaeology is primarily a research-based discipline based on investigation, survey and preservation planning rather than creativity. The implication in this chapter, and others in this book, is that landscape architecture is weak on theory and there is both a lack of theorising and conceptualisation by landscape architects of their practice.

Some of the disciplines feel that there is considerable collaboration and crossover already with landscape architects e.g. forestry (Bento and Lopes, Chapter 8) and cultural geography (Daniels, Chapter 12). Some disciplines appear still to have quite a narrow view of what landscape architecture covers; much narrower than is recognised by the discipline itself. This view tends to be from the more traditional disciplines rather than the more recently emerging discipline areas. There is still a view from the outside that landscape architecture concentrates on the 'pictures' painted – the visual and subjective aspects of landscape – and there is a continuing misconception that the materials that landscape architects predominantly deal with are planting and 'greenery' when much if not most of their work involves hard surfaces, engineering matters, consideration of and design of aspects of grey infra-

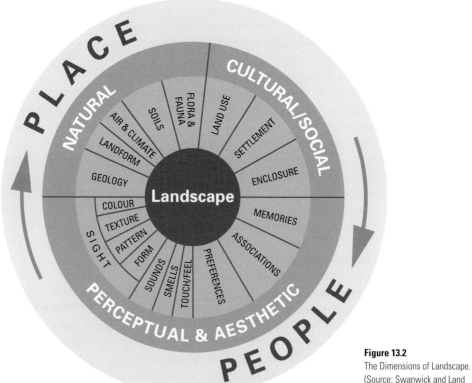

**Figure 13.2**
The Dimensions of Landscape.
(Source: Swanwick and Land Use Consultants, 2002)

structure, etc. The implication is that landscape architects are not very good at publicising what they can and actually do.

While some views seemed to consider that landscape architects presently practise predominantly on one side or the other of the arts–natural sciences divide, the view from cultural anthropology (Rotenberg, Chapter 10) is that landscape designers manage to mediate between these two spheres. Bunce (Chapter 6) suggests that landscape architects could help landscape ecologists apply landscape science while landscape ecologists can help landscape architects understand the science. Education is the critical area here since, as is suggested, the application of landscape ecological knowledge often does not appear to be occurring within landscape designs and plans at present.

The concentration in Chapter 1 (Jormakka, architecture) indicates that traditional garden design is still seen as an important source of knowledge, particularly in relation to the style of design projects. Most landscape architects, while acknowledging the historical importance of garden design in the development of contemporary designed landscapes, would also probably regard the study of such tradition as only one part of a much wider area of studies. The study of garden design (in its broadest sense) can provide a doorway into a wide range of aesthetic and historical studies, the psychology of perception, theoretical underpinnings of aesthetics and social and economic management of the landscape. However there are problems with the association of the work of contemporary landscape architects with that of garden design, even in its widest sense, because it can severely limit the understanding of the scope of work presently carried out by practitioners by the general public in particular, but also by potential clients, policy makers, politicians etc.

Planners have for some time (if not always) theorised on new concepts and formats for cities (see Venturi, Chapter 11). While urban processes and fabric is seen to be in a state of flux, the view from the planning profession is that there would seem to be a role for landscape architects in helping to stabilise the material world of the city, but these opportunities need to be more actively taken up. The suggestion is that landscape professionals need to innovate, to develop agility of intellect and new knowledge relating to a range of design aspects including mitigation and adaptation for actual and perceived threats, the creation of new identities for new cultures emerging in the city, the potential of new technologies and the demand for alternative forms of infrastructure.

It was surprising in examining these chapters that two key considerations in the work of landscape architects were given scant mention, or perhaps landscape architects do not provide enough focus upon: sustainability and public participation. Ten years ago it was recognised that many landscape architects had much to learn with regard to the incorporation of sustainability concerns, including gaining skills for working with communities within practice (see Roe, 2000; Roe and Rowe, 2000). The view from planning (Venturi, Chapter 11) is that landscape architects should not change tools and materials, but change attitudes and visions – perhaps theories and skills for using the tools and the materials – and employ a change of focus to take the longer view, producing designs that mitigate rather than adapt; 'preventing' or providing risk avoidance strategies and the starting point for future building design rather than solving problems post-construction. The inference is that landscape architects tend to take a short-term view in design expression. Åsdam (fine art, Chapter 5) sees their focus primarily on the site level and suggests that further collaboration would help landscape architects to gain insights by 'asking the right questions in relation to the right site'.

The view expressed by Rotenberg (cultural anthropology, Chapter 10) suggests that what landscape architects do is to extend 'control' over the natural world through their work. The more 'green' view would be that landscape architects endeavour to work *with* nature in a more William Robinson style, grappling with concepts such as managed retreat, resilience and succession rather than reproducing a kind of Le Nôtre/Louis XIV approach to the control of landscape. As identified in these chapters there are still many skills to be learned and theoretical concepts to be grappled with relating to more sustainable practice and this dilemma between control and change is fundamental. New practical and theoretical approaches to tackle such dilemmas could emerge from learning from a range of disciplines.

## Communication and language

The different language of disciplines is commonly regarded as a problem area for interdisciplinary working (Hansson, 1999; Tress *et al.*, 2006). Anthropologists (Rotenberg, Chapter 10) communicate using words in a 'short-hand' way understandable only to the discipline. Thus language is seen to be important in communication not only between discipline areas but also within disciplines themselves in project development and in research. Lack of common terminology may be an obstacle to cross-disciplinary work in sharing information (Florgård, 2007).

Communication is not just about the terms and jargon used, but about the concepts and meanings behind the words. The issue of language and interpretation are areas of some considerable academic study (e.g. Benson, 2004; Campbell, 2003; Eco, 1992, 1990; Fisher, 2003; Longatti and Dalang, 2007; Roe *et al.*, 2008). It is generally acknowledged that the use of words, and their definitions and uses, matter. Since language is seen as an indicator of conceptual understandings it is important in interdisciplinary work for participants to understand both the language and the concepts behind language used. Sometime there is a call for the development of a 'common language' between different disciplines, but this is problematic and perhaps Benson's (2004) suggestions to use clear language and more explicit approaches to defining meanings in relation to landscape is most helpful in reducing obfuscation in terms, reduce distrust between experts and promote a more inclusive approach to addressing landscape. Communication – particularly through mediation and negotiation – and management are regarded as key skills in the education of landscape architects (Venturi, Chapter 11), as is the importance of facilitating the communication of ordinary people with each other through the way landscape is designed and structured.

Thus three key suggestions emerge, that landscape architects need:

1  a greater understanding of the way landscape language is used in order to work with different disciplines

2   to build up skills for conversations with other disciplines
3   a greater recognition of the way design can foster communication in the landscape.

## Theoretical underpinning

Landscape architecture is sometimes criticised for not having 'a theory'. However, like many disciplines in this book, the sources of theory and value for landscape architecture are broad and theory itself is continually being added to as the scope of work changes. Ian Thompson (1999) provides a useful way of categorising key value systems into three main areas: ecology, community, delight (Figure 13.3) which broadly cover the science, social science and arts-based sources of theory for landscape architecture. Understanding the key theoretical sources of other disciplines can also provide considerable insights to those disciplines.

Landscape architects generally do not discuss theory as much as some other disciplines, but if theory is translated as 'ways of seeing' or 'bases for working' then it is perhaps easier to understand the breadth of landscape architecture theory that landscape practitioners either consciously or unconsciously use in practice. However, the examination of new areas of theory can open up 'restricted territories', or areas of understanding that landscape architects tend to avoid. Theory can emerge from discipline-based or curiosity-driven inquiry, but perhaps as primarily practical or action-based animals, the development of landscape architectural theory and methods can be seen to emerge from the problems themselves rather than the reverse (Brewer, 1999). The important point perhaps is that theories should be approached critically and not swallowed whole. Theoretical development can open

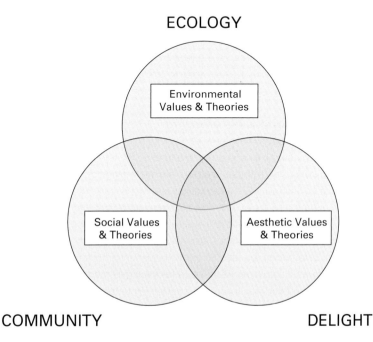

**Figure 13.3**
Overlapping fields of value and sources of theory in landscape architecture. (Source: Based on Thompson, 1999)

up new ideas and a basis for working. Interdisciplinary working has been shown to stimulate the flow of ideas and techniques between cultures and be exciting and challenging (Phillipson et al., 2009).

Landscape ecology theory has for some time had a considerable influence on landscape architectural thinking in many countries. Bunce (Chapter 6) regards a good understanding of landscape ecology as essential for landscape architects. However, the take up of landscape ecological theory is – like many landscapes – somewhat fragmented and a little ecological knowledge interpreted poorly can cause more problems than it solves (Beunen and Hagens, 2010; Grose, 2010).

Cultural geographers tend to have a strong theoretical base. The emphasis is on examining in-depth ways of looking at landscape and, importantly, ways of seeing with landscape – that is taking landscape as the starting point for the way you look at the world (Daniels, Chaper 12). Much landscape architectural knowledge is grounded primarily in an experience of working *with* the physical landscape (experience, reflection and action) and with those stakeholders who use it and manipulate it, such as developers, managers, landowners and policy-makers. Landscape architects work on what might be labelled an 'applied grounded action theory' whereas the knowledge and understanding of cultural geographers is through a more academic direction: research, evaluation and reflection (see Figure 13.4).

Ipsen (sociology, Chapter 3) suggests that key areas of understanding for landscape architecture are the ongoing understanding of systems, change, impact and particularly landscape consciousness. The theory behind the relationship of people with the landscape (cognitive, aesthetic and emotional) and the practical effects that people's landscape consciousness has on the material landscape are important. This is because the consciousness of the landscape defines how it is valued by people and therefore how they view change or potential change and, for example, whether they will want to protect the existing state of landscape or support the creation of new laws that will change the landscape. Understanding the key theoretical perspectives and the value systems by which disciplines work provides a really useful understanding of their working practice as well as the differences in educational approaches.

## Ways of working and ways of seeing

Collaborating with other disciplines and crossing into new areas of study requires the acquisition of new skills and an understanding of new methods for working. Highlighted in particular in these chapters are technological skills e.g. the many possibilities of working with GIS and satellite imagery, but also economic calculation and using economic and social data. The key here is that while landscape architects cannot possibly master all the different skills of all the potential disciplines with which they may be working, it is important to understand where such skills are needed, how they can help inform practice and provide planning, design and management solutions with a much stronger justification for decisions made. Skills may need to be acquired or improved but there is a caution (Rotenberg, Chapter 10) against shallow learning of skills and highlighting the need for a better understanding of discipline approaches and theory.

**Figure 13.4**
Knowledge building in Landscape Architecture and Cultural Geography.

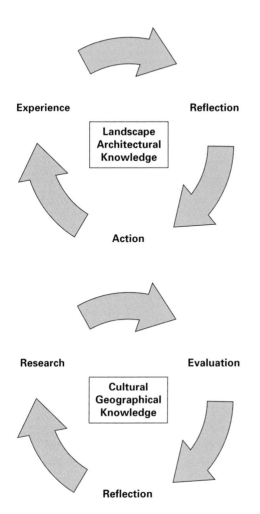

Collaborating with landscape archaeologists and historical geographers could help understand previous change and time depth (Kleefeld and Schenk, Chapter 7; Fairclough, Chapter 4). So with the future envisioning skills of landscape architects, collaboration could provide a fuller picture of the temporality of problems by travelling backwards and forwards in time using and creating scenarios to explore possibilities. Understanding both the fieldwork techniques and scenario-building/ design skills may help the development of innovative tools.

Professional practice increasingly demands robustness in working methods, a point emphasised in these chapters, particularly in relation to survey methods and in relation to ecological planning. Design decisions need to consider a range of impacts, and collaboration with landscape ecologists could improve understandings of the ecological impact a design might have and avoid the 'greenwash' of the past where ecological assumptions (e.g. concerning the benefits of corridors) were made. Landscape ecology has already taught much about landscape processes and interaction, but there is still a need to think in terms of processes and dynamics

(Sarlöv-Herlin, 2004) instead of landscapes as areas where 'a design' is an apparently static and achievable norm.

A point emphasised in these chapters is that for landscape architects it is important to define the problem or questions in terms of landscape *first* then define what areas of knowledge are required in order to understand and solve a problem. This may seem straightforward, but in practice the solution to research and design problems may require another focus (e.g. economic and social); however, landscape architects' expertise is in providing the 'landscape view'. Starting with the landscape questions will help define whether additional knowledge/understanding is needed and perhaps reduce the danger of being primarily responsive to financial requirements.

There is a perception in these chapters that landscape architects do not adequately address issues of scale and 'joined-up thinking' and have a tendency for the oversimplification of problems and issues that need to be considered to create meaningful spaces. Interdisciplinary work could help provide new techniques that would help deal with the confusing complexity of information related to landscape, particularly at the larger scale. The suggestion is that it is important to assess the capabilities of the site/problem in much broader terms than is often provided by the brief. Åsdam (fine art, Chapter 5) suggests that the 'ways of seeing' of the artist can help reveal the potentials of 'forbidden' spaces. Landscape architects are generally concerned with creating pleasant places, but artists' objectives may be more about provoking reaction. Collaboration could help landscape architects provide designs that challenge both clients and the public to think about and use landscape in new ways.

As is suggested by Venturi (planning, Chapter 11) there is a need for a change in perspective on all sides if disciplines are to work together well. Examples of new possible approaches that are already emerging and attracting cross-disciplinary attention are concepts of resilience and 'wicked' problems. Representation of the landscape has been a particular skill of landscape architects, and is seen in these chapters as a key way landscape architects can push forward thinking, such as the development of 3D and 4D representations which incorporate flows and processes and are not static representations of landscape. As the ELC emphasises, landscape is dynamic in time and space, in concept and in reality. Although there are many tools and techniques available, landscape architects should, as part artists, part technicians, be at the forefront of devising new ways of representing the landscape, its complexities, the speed and dynamics of change and what is foreseeable and what is not.

## *Knowledge building*

Knowledge exchange can be one of the main benefits of interdisciplinary working. Landscape architects for example could learn much from the strong research and knowledge base of a number of the disciplines in this book, while they could teach about participatory practices and thinking, about bringing together overlapping ideas about landscapes, about the importance of inspiration, aesthetics and creativity and the way it potentially affects the way people live or can live their lives. In some cultures the separation in landscape studies between 'culture' and 'nature' or arts and science subjects has not been apparent (see Taylor, 2009) but in European cultures this has resulted in real difficulties in the research and teaching traditions in building knowledge that integrates arts and science subjects. Even within cognate areas (e.g. landscape architecture and architecture) there are considerable differences between teaching methods, philosophies, attitudes and approaches to the world.

Taking up opportunities for 'chance encounters' with knowledge is emphasised by Daniels (Chapter 12) to build a 'new antiquarian' approach; this could be through wandering around a library or keeping notebooks such as those used by Laurence Halprin. Cultural geographers see that effective fieldwork demands both 'a tool kit of systematic techniques' for site recording and survey and 'the cultivation of a receptive and empathetic outlook' which allows for engagement with people as well as the locality or site (Daniels, Chapter 12). 'Layering' of knowledge is a technique used by a number of disciplines such as historical geography (Kleefeld and Schenk, Chapter 7) where the emphasis is on creating databases which can be manipulated to protect landscapes and understand human processes.

This book recognises the need for landscape architects to consider their roles in a rapidly changing landscape context. The roles of landscape architects such as mediators, communicators, facilitators as well as those required traditionally as a designer, planner and manager, have been recognised for some time (Roe, 2000). Building awareness of where new knowledge and new skills are needed through engagement with other disciplines will help landscape architects not only to produce long-lasting, beautiful, economically and ecologically sustainable and appropriate designs, but help ensure the sustainability of the profession itself.

## Landscape architecture as an academic endeavour: research, education and training

The LE:NOTRE project has over the last 11 years allowed for considerable reflection throughout Europe on the ways that landscape architecture is taught and has, I believe, gone much further than many professions in this respect by (i) looking at core competencies for the professional expertise of landscape architecture across Europe (ii) having an on-going and detailed conversation on all areas of professional practice (iii) providing opportunities to exchange and build knowledge within the profession, and (iv) reaching out to cognate discipline areas through expert advisors. The 'cake' of discipline areas relating to landscape can be cut in many different ways, but if landscape architects are to respond to external professions, it may mean that new ways of structuring knowledge about landscape is also needed. As new discipline areas and knowledge develops, so landscape architecture training must respond.

These chapters highlight the need to engage with a range of theory and practice emerging from other disciplines. Plant knowledge is an area that is identified as one where landscape architects need to pay more attention (Schmidt, Chapter 2). It is perceived that the way landscape architects are taught using a limited palette of plants that are then commonly used in practice can create monotonous or poor planting design. Landscape architects still need to be more exploratory and innovative in the way they gather plant knowledge and extend their palette.

Although there is a comment that landscape architectural education does not produce 'strong' design styles (Jormakka, Chapter 1) this can be seen in both positive and negative terms. While it may reduce conflict when working with other disciplines, it does little to help the public profile of landscape design.

Landscape architecture is a composite discipline; it takes from the arts, science and social science methodologies. Increasingly students need to learn not only how to gain design inspiration and demonstrate design technologies, and soft and hard materials manipulation and management, but also demonstrate social science skills such as participatory working, focus groups and observational skills. Add to these the ability to read aerial photographs, historical maps, carry out computer-based analysis and manipulation, ecological survey, policy analysis, site supervision and communication with clients – and you have the basis for a landscape professional!

It would seem important for landscape architectural teaching to embrace an outlook that fosters an ability to view the landscape through different lenses if cross-disciplinary working is to be successful. It is acknowledged that some study programmes already include courses which are jointly taken by a number of disciplines and that this is a good start in interdisciplinarity. Building student work on areas such as landscape assessment (Price, Chapter 9), which potentially allows many disciplines an input, would be a good starting point because although working next to or with other disciplines is important (Bruce et al., 2004) it is recognised in these chapters that building understanding and interdisciplinarity potential is about devising programmes that challenge students not only to *do* things in different ways, but also to *think* in different ways through more problem-orientated approaches. This requires 'cutting across traditional discipline-based academic structures and systems of reward and resource allocation that are found in most universities' (Ibid, p.459).

## Final observations

The collection of chapters and the disciplines represented in this book are just a few of those that landscape architects come into contact with in their professional work on a daily basis. However by examining the way that these disciplines see their own relationship to landscape and in particular to landscape architecture, it is possible to gain a better understanding of where landscape architecture as a discipline sits within the professional and academic world. It is accepted that exchange of information and interpretation can be an important step to greater understanding between disciplines. Many landscape architects admit to finding it difficult in describing what they do, and also to being irritated by the common belief that 'designing gardens' is what landscape architecture is about. However it is up to the profession to do something about this. This is not a novel idea, but a third perception seems to be arising, which is that other disciplines seem to be taking over the 'ground' of landscape architecture. The professions in general have been very protective of their areas of focus, indeed professional codes of conduct and training were established to help the professions define their focus and protect their control over that area of work.

Unfortunately this also means that landscape architecture as a profession has been slow to take up the opportunities that are arising from present environmental crises and demands and is rapidly losing ground to other more active and innovative disciplines which are straying outside their traditional discipline areas to develop new knowledge and new composite skills.

Academic disciplines are no better; the academic systems in European universities still discriminate against the development of interdisciplinary research and development of cross-disciplinary understandings in research and teaching (see Bruce *et al.*, 2004). Professional work is constantly changing, and as once professions developed such ways of working, now these need to be reassessed as the boundaries of professional work are becoming more blurred. The ELC approach to considering the landscape as a holistic system rather than the previous 'action for landscape' approach provides the basis for widening collaboration and innovation in the way landscape-related subjects are taught in universities (Matthews and Selman, 2006; Brewer, 1999).

As landscape change becomes faster and the demands humans place on natural resources grows it seems that we are facing ever more complex and urgent challenges which emanate both from human actions and natural events (Roe, 2009). But what does 'challenge' really mean for the landscape professions? Is it really possible to meet these challenges? Some believe that advances in technology will allow us to assuage the challenges we place upon the landscape, but at what price? And are we willing to pay the price? How can landscape architects respond to the demands of such challenges?

I do not pretend to agree with everything the authors say in their chapters. My own recent experience suggests that good interdisciplinary working is as rare as good commentary on such working which provides specific recommendations for ways forward. So this book is spearheading an area which is poorly represented by paving the way for better understanding through well-considered views of landscape architecture with the aim of further self-examination and development of the profession and the development of cross-disciplinary collaboration. The following summarises the clues given in this book as to possible future routes for the profession of landscape architecture:

1. Strengthen the basis for working through engagement with other disciplines and in interdisciplinary investigation: consider language(s), exchange methods and methodologies, theory and discourses.
2. Strengthen research sensibilities, cultures, traditions and agendas (see Benson, 1998) and the way practitioners employ research findings to inform design, planning and management decisions.
3. Develop new technological skills which aid interdisciplinary working.
4. Improve communication skills, because learning to converse with other disciplines is important.
5. 'Be comfortable in your own skin' or recognise and build upon the potentially instrumental and critical role that landscape architecture can play in the determination of future landscapes; in the way that landscape architects can, more

than any other profession concerned with landscape, provide a holistic view which crosses the arts-science boundaries and, in particular, focuses on the material expression of communities' desires, needs, identity and wants related to the landscape.
6. Recognise that landscape architects have something valuable to contribute in both theoretical and practical discourse. These chapters showed that there is much overlap in the professional areas of interest, and other disciplines are showing some awareness of the skills that landscape architects have and that the work of landscape architects does help inspire others.
7. Embrace and celebrate the wide range of influences on landscape architecture and recognise that within the profession there are likely to be specialist areas that particular individuals and practices will wish to focus upon; be open and receptive to new ideas and generous with those within the profession (as well as within other professions) who hold an alternative view of the world.
8. Generate a more proactive approach to establishing cross-disciplinary fora; in leading research, in being political and taking a view. Consider the significance of landscape as a potentially unifying force in policy, practice and research (see ESF/Cost, 2010).
9. Challenge yourself: the focus group/workshop approach to the structure of this book aims to challenge and produce new insights. As professionals, landscape architects profess to have a special knowledge of landscape, so it is important that this is the case and that it includes a strong theoretical understanding behind the work; landscape architects need to embrace and engage more fully with theory but also be prepared and have the confidence to challenge others.
10. Develop intellectual agility and recognise the need for changing perspectives to address the enormous potential change in the scope of work in, for example, cities, infrastructure, cultural identity, and in relation to climate change.
11. Create new opportunities for new interactions with landscape and for professionals themselves to have new 'conversations' with landscape. Landscape architects are good at this – they spend a lot of time 'in the landscape', perhaps more so in the past than today; being stuck at the computer screen, the drawing board or in the meeting room should not be a substitute for real world experience.

## Conclusions

These chapters indicate a considerable regard for landscape architects and although some undoubtedly indicate a partial understanding of the scope of landscape architectural work in practice and research presently being carried out, there is also much good understanding of the potential overlap areas and a considerable thirst for further discussion.

It has always seemed to me that landscape architecture is more than a job; it is a philosophy, a way of life, or a way of living. Landscape architects are nice people, and long may it stay this way. Landscape architects have in the past been criticised for not being forceful enough or political enough and while there are arguments to support the need for more and better involvement of landscape architects in political and policy-making spheres, there are many different ways of doing this

**Crossing the boundaries?**

**Figure 13.5**
A summer moment, St Abbs Head, Scotland.

which do not necessarily entail a total 'change of skin'. In a recent conversation with an architect it was suggested to me that architecture had created no new valuable or significant 'places' for people over the last 5–10 years compared to those of the past. I would argue that recent and contemporary landscape design has in fact produced some really wonderful places; of course also some very run-of-the-mill ones, which will no doubt be forgotten. But that perhaps is the great thing about landscape – it changes. Whether we intervene directly through design or through management, or whether we leave it to natural forces and chance impacts, change will occur. Much of what attracts us to landscape is transient and lasts only in the memory – such as the effects that the sun has at a particular time of year on a particular patch of flowers (Figure 13.5) or a particular May Day celebration in a small traditional village green in the countryside where children dance around a maypole. However just as much of what attracts us is about longer-term change: it is the physical result of the interactions between the various processes of change over time, the layering of present and past and the possibility of future that makes landscape significant to people. Change is often portrayed as a bad thing, but landscape architects can provide an alternative reality and in particular can help people to look at landscape afresh, encourage the enjoyment of change and help harness change that enriches both the landscape and people's lives. This ability to work with the dynamic landscape to realise the visions of individuals and communities is perhaps landscape architects' greatest professional asset. Now is the time to reassess how the acquisition of new theories, skills and approaches can ensure the profession responds to the great landscape challenges of our time.

## Bibliography

Benson, J. F. (1998) 'On Research, Scholarship and Design in Landscape Architecture', *Landscape Research*, 23(2): 198–204.

Benson, J. F. (2004*)* 'A Logodaedalic Adventure in Landscape Sensitivity and Landscape Capacity', *Proceedings of a CCN Workshop on Landscape Capacity and Sensitivity* 27 January 2004.

Beunen, R. and Hagens, J.E. (2010) 'Use of the Ecological Networks Concept in Spatial Planning Practices', *Landscape Research*, 34(5): 563–580.

Bhatia, V. K. (2002) 'Professional Discourse: Towards a Multi-dimensional Approach and Shared Practice', In: Candlin, C.N. (Ed.) *Research and Practice in Professional Discourse* (Hong Kong, City University of Hong Kong Press) pp. 39–60.

Brewer, G. D. (1999) 'The challenges of interdisciplinarity', *Policy Sciences*, 32: 327–337.

Brewer, G. D. and Lövgren, K. (1999) 'The theory and practice of interdisciplinary work', *Policy Sciences*, 32: 315–317.

Bruce, A., Lyall, C., Tait, J., and Williams, R., (2004) 'Interdisciplinary integration in Europe: the case of the Fifth Framework programme', *Futures*, 36: 457–470.

Campbell, H. (2003) 'Editorial: Talking the Same Words but Speaking Different Languages: The Need for More Meaningful Dialogue', *Planning Theory and Practice*, 4(4): 389–392.

Eco, U. (1990) *The Limits of Interpretation* (Bloomington & Indianapolis: Indiana University Press).

Eco, U. (1992) *Interpretation and Over Interpretation* (Cambridge, CUP).

ESF/COST (2010) 'Landscape in a Changing World: Bridging Divides, Integrating Disciplines, Serving Society', *European Science Foundation/COST Science Policy Briefing Note 41*. Available at: http://www.esf.org/publications/science-policy-briefings.html (accessed 7.4.11).

Fischer, F. (2003) *Reframing Public Policy: Discursive Politics and Deliberative Practices* (Oxford, OUP).

Florgård, C. (2007) 'Preserved and Remnant Natural Vegetation in Cities: A Geographically Divided Field of Research', (Review Paper), *Landscape Research*, 32(1): 79–94.

Grose, M. (2010) 'Small Decisions in Suburban Open Spaces; Ecological Perspectives from a Hotspot of Global Biodiversity Concerning Knowledge Flows between Disciplinary Territories', *Landscape Research*, 35(1): 47–62.

Gould, S. Jay (2004) *The Hedgehog, the Fox and the Magister's Pox* (London, Vintage Books).

Hansson, B. (1999) 'Interdisciplinarity: For what purpose?' *Policy Sciences* 32: 339–343.

Karlqvist, A., (1999) 'Going beyond disciplines, the meanings of interdisciplinarity', *Policy Sciences*, 32: 379–383.

Longatti, P. and Dalang, T., (2007) 'The Meaning of "Landscape" – An Exegesis of Swiss Government Texts', in: Kienast, F. Wildi, O., and Ghosh, S. (Eds) *A Changing World, Challenges for Landscape Research*, (Dordrecht, Springer) Landscape Series, 8, pp. 35–46.

Marzano, M., Carss, D. and Bell, S. (2006) 'Working to Make Interdisciplinarity Work: Investing in Communication and Interpersonal Relationships', *Journal of Agricultural Economics*, 57(2): 165–184.

Matthews, R. and Selman, P. (2006) 'Landscape as a Focus for Integrating Human and Environmental Processes', *Journal of Agricultural Economics*, 57(2): 199–212.

Phillipson, J., Lowe, P. and Bullock, J. M. (2009) 'Navigating the Social Sciences: Interdisciplinarity and Ecology', *Journal of Applied Ecology*, 46: 261–264.

Roe, M. H. (2000) 'The Social Dimensions of Landscape Sustainability', In: Benson, J. F. and Roe, M. H. (Eds) *Landscapes and Sustainability* (Oxford: Routledge)

Roe, M. H. (2009) 'Editorial: The Times They Are A-Changin', *Landscape Research*, 34(1): 1–6.

Roe, M. H., Jones, C. and Mell, I. C. (2008) *Research to Support the Implementation of the European Landscape Convention in England*. Contract No. PYT02/10/1.16.

Roe, M. H. and Rowe, A. M. (2000) 'The Community and the Landscape Professional', in: J. F. Benson and M. H. Roe (Eds) *Landscape and Sustainability* (Routledge, Oxford).

Sarlöv-Herlin, I. (2004). 'New Challenges in Spatial Planning, Landscapes', *Landscape Research*, 29(4): 399–411.

Snow, C.P. (2008, orig. 1959) *The Two Cultures* (from the Rede Lecture, and *The Two Cultures: A Second Look*) (Cambridge University Press, Cambridge).

Swanwick, C and Land Use Consultants (2002) *Landscape Character Assessment Guidance for England and Scotland* (Countryside Agency, Cheltenham and Scottish Natural Heritage: Edinburgh).

Taylor, K. (2009) 'Cultural Landscapes and Asia: Reconciling International and Southeast Asian Regional Values', *Landscape Research* 34(1): 7–32.

Thompson, I. H. (1999) *Ecology, Community, Delight: Sources of Values in Landscape Architecture* (Spon, London).

Tress, B., Tress, G. and Fry, G. (2006) 'Defining concepts and the process of knowledge production in integrative research'. In: Tress, B., Tress, G., Fry, G. and Opdam, P. (Eds) *From Landscape Research to Landscape Planning: Aspects of Integration, Education and Application* (Springer, Dordrecht), pp.13–26.

# Name Index

Aalto, Alvar 34
Abelson, P. 226
Aglietta, Michel 69, 70–1
Alberti, Leon Battista 22
Aldred, A. 103
Alonso, W. 10
Altschuler, B. 125
Anderson, K. 288
Angelstam, P. 139
Apostol, D. 133, 137, 179, 182
Appleton, Jay 281
Aristotle 27, 29, 31
Armitage, P. 139
Arnaud, J.-F. 134
Arnould, P. 175
Åsdam, Knut 9, 117–29, 304, 308
Ashmore, W. 96
Aston, M. 85
Aubrey, John 85
Audot, Louis-Eustache *33*
Augé, Mark 252, 253
Austen, Jane 292–3

Baccini, A. 80
Bachelard, G. 123
Baltrusaitis, Jurgis 17
Banham, Reyner 279
Barabesi, L. 80
Bärtels, A. 52
Bartels, C. 88
Basso, K. 255
Bataille, Georges 21, 22
Baudrillard, J. 68
Baudry, Jacques 134, 136, 140
Baxandall, Michael 21
Becker, G. 218
Behrens, Peter 34
Bell, S. 133, 137, 179, 182
Bellamy, P.E. 134

Bender, B. 85
Benson, J.F. 304
Bento, João 10
Berger, John 288
Bergin, J. 224
Berman, M. 65
Beuys, Joseph 125
Bhatia, V.K. 301
Bintliff, , J. 85
Birks, H.H. 139
Bloch-Petersen, M. 142
Bloom, A. 53
Boas, Franz 234, 237, 238
Boitard, Pierre 15
Bonnefoy, Yves 17
Borer, A. 125
Bosc, Louis Augustin Guillaume 44
Bourdieu, Pierre 35, 247
Bowden, M. 85
Bowers, J.K. 210
Brandt, Jesper 145
Brenner, N. 253
Brewer, G.D. 299, 305
Brockman, J. 6
Brown, Capability 210
Buczaki, S. 50, 52
Bull, Catherin 293
Bunce, Bob 9, 30, 131–47, 135, 137, 139, 142, 143, 144, 303
Burckhardt, Lucius 60, 66
Burel, F. 136, 140
Burggraaff, Peter 160, 164
Burnet, Thomas 28–31, *28*

Castells, M. 75
Castro, P.V. 85
Chen, H. 80
Cheshire, P.C. 210
Choumert, J. 224

316

# Name Index

Christaller, W. 210
Clark, J. 88
Clawson, M. 224
Clements, Frederic 22
Coolidge, M. 121
Cooper, A. 144
Corner, James 22, 281, 287
Cosgrove, Denis 277, 279, 285–6, 291, *292*
Cowell, B. 282
Cowley, Abraham 19
Craik, Kenneth 281
Crawford, O.G.S. 85
Cronon, William 289
Cross, N. 6
Crouch, D. 287
Crowe, Sylvia 10, 198, *215*

Daniels, Stephen 11, 276–93, 309
d'Alambert, J. 19
de Quincy, Antoine Chrysostome Quatremère 18–19
de Sousa Santos, B. 248
de Ville, Antoine 17–18
Deleuze, G. 123
Derrida, Jacques 22
Diderot, Denis 19, 27
Dix, Andreas 158
Douglas, David 44
Dramstad, W.E. 133, 134
Dreslerová, D. 88
Dujesiefken, D. 50
Duncan, J. 280

Elias, Norbert 69
Elliott, P. 281
Evans, A.W. 210

Fairclough, Graham 9, 83–111, 302, 307
Feld, S. 256
Finke, Peter 67
Flaubert, Gustave 15–16
Flogård, C. 304
Ford, L. 246
Forman, Richard 133, 134, 137
Fortune, Robert 44
Foucault, Michel 16, 119, 123, 249, 250, 251
Fox, Cyril 85, 103
Friedman, Milton 200
Friedman, T. 120
Fry, G. 133, 134, 136
Fulton, Hamish 119, 120, *120*

Gandy, M. 280
Gehl, J. 248
Gencsi, L. 45
Giddens, Anthony 69–70
Girouard, Mark 279
Gobakken, T. 179
Godron, M. 133
Gold, J. 285
Goldsworthy, Anthony 120
Gombrich, Ernst 34
Gontard, Carl von 17
Goodey, Brian 281
Gorer, R. 53
Gradmann, Robert 152
Graham, B. 90
Graham, Dan 123
Grahn, P. 219
Gramsci, A. 244
Grant, W. 120
Gray, Thomas 26
Gröning, G. 60
Gross, W. 42, 50
Guattari, F. 123
Gunzelmann, Thomas 164

Habermas, Jürgen 69, 70
Hackworth, J. 253
Haines-Young, R.H. 139, 142, 144
Hall, Sir James 19, *20*
Halprin, Laurence 309
Hansson, B. 301, 304
Haraway, Donna 32
Harlow, W.M. 51
Harms, W.B. 139
Harris, D. 281
Harris, P. 120
Harrison, R. 107
Harshaw, H.W. 204
Harvey, David 244
Hauschild, E.I. 256
Hawkins, C.J. 209
Hegel, G.W.F. 18
Heidegger, Martin 36
Heinz, W. 52
Helliwell, D.R. 223
Henry, M.S. 226
Herlyn, U. 60
Herzfeld, M. 253
Hidding, M. 88
Hinsley, S.A. 134, 137, 140
Hirsch, E. 239

# Name Index

Hoffmann, Josef 34–5
Holston, J. 254
Hoskins, W.G. 85, 288–9
Howard, D.C. 135
Howard, Ebenezer 256
Howard, P. 90
Hunt, John Dixon 16, 17

Ihse, M. 142, 143
Iliev, I. 51
Ipsen, Detlev 8, 60–80, 306
Irwin, E. 211
Isard, W. 210
Isenstadt, S. 286

Jackson, J.B. 18, 281
Jacobs, J. 286
Jevons, W.S. 200
Johnson, Matthew 86, 288
Johnson, Dr. Samuel 36
Jones, M. 88
Jones, P. 220
Jongman, Rob 145
Jormakka, Kari 8, 15–38
Judd, Donald 122

Kahneman, D. 225
Kaufmann, S. 60
Kelty, M.J. 180
Kent, William 19
Keynes, John Maynard 200, 211
Kitaibel, Paul 44
Kleefeld, Klaus-Dieter 10, 150–71, 307
Klinka, K. 180
Knapp, B.A. 96
Knetsch, J.L. 224, 225
Koch, Karl Heinrich 44
Kondratiev, N.D. 70
Konijnendijk, C. 50
Kordes, G. 53
Kraynak, J. 122
Kruse, A. 88
Kuna, M. 88
Kushlin, A.V. 137

Labov, W. 240
Lacan, J. 126
Lambert, D. 280
Lancester, R. 44, 52
Langdon, D. 217
Langhans, Carl Gotthard 17

Langley, Batty *15*
Larmessin, Nicolas de *23*
Larsson, L. 87
Laugier, Marc-Antoine 19, 26, 27
Le Corbusier 35
Lefebvre, Henri 74, 244–5
Leland, John 85
Leopardi, Giacomo 28
Lewis, David 36
Lifran, R. 206
Linde, C. 240
Linnaeus, C. 242
Linné, Carl 22
Lipsey, R.G. 200
Lipsky, Z. 142
Lock, G. 96
Long, Richard 119
Lopes, Domingos 10
Lorimer, H. 280
Loudon, J.C. *26*, 43
Lövgren, K. 299
Lowenthal, D. 281
Lutik, J. 224

McCann, T. 144
McDonogh, Gary 246
McHarg, Ian 6
Malthus, T.R. 209
Marshall, A. 200
Marx, Karl 65, 209
Marzano, M. 299
Matless, D. 279, 287
Meinig, D.W. 287
Metzger, M.J. 143
Milanova, E.V. 137
Mishan, E.J. 2
Moggridge, Hal 281
Møller, P.G. 88
Molyneaux, B. 96
Muir, R. 96

Næsset, E. 179
Nauman, Bruce 122
Naveh, Zev 134
Nietzsche, Friedrich 36
Nilsson, K. 50, 220
Nitz, Hans-Jürgen 158

O'Hanlon, M. 239
Olwig, Kenneth 88, 288
Opdam, P.F.M. 135

# Name Index

Orejas, A. 88
Oria, J.A. 189
Orians, Gordon H. 281
Oueslati, W. 206

Pardal, S. 177
Pauwels, R. 136
Payne, B.R. 224
Pearce, D.W. 207, 209
Pearson, M. 279, 287
Pedroli, B. 137
Penning-Roswell, E. 281
Penrose, S. 107
Petit, S. 141
Phillipson, J. 301, 306
Piper, Adrian 125
Pliny 32
Ponte, Alessandro 17
Pope, Alexander 18
Price, Colin 10, 197–229, 301
Probocskai, E. 43
Pückler, Fürst 210, 217
Pungetti, G. 88

Radó, D. 50
Randall, A. 221, 222
Regato-Pajares, P. 140
Repton, Humphry 279, *279*
Revill, G. 285
Reynolds, Sir Joshua 27
Ricardo, D. 209
Roberts, B.K. 87
Robinson, J. 202
Rodgers, C.P. 284
Roe, Maggie 11, 299–313
Rotenberg, Robert 10, 233–57, 304
Rousseau, Jean-Jacques 20–1
Rowe, A.M. 303
Rowe, Colin 35
Rowley, T. 85
Ruggles, D. 281
Rutheiser, C. 252

Sahlins, M. 241
Samuelson, P.A. 231
Sapir, Edward 237–8
Särlov-Herlin, I. 133, 136, 308
Schenk, Winfried 10, 150–71, 307
Schlüter, Otto 152
Schmidt, Gabor 8, 41–57
Schofield, J. 107

Schreber, D.G.M. 256
Schreiber, D. 52
Schwartz, T. 238
Seneta, W. 51
Serota, N. 122
Shanks, M. 287
Shaw, George Bernard 33
Sheail, J. 139, 143
Sheppard, S.R.J. 204
Simmel, Georg 60
Simons, S. 121
Smirnov, Michael 44
Smith, Adam 199, 205
Smithson, Robert 119
Snow, C.P. 3, 5
Soja, Edward 244
Sorkin, M. 252
Sperber, D. 238
Spirn, Ann Whiston 289
Stigler, G. 218
Stigsdotter, U.A. 219
Stilgoe, John 289, 290
Strom, S. 224
Sukopp, H. 65, 75
Sullivan, W.C. 219
Summerson, John 279
Swyngedouw, E. 253

Taylor, C.C. 85
Taylor, K. 309
Teyssot, Georges 18, 19
Thacker, Christopher 27
Theodore, N. 253
Thomas, A. 229
Thomas, Keith 242
Thompson, Ian 6, 305
Tilley, C. 85
Tóth, I. 45
Tradescant, John (the Younger) 44
Travers, M. 224
Tress, B. 299, 304
Trieb, M. 281
Tschumi, Bernard 21
Turner, .K. 207
Tyrväinen, L. 221

Ulrich, R.S. 220
Upton, Dell 280–1
Utzon, Jørn 34

van de Velde, Henry 34
van der Rohe, Mies 22

**Name Index**

van Londen, H. 83
Vancsura, R. 45
Venturi, Marco 11, 259–75, 308
Verboom, J. 136
Vermeulen, H.J.W. 135
Viollet-le-Duc, E.E. 235
Vitruvius 19
Von Droster zu Hülshof, B. 159
Von Thünen, J.-H. 74, 210

Walpole, Horace 19
Webb, N.R. 134
Webber, M. 252
Weller, Richard 22

Whitby, M. 210
Whyte, W.H. 248
Wiens, J.A. 140
Wilde, Oscar 31–2
Williams, Raymond 287–8
Witcher, R.E. 283
Wittgenstein, Ludwig 35
Wrathmell, S. 87
Wu, J. 211
Wylie, John 278, 288

Zavala, M.A. 189
Zemach, Eddy M. 32
Zukin, S. 252

# Subject Index

Page numbers in *italics* denotes an illustration/table/figure

abandoned settlements 152, 156, 167
abandonment, agricultural *138*, 141
absolute space 245
accessibility 239
accounting 211, 216–18
aerial photography 85, 88, 100, 103, 143–4
aesthetics 31–2, 34, 67, 77, *78*, 96, 137, 197–8, 202, 233; and art 117–18, 119; ecological 133, 204, 207; and economics 204, 208, 211, 213, 214–15, 216–17, 220–1, 224–5, 226; and forests 177, 188; of space 74; and woody plants 50
afforestation 10, 178–9, 191
Agri-Environmental Schemes 141, 210
agriculture: mechanization in 72, 73
agro-forestry 191
allotment garden movement 256
Almeria (Spain) *143*
Alps 67, 68, 80
Amazon rain forests 77
analytical survey 104
anthropogenetic geomorphology 155
anthropology 234, 271, 304 *see also* cultural anthropology
applied historical geography 150, 151, 153, 158–9
arboretum 43, 55
arboricultura 42
arboriculture 42, 50, 54
archaeology 9, 83–4, 302; history of landscape study within 85; interdisciplinarity of 90; scope and methods of 83–4; as study of change in history 94; theory within 93–4; *see also* landscape archaeology
archive: landscape as an 155–8
area elements 169

art 9, 117–29, 155, 301;cross-disciplinarity and cross-media approach 127–8; engagement with space/place 122–4; land 9, 119–20; and performance 125; and political activism 118; researching into goal and context of work of 128–9; and viewer/user notion 124–5
Arts and Humanities Research Council, Landscape and Environment programme 282, 283
atopia 247, 252–4
Austria 139
autonomy: and landscape architecture 128, 129
autotopia 247, 254
avant-garde 34, 35

backtracking 154, 159
banal 34–5
baroque palace garden 244
Bavaria 159
Berlin 274
Bible 31
Biblical Paradise 19–20
biodiversity 139–40, 141, *142*, 189
Blindcrake *86*
Boasian School 234
*bocage* landscape 136, 139, 140
botanical garden 43
Branitz 217, *217*
bridges: and tolls 206
brownfield sites 272, 274
buses 252

capital 201–2
capitalism 65, 70–1, 100, 207
cartography *see* maps
cemeteries 250
Center for Land Use Interpretation (CLUI) 121–2
Château de Villandry 21

321

## Subject Index

cinematic 127
Cistercians 158
city/cities 8, 19, 262, 264, 303; in ancient world 244; and consolidation 267; criticism of 67; features of innovation 267–70; Fordist 65; industrial 262; nd landscape 60–1; and redevelopment 262; seen as obstacle to general mobility 264; shrinking 80, 262, 264; transformations 262–4
city-as-form 265
civilization, theory of 69
classes *see* social classes
Claude glass 26–7
Cleve (Germany) 164
climate research, historical 157
Clun (Shropshire) *142*
colonialization 43, 73
Common Agricultural Policy (CAP) 141, 144
Common Ground 287
commons: privatization of 69
communication 304–5
communicative action 69, 70
connectivity 132, 134, 135, 136
conservation movement 92
consumption 203
contingent valuation method (CVM) 221–2, 224
corridors 132, 134, 135, 136, 139, 145, 307
COST 5, 189–90
cost-benefit analysis 209
Council of Tree and Landscape Appraisers (CTLA) 219, 223–4
Countryside Information System 146
countryside policy 145–6
cross-disciplinary working 127–8, 299, 299–310
cultural anthropology 10, 233–57, 303; areas of basic knowledge 239–54; and atopia 247, 252–4; and autotopia 247, 254; and ethnography 234, 254–5; and ethnology 234, 255–6; and heterotopias 247, 248–52; and orthotopias 247, 248, 249, 253; research approaches 254–6; social production of space 243–6; and spatial discourse 241–3; and spatial practices 246–7
cultural geography 11, 276–93, 301, 302, 309; and cartography 286–7; development of 277; and fieldwork 290–3, 309; and landscape 277–9; and landscape architecture 279–81; landscape in pictures 285–7; and libraries 289–90; and living landscapes 282–5; as a scholarly field 285; and theory 287–9, 306
cultural landscape 88, 110, 139, 150, 152, 155, 158–70, 280; area elements 169; definition 160–1; as form of economic capital 163; function of elements 169–70; grades of future development 162–3; and land register 166–7; and land use maps 167, 168; linear elements 169; model for respecting and preserving *162*; point elements 169; research and analysis 164–70, *165*; use of maps 167–70
cultural landscape change map 167, 168, 170
cultural landscape conservation 159–60
cultural landscape inventories 159, 164
cultural landscape structure maps 170
culture(s) 62, 256; contradictory trends in development of 238; distributive quality of 237–9; interpretative approaches 236–7; in landscape archaeology 102–3; mediation theories of 235–6; as a population of meanings 238
Cumbria 139
Czech Republic 136

'dark woods' 158
Darwinian evolution 30
demand 203–5, 218–19
demographics 61
dendrology 8, 41–57; applied 52; and arboriculture 42, 50, 54; basic concepts of 45–6; botanical aspect 45; definition 41; general 51; hardiness zones 52–3; history of as an independent subject 43–4; horticultural 45; intersection and collaboration between landscape architecture and 47–50, 54–6; landscape-oriented 45; main areas of knowledge of use to landscape architects 51–4; practical aspect 45; role of woody plants in the landscape 47–50; selection according to ornamental/function value 54; selection to the function of landscape type 53; teaching of 46–7, 57; *see also* silviculture
Derwent Valley (Derbyshire) 293
description of landscapes 136–9, *141*, 146
Desert Research Station *126*
Design Argument 29–31
diminishing marginal utility 204
discounting 209, 228
disembedding 69–70

# Subject Index

dispersal 134, 135, 146
distribution 209–10
documentation of landscape history 153–5
Downton (Herefordshire) *91*
DPSIR Framework 141
DRIFT (Daventry International Rail-Freight Terminal) *107*
Durham Cathedral *225*
Dutch elm disease 220
dystopia 247

EAA (European Association of Archaeologists) 84, 88, 94, 105, 110
EAC (European Archaeological Council) 105, 111
Eastern Europe 261
ECLAS (European Council of Landscape Architecture Schools) 2, 4, 24, 25
ecological aesthetic 133, 204, 207
Ecological Main Network 136
ecological paradigm 23
ecology 67, 131 *see also* landscape ecology
economics 10, 197–229; accounting and costing 211, 216–18; and demand 203–5, 218–19; and diminishing marginal utility 204; and distribution 209–10; and factors of production 200–2; and government intervention 210, 211; and income distribution 209–10; integrated approaches 226–7; and investment 207–9; macroeconomics 200, 211; and markets 205–7, 212; need for awareness of economic arguments 214–16; normative 211–12; origins of formal 199–200; positive 197, 211; premia 225, 227; and private/public goods 206–7; pushing out economic understandings 216; research methods and approaches 213–28; and scale economies 203; spatial 210–11; and supply 200–3, 204; and sustainability debate 227–8; and taste 204–5, 218–19; teaching of 211–13; and 'theory of the firm' 202; valuation methods 219–25, 226
ecosystems 22
ecotourism 192
education: forestry 185–6, *186*, 187–8, 192–3; landscape architecture 270–2, 309–10; landscape ecology 144–5, 303
Ellerstadt (Germany) 72
elm trees 220
emptiness: construction of 246

enclosures 100
*Encyclopédie* 19
energy 191
Environmental Design Research Association (EDRA) 257
environmental history 151, 152, 153, 158, 171, 289
Environmental Impact Assessment 146
environmental issues 299
environmental strata 143
environmental study, science-based 104
Eskdale (Cumbria) 283–4, *284*
ethnography 234, 254–5
ethnology 234, 255–6
Eucalyptus species 179
Europe: forests in 175
European Archaeological Council *see* EAC
European Association of Archaeologists *see* EAA
European Council of Landscape Architecture Schools *see* ECLAS
European Landscape Convention (ELC) 1–2, 4, 83, 86, 90, 93, 108, 109, 171, 197, 274, 299, 308, 311
European Landscape Dynamics 80
European Science Foundation (ESF) 5
European Union: COST programme 5, 189–90; Thematic Network Projects 2
excavations 100–1, 104
experts/expertise 33–5
externalities 206–7
extrinsic values 27

families 272
FAO 190
Faro Convention 93
Federal Regional Planning Act 159–60
feminism 125
feudal landscapes 157–8
field walking 85
fieldwork: and cultural geography 290–3, 309; long-term 254–5
Finland 174
Fire Paradox 189
Fordism 71–3
Fordist city 65
forest clearings 155
forest fires 180, 183, 189
forest inventories 186–7
Forest Landscape Ecology Lab (FLEL) 188–9
forest management 174, 175–7, 178–83, 190;

323

## Subject Index

composition of forest 179–80; and fertilization 181; mixed forest stands 179–80; plantation, sowing and natural regeneration 180; pruning and thinning 181; roads and tracks 182–3, *182*; scientific 176; selection of sites and reforestation 178–9; selection of species 179; silvicultural models 183–4; site preparation 180; stand tending 181; sustainable 175, 178, 192; weeding 181
forestry/forests 10, 77, 173–93, 216, 227, 302; basic concepts of 175–83; benefits of 192; definition 42; distribution of 174–5, *174*; and education 185–6, *186*, 187–8, 192–3; and EU COST programmes 189–90; functions of 177, 183–4; and historical geography 158; importance of 175, 192; innovative challenges in research 188–90; and landscape architecture 173–4, 187; opportunities for developing collaboration 190–1; over-exploitation of 176; threats against 192; timber harvesting 181–2; urban 42, 192
fountains 17
freshwater ecology 139

garden archaeology 87
*Garden History* 281
gardens/garden design 14–19, 245–6, 303; Biblical Paradise as first 19–21; botanical 43; and contrariety 18; Flaubert's categories of 15–16; as fluid signifiers without a signified 19; following pictorial models 26–7; as heterotopias 16–17; and illusion 18; labyrinths 21–2; origin of name 21; as a technological artifact 27; undermining authority of architecture 17
gaze 126
GB Countryside Survey 139, 142
gelding 32
genetic diversity 140
Geographical Information Systems (GIS) 136, 157, 177, 186
geography 276, 277, 281 *see also* cultural geography
geophysics 85
Germany 76, 152; and 'death of the forest' 77
Gestalt 75
globalization 52, 141, 262
Gothenberg Agreement 140
government intervention 211

grasslands *138*, 140
Greeks, ancient 21
Grizedale Forest (Cumbria) 120

Hadrian's Wall 282–3, *283*, 284
hardiness zones 52–3, *53*
health benefits 219–20
heaths 167
Hedgerow Protection Act 144
hedgerows 134, 135, 136, 139, 140, 144, 146, 155
hedonic house price method 224–5
hegemony 244
heritage management 90, 92
heterotopias 16–17, 247, 248–52
Historic Landscape Characterisation (HLC) 98–9, 100, 103
historic maps 100, 104
historical geography 10, 150–71, 307, 309; applied 150, 151, 153, 158–9; basic concepts of 153–63; as a bridge 171; climate research 157; and cultural landscape *see* cultural landscape; documentation of landscape history 153–5; and feudal landscape research 157–8; and forests 158; history and development of 151–3; and history of water and air 158; landscape as an archive 155–8; principal objectives 151; and scientific methods 156; use of maps 154
historical relicts 155
'historico-genetical' geography 150–1, 152, 171
Holy See garden (Potsdam) 17
homogenization 11, 252, 260
horticultural dendrology 45

IALE (International Association for Landscape Ecology) 132, 145, 146, 147
IFLA (International Federation of Landscape Architects) 132, 147, 274
images: and cultural geography 285–7
income distribution: inequality of 206, 209–10
income levels 204
industrial city 262
innovation: features of 267–70
Institute for Archaeologists (IfA) 111
interdisciplinarity 61–8, *62*, 83, 90–1, 109, 261, 299–300, 304, 308
International Association for Landscape Ecology *see* IALE

## Subject Index

International Association for People-Environment Studies (IAPS) 257
International Construction Exhibition Emscherpark 68
International Federation of Landscape Architects *see* IFLA
International Society of Arboriculture 42
interpretation: and culture 236–7; and landscape 121–2
intrinsic values 27
investment 207–9
island biogeography theory 134
isolation 134–5

kitsch 34, 35
knowledge building *307*, 308–9
Koper (Slovenia) *263*

labour 200, 200–1
labyrinths 21–2
land 200; origin of word 26
land art 9, 119–20
land register 166–7
land surveys 167
land use maps 167, 168
landscape(s): analysis of 64–5; as a cultural creation 26, 27; and cultural geography 277–9; definition of 197; dimensions of *302*; feudal 157–8; horizontal approach 278; images of 284–6; interdisciplinary concept of 61–8, *62*; living 282–5; mental 69, 76–9, 86; modernization and dynamics of 65–6; origin of word 25–6, 155, *156*; perception of 73, 76–8, 79; regressive 67–9; sociology of 60–80; as space 74–5; and theory 287–9; transitional 65–8, *68*;; vertical approach 277
landscape archaeology 83–111, 301, 302; and action research 108; archaeology at landscape scale 97–8; archaeology of past landscapes 99–100; archaeology of present day landscape character 98–9; branches of 96–100; collaborative research opportunities 102–4; cultures in 102–3; definition of landscape 92–3; diversity of approaches 87–8; focus on theory 93–4; and heritage management 90; interdisciplinarity of 90–1; and knowledge research 105–7; and landscape architecture 82–101, 91–2; and landscape character 84, 87, 95, 97, 98, 108; and landscape ecology 101; and landscape history 88–9, 110;

methods and techniques 100–1, 103–5; nature of 85–6; and past in the present landscape 94, 98–9, 108; and politics 94; scale and spatial patterns 96; subjectivity versus scientism 93
landscape architecture: as an academic endeavour 309–10; complexity of 25; defining 24–5, 90–1; misconceptions about 302–3, 310; need for acquisition of new skills and new methods of working 306–7; possible future roles for 311–12; task of 1; and theory 8, 15–38, 302, 305–6, *305*
landscape assessment 133, 137, 226, 310
landscape biography 104
landscape change 67, 73, 95, 103, 109–10, 142–4, 161, 166, 198, 210, 311
landscape character 84, 87, 95, 97, 98, 108 *see also* Historic Landscape Characterisation
Landscape Character Assessment (LCA) 98, 103
landscape consciousness 77–9, *78*, 306
landscape description 136–9, *141*, 146
landscape ecology 9, 101, 104, 131–47, 301, 303, 307; areas of basic knowledge 133–44; and biodiversity 139–40; collaboration and overlap with landscape architecture 132–3, 145–7, 303, 307–8; and countryside policy 145–6; definition 131; description and classification of landscapes 136–9, *141*, 146; development of as a discipline 131; and forest planning 177; and landscape function *138*, 140–1, 145; landscape structure and pattern 134–6, *135*, 145; main principles of 132; monitoring and assessment of change 142–4, *143*; and restoration techniques 146; teaching methods 144–5, 303; and theory 306
landscape function *138*, 140–1, 145
landscape history 88–9, 110; documentation of 153–5
landscape painting 155
landscape patches 134
landscape patterns 136, 188, 264–7
Landscape Research Group 281
landscape scale 9
landscape schools 274
landscape structure 134–6, 145
landscape urbanism 22–3
language 304–5
Languedoc (France) *135*

325

# Subject Index

LE: NOTRE Project 2–3, 3, 4, 56, 90–1, 110, 293, 309
libraries 289–90
Life EECONET initiative 136
lifestyles 73, 76
lifeworld: colonization of 70
lighthouses 206
Lincoln Cathedral 221
Lincoln Park (Chicago) 247
line walks 119–20
linear elements 169
linear features 135, 136, *142*
lived space 245
living landscapes 282–5
locality 240
long-term change, study of 100, 103, 106–7

macroeconomics 200, 211
Malahide Conference 140
manorial registers 153, 154
maps 286–7; and cultural geography 286–7; and cultural landscape research 167–70; historic 100, 104; and historical geography 154; land use 167, 168; parish 287; topographical 167–8, 170
markets 205–7, 212
Marx, Karl 201–2
mediation: and culture 235–6
medieval archaeology 87
medieval city 65
Megacity 62, 65
megaurban landscapes 9
mental landscape 69, 76–9, 86
meta-populations 134
migration 9, 73, 272, 280
modernization: and communicative action theory 70; and control over nature 69 and disembedding 69–70; driving forces behind 65; and dynamics of landscape 65–6; and landscape planning/architecture 69–73; and regulation theory 70–1
monitoring of landscape 142–4, *143*
morphogenetic approach 151, 152
motorway verges 133
Multiple Use Sustained Yield Act (1960) 176
music 34

narrative place 127
national parks 176, 226–7
National Trust 221

natural resource economics 197, 198, 200, 208, 210
nature 50, 61, 62–3, 235, 250; Aristotelian conception of 27, 29; Burnet's conception of 28–9, 30–1; as a contingent construction 28, 32–3; control of and modernization 69; and culture 235; and good 27–8; as mother of all architecture 19; organicist theories 22, 30–1; Wilde on 31–2
nature reserves 136
Nature 2000 Network 140
natural sciences: and landscape architecture 6–7
Navy Target 103A *121*
Netherlands 88, 104, 136
network analysis 80
New York City apartments 240
nitrogen-fixing plants 180
non-places (atopia) 247, 252–4
Nottingham University 292
nurseries 42–3, 52

ordinary places (orthotopia) 247, 248, 249, 253
Ordnance Survey 167
organicism 22, 30–1
orthotopia (ordinary places) 247, 248, 249, 253

palaeoecological investigations 156
palimpsest 94, 98, 100, 104, 109, 155, 260, 290
panorama 71
paradise 18, 19–21, 73
parish maps 287
parking 253
parks 124, 251; urban 250, 251–2
past landscapes, archaeology of 99–100
perception of landscape 73, 76–8, 79, 105
performative 126–7
peri-urban landscapes 9
Permanent European Conference for the Study of Rural Landscape (PECSRL) 152
philosophy of architecture 35–6
phosphate analysis 156
pinetum 43
place(s) 118–19, 241–2, 246–7; atopia 247, 252–4; autotopia 247, 254; heterotopia 247, 248–52; narrative 127; orthotopia 247, 248, 249, 253; and space 240; and spatial practices 246–7; and visual arts 122–3
place-relatedness 77
planning, regional *see* regional planning
plant colonisation 135

## Subject Index

'plant hunters' 43–4
plant knowledge 42, 54, 242, 309–10
point elements 169
politics of landscape 94
Poprad (Slovakia) *138*
positive economics 197, 211
premia 225, 227
preservation 162
preventative silviculture 183
private goods 206, 207
production, factors of 200–2
property ownership 76
pruning: and forests 181
Prussian land surveys 167
public goods 206–7

quality: and quantity 264–5
queer theory 125

railways 71
raw materials 201
redevelopment 262
regeneration 272, 274–5; forest 176, 177, 179, 180
regional planning 259–75, 303; and families 272–3; features of innovation 267–70; forms of transformations 262–4; landscape patterns 264–7; and mobility 273–4; research fields 272–4; subjects of landscape architecture planning 70–2
regional studies 100, 103
regressive landscape 67–8
regulation theory 69, 70–3
representation systems 270–1, 277, 279
representational space 240, 245–6
rescue excavations 85, 92
residential areas 266
Ring of Brodgar (Orkney) *201*
Rio Declaration 140
roads 265–6
Rovigo (Italy) *273*

St Abbs Head (Scotland) *313*
Salisbury Cathedral 221
sampling 143
scale economies 203
Scampton (Lincolnshire), military airbase at *95*
scenarios of change 47
Science Policy Briefing 'Landscape in a Changing World' 5, 6, 9, 105, 111
Scillies (UK) *99*

settlements 152, 154–5; abandoned 152, 156, 167
Seville cathedral 217
shelter belts 48–9
shopping centre 253–4
shrubs *see* trees/shrubs
silviculture 45, 173, 183–4
sites 240
Sitka spruce 179
Slovak Academy of Sciences 131
Snowdonia National Park (Wales) 229
social classes 73, 76
social fields: Bourdieu's theory of 35
social strata 76
soil science 155
*Sorbus* 45–6
space 118, 150; absolute 245; landscape as 74–5; lived 245; and place 240; political analysis of 245; representational 240, 245–6; social production of 243–6; sociology of 68, 73, 74–5; and visual arts 122–3
space flows 75
spacing 75
spatial discourse 241–3
spatial economics 210–11
spatial practices 246–7
Spiral Jetty 119
squatter development 254
stand tending (forestry) 181
Stanezice (Slovenia) *266*
Stowe garden 17
supply 200–3, 204
survey, analytical 104
sustainability 75, 239, 303–4; and economics 227–8; and forest management 175, 178, 192
sustainable forest management 175, 178, 192
Swedish 'Ystad Project' 87
Switzerland 159
synchronicity 67

taste 26, 34–5, 124–5, 204–5, 218, 218–19, 226
teaching methods *see* education
technological skills 306
theme parks 251
theory: collaboration on 101, 104–5; and cultural geography 287–9, 306; and landscape 287–9; and landscape archaeology 93–4; and landscape architecture 8, 15–38, 302, 305–6, *305*; and landscape ecology 306

**327**

## Subject Index

'third culture' 6–7
Thünen circles 74
timber harvesting 181–2
time 150
time-cycles 74
topographical maps 167–8, 170
totality 22
tourism 163, 272
townscape 106, 110
traffic 73
transhumance 141
transitional landscape 65–8, *68*
travel cost method 224
tree valuation 219
trees/shrubs 8, 47–50; aesthetic role 50; and arboriculture 42; ecological role 49–50; functional role 48–9; greening in industrial areas 49; importance of 173; maintenance of *see* arboriculture; psychological role 50; shelter belts and wind breaks 48–9; as therapeutic 220; *see also* dendrology; forestry/forests
Trsteno Arboretum (Croatia) 43
'two cultures' divide 1, 3, 5, 6
*Two Cultures, The* (Snow) 6

UNESCO 96
urban ecology 139, 147
urban forests 42, 192
urban landscape 65, 75, 110, 127

urban parks 250, 251–2
urban planning 262
urbanization 11, 75, 106, 143, 259
utopia 16, 247

valuation 219–25, 226
Vaux-le-Vicomte (Le Nôtre) 17
Veneto 291, *292*
Versailles *205*
vertical approach: and landscape 277
viewer/user 124–5
Villa d'Este (Tivoli) 17
Villa Garzoni garden (Gollodi) 17
Villa Lante garden (Bagnaia) 17
voluntary subscriptions 221

water 158
weeding: and forests 181
willow cathedral *20*
wind breaks: and trees/shrubs 48–9
wind turbines 202
wine-growing 72
woodlands 135, 136, 144, 158, 220
woody plants *see* trees/shrubs
Working Group for Applied Historical Geography 153
World Archaeological Congress 88
World Forestry Congress, 5th 176
Wylfa Nuclear Power Station (Anglesey) 215